THE PHENOMENOLOGICAL REALISM
OF THE POSSIBLE WORLDS

ANALECTA HUSSERLIANA

THE YEARBOOK OF PHENOMENOLOGICAL RESEARCH

VOLUME III

Editor:

Anna-Teresa Tymieniecka

THE PHENOMENOLOGICAL REALISM OF THE POSSIBLE WORLDS

*The 'A Priori', Activity and Passivity of Consciousness,
Phenomenology and Nature*

PAPERS AND DEBATE OF THE SECOND INTERNATIONAL
CONFERENCE HELD BY THE INTERNATIONAL HUSSERL
AND PHENOMENOLOGICAL RESEARCH SOCIETY
NEW YORK, N.Y., SEPTEMBER 4–9, 1972

Edited by

ANNA-TERESA TYMIENIECKA

D. REIDEL PUBLISHING COMPANY

DORDRECHT-HOLLAND / BOSTON-U.S.A.

Library of Congress Catalog Card Number 78–25369

ISBN 90 277 0426 0

Published by D. Reidel Publishing Company,
P.O. Box 17, Dordrecht, Holland

Sold and distributed in the U.S.A., Canada, and Mexico
by D. Reidel Publishing Company, Inc.
306 Dartmouth Street, Boston,
Mass. 02116, U.S.A.

Printed in The Netherlands by D. Reidel, Dordrecht

TABLE OF CONTENTS

PART II / ACTIVITY AND PASSIVITY OF CONSCIOUSNESS

PART III / PHENOMENOLOGY AND NATURE

COMPLEMENTARY ESSAYS

ACKNOWLEDGEMENTS

The Debate of the Conference was prepared
for publication by

PROFESSOR D. N. K. DARNOI

of Monmouth College, N.J.

I express my thanks to my assistant, Mr Shahrokh Haghighi, and to
Louis Tymieniecki-Houthakker for their editorial help in preparing this
volume for publication. Professor Augustin Riska and Mr Robert Green-
burg have contributed to it by tape-recording the debate of the Conference. Most of all, however, thanks are due to Professor Sebastian A.
Matczak for his generous efforts in the organisation of the Conference and for securing the funds for its support.

I am greatly indebted to Professor D. N. K. Darnoi for his dedicated
and most expert collaboration and to Monmouth College, New Jersey,
for contributing toward editorial expenses.

A.-T. Tymieniecka

INAUGURAL LECTURE

Dedicated to Monsieur Jean Wahl

yes because
must seem salient to the theorist.

eh what's this? is not this
exactly where P/G come in?
eh? emotion/passion etc?

ANNA-TERESA TYMIENIECKA

IMAGINATIO CREATRIX

The 'Creative' versus the 'Constitutive' Function of Man,
and the 'Possible Worlds'

I. INTRODUCTION

Every serious philosophical quest reaches sooner or later the fundamental question of *the origin of the human world and man's role in it*. Philosophy since Descartes adopts a framework of inquiry in which the factors and sources of the world's origin are sought chiefly in the nature and role of human consciousness; the recognition of the nature and role of consciousness seems to be the major accomplishment and progress of modern philosophy over the past. The two major and lasting contributions to this effect, as well as the two major treatments of the problem of the origin of the world, have been offered, by Kant in modern times and by Husserl in the contemporary period. And yet, both of them, in spite of the wealth of detailed analysis which they have left as a lasting heritage for philosophical scholarship, seem to have failed in the adequate formulation and treatment of this problem. I see the reason for their failure in their main assumptions, which seem to consist, firstly, in a tacit acceptance of the Cartesian conviction about the *absolute sovereignty of the logical reason* over other dimensions of human functioning, which stretches in gradation of intelligibility from the blind organic operations, the impulses through the affective and sensory levels and the whole dynamic dimension of 'passions', to the highest rational operations and transcending-élan of the spirit.[1] Secondly, *both Kant and Husserl have downgraded the structurizing role of passions by relegating it as the 'empirical soul' to the brute functioning of organism*. Therefore when at the culminating point of their quest they were led to search after the crucial clues to the sources and factors of the origin of the human world in the empirical realm of man's functioning, Husserl, in spite of his ever renewed efforts in his last works, never broke through the screen of reason he had established at the start, whereas Kant broke through by recourse to the *Einbildungskraft* and the role it plays with respect to the Genius, but could not handle it satisfactorily since his exit was already

Tymieniecka (ed.), Analecta Husserliana, Vol. III, 3–41. *All Rights Reserved.*
Copyright © 1974 by D. Reidel Publishing Company, Dordrecht-Holland.

blocked by the constitutive apparatus he had laid down. They were, albeit in different ways, caught into (a) the absolute correlativity: rational consciousness-human world, and (b) the necessity to equate the actual existing world with the potentialities of constitutive consciousness. Thus coerced into the position of a transcendental idealism, meaning the human world as it is, being absolutely dependent in its existence upon the rational consciousness, no other world than this one is possible. (Husserl insisted that the idea of a different world than the actual is absurd.)

My own work, in its slow progress which has been punctuated by various critical treatments of Husserl's transcendentalism, especially of the 'absolute status' that the transcendental consciousness as an agency of intentionality assumed in the classic phenomenology as the exclusive agency, leads me now to a *radical challenge* to this classic and current assumption of the sovereignty of reason in the functions which lead to the origin of the human world.

But in order to inquire into the factors and sources of the origin of the human world, should we not in the first place start by the fundamental statements *about the nature of both of them*? In fact, in an unprejudiced analysis – not restricted arbitrarily to the prism of conscious operations – the world appears as a system of things, beings and events intimately interwoven through the processes they generate; a *relatively stable system* caught in a dynamic progress of structural unfolding onwards. The cornerstones of this progress appear to be irruptions of *original and unpredictably novel entities*, fruits of man's inventive genius. Human being appears as the major dynamic factor within this system not only with respect to the progress onwards but also at the level of introducing the essential and basic level of *meaning* through the structurizing work of his consciousness. In order to account for the origin of the human world he have to account for both.

In fact, it seems to have been the oversight at the start of this double perspective upon the human world that explains why the otherwise incontestable contributions of genius towards the elucidation of the issues concerning the origin of the *life-world* have been vitiated in their scope. Indeed, neither Husserl nor Kant raises the question from having marveled about the nature of human reality, or from having taken into consideration the interplay of the two factors: conservation and renovation; both of them come to it indirectly from the question of the foundation of

I come to it through music, science, art, etc *but not to validate science, but my first inclinations were to invalidate*

the validity of science (identifying science with a higher level of the constitutive function). The basic meaning-giving and structurizing agency
of man is what has been called 'phenomenological or transcendental
constitution'. However, can the transcendental constitution, as it appears
in the phenomenological analysis both Husserlian and post-Husserlian, *No, neither*
account for *original, radically novel* meaningful structures introduced *can I.*
into the human world as the fruit of man's inventive genius? Is, as
phenomenology claims, the constitutive system of the consciousness to *rationality*
be identified with the whole extent of man's structurizing virtualities? The
classic and current phenomenology answers these questions in the affirmative. On the contrary, in what follows, I propose to deny to the constitutive consciousness these universally accepted prerogatives by distinguishing within the complete *human functioning system*, conjoined
with the constitutive function and yet autonomous with respect to it,
another structurizing and productive function of man: *creative function.*[2]

Unraveling the ways in which it works, its progress and its sources,
we propose, that:

(1) the constitutive activity of consciousness is not the one and only
system of man's functioning, but merely one of the possible conscious
functions;

(2) *the creative function* being another's, I submit that the analysis of
its workings reaches deeper into the nature of man's complete functioning, bringing to light: (a) various hidden factors of human functioning
instrumental in the origin of the human word, dispelling the traditional
division into 'faculties', and (b) revindicating the basic role of the impulsive, emotive, and affective dimension of passions, (c) showing the
specific type of *orchestration* of human virtualities into the *creative
function*, establish: (i) the *creative imagination* as the agency of the *a
priori* in the 'creative freedom' as well as (ii) the *plurifunctionality of
human consciousness as the source of possible worlds.*

Thus an opening appears towards *phenomenological realism.*

radical part of departure seeing utility dependent.

Is that phil to me?

II. CREATIVE VERSUS THE CONSTITUTIVE FUNCTION OF MAN

*Differentiation of the Two Functions with Reference to the Modal
Opposites: 'Activity and Passivity of Human Functioning'*

As we have already pointed out, no serious philosophical quest can bypass

I had mode emotive = aesthetic.

the concern with the origin of human reality. In both of its classic phases, ontological and transcendental, this concern is their focal point of interest. A wealth of analysis has thrown light upon many puzzling questions concerning the sources, principles and vehicles of the establishment of order upon which human reality and life itself depend.

In our proposed attempt to diversify the human functioning towards distinguishing its specific means and ends oriented by the concern of finding how both, perdurance and novelty – routine and invention – present in the human world, may be accounted for, we will encounter invariably some classic pairs of principles standing in opposition, which, bearing upon the ways of proceeding proper to types of functioning, divide them along their opposed respective lines. It appears that, first of all, asking the question about the conditions of unprecedented and novel instances emerging within the world routine, e.g., works of great art and invention, we find that disentangling the array of problems which is thereby thrown open to be treated, they fall naturally into a division along the line of *passivity/activity* of human functions. In the classic approach to our issue we find this demarcation principle applied already at the incipient stage of cognition, that is perception.

Indeed, phenomenology shares the Lockean-Kantian conviction that it is the transcendent perception, or sensory experience, that is the original, primeval source from which surge the germinal elements of human reality, and from which take off, as Locke so firmly states, all man's cognitive faculties in their respective modes.

This crucial role attributed to perception stems in the Lockean tradition from emphasizing in the sensory perception the particular vividness of the presence of the cognitive object within the cognitive act, which Husserl calls 'bodily selfhood', and assuming it as the strongest mode of evidence. Since upon this analysis only perception – transcendent as well as immanent – seems the only one, from among all the cognitive instances, to be endowed with the 'selfgiven' presence of its object, perception is accepted as the basic mode of cognition, 'original' in Husserl's terms, laying the foundation and outlining the construction of the edifice of human reality for all the others. In both, the Lockean-Kantian and the phenomenological conviction – although in the latter for more nuanced reasons – the cognitive-constitutive flow of experience, intermingling present sensation with remembrance, imagination and expectation, the flow through which – as it seems – we *nolens volens* exist by perduring from instant to instant without being in a position to choose it, decide upon it, or as it is, without even being surprised by it, is understood as our 'submission' to it. Kant, laying stress upon the singular sense-perception as an occurrence with a quality and content of its own, calls this special state of our being in perception 'receptivity'. Husserl, approaching experience from the processional nature of perception and the 'involuntary', and yet 'indispensable' nature of its flow, calls it 'passive genesis'. Indeed, we never stop to sense, feel, remember, be aware of something, etc., unless we leave the 'normal' state of mind. Zen

For Kant, although these foundations laid by experience for the origin of the world are at the first level, that is of the 'manifold of experience', 'received' by the cognizing subject without his willing or selecting yet in their forming and ordering, in the *a priori* forms of space and time with reference to the categorial schemata, – which corresponds to the Husserlian process of 'passive genesis' – without which they would be neither cognitive nor constructive of our reality, an 'activity' of the subject is already involved. In the Husserlian conception of the original experience of perception, or as Husserl calls it *'Urstiftung der Realität'*, the manifold, in spite of the change of emphasis from the Kantian instantaneity of structuration of the manifold (all happening at once), to the processional unfolding of the perceptual process, consisting of *a series of perceptual glimpses* put into motion and oriented by the perceptual development, 'receptivity' takes a strictly 'passive' turn.

Furthermore,[3] the 'passive genesis', as the complete set up of all the perceptual forms operates by a 'spontaneous' – here it means 'involuntary' – unfolding of a perceptual *series of individual glimpses*, carried on by their anticipatory-fulfillment structure. The content and the, so to speak, hooks of their progress, as well as their points of reference, being already preestablished, guarantee, on the one hand, the continuity of the structurizing constitution of the object and on the other hand, statement and correction of possible errors. Human reality emerges evidently from what, after Husserl, we call appropriately 'passive genesis' or 'synthesis', from this basic flow of experience carrying us willy nilly onwards. The so-called 'passivity', which means the pre-installed inevitability of its progress, is, in fact, the expression of its way of organization. Its 'spontaneous' unrolling, far from being a blind outburst, is on the contrary preestablished by rules, principles, blueprints preparing progressively appropriate means as the vehicles of their accomplishment. Due to the preestablished principles and norms for its universal progress and to the fact that this progress established laboriously, as Bergson has shown, in the backyard of nature becomes thoroughly *automatized* in the forefront of its operations, which is the intentional consciousness, we accomplish most complex and intricate operations of cognitive-constitutive nature effortlessly and involuntarily. To this mechanized apparatus carrying along established universal function of man, we owe an enormously complex and rich intersubjective world. From this anonymous and universally shared ground of primordial experience we raise to deliberation and selection only at the level of theoretical thinking, reflection and judgment; there it is that, according to Husserl, consciousness becomes 'active'.

However, does this mean that at the level of 'active consciousness' we are freed from the preestablished pattern of the structurization laid down by original perception? Would there be a disruption in the continuity of our constitutive system, the basic functions following a set of directives, the others oriented by individual whim? If that were the case, how then could the world, at the more advanced level of construction to which the higher intellectual activities of man are essential, achieve its unity and homogeneous nature, instead of falling apart into chaos?

In fact, as we have already pointed out elsewhere,[4] the originary experience being the motor for the emergence of all conscious functioning and the lawgiver delineating their scheme of activity according to its own preestablished regulations and structurizing principles, the constitutive system at all its levels holds together following *the same trend and subservient to the same telos*. This trend is clear. Although recognition of intentionality as the essential nature of consciousness seemed to introduce a 'neutral' medium, that is, unprejudiced with respect to the stand upon the great metaphysical questions concerning the nature and origin of the world, e.g., idealism/realism issue, with all the detailed work made in the field, it becomes obvious that intentionality as understood by Husserl and the classic phenomenology means the vehicle for a specifically rigorously rational structurizing level of the constitution of the world. In fact, Husserl himself admits the basic point he makes in interpreting both originary experience and further constitutive levels of intentionality as he sees it, writing:

Die Wahrnemung und ihre parallelen Bewusstseisweisen der Anschauung sind aber die ersten Grundgestalten des Bewusstseins, die für den Aufbau des spezifisch logischen Bewusstseins in Frage kommen, sie sind erste Grundlagen im logischen Bau, die gelegt und verstanden werden müssen. Wir schweifen also nicht etwa ab, sondern wir sind dabei schon Logiker, ohne es zu wissen.[5]

Clearly, intentionality is interpreted by the classic phenomenology as the vehicle of the exclusively *logically rational structurization*, its operations following the outline, preestablished regulations and principles of the 'passive', that is, effortless and involuntary unfolding of the originary experience.[6] Its regulative principles, its aim, will be treated later on. The question arises: "How could the constitutive system predelineated by universally implemented rules, regulations and principles – the implementation of which is automated in ever repeated routine operations – account for the emergence of activities which are novel and different from those of the routine constitution?" How could the constitutive genetic system, as understood by the classic phenomenology account for the origin of entities, obvious fruits of man's conscious efforts which are *original* and *unprecedented* with respect to the established world? That is, how could the constitutive trend of our functioning account for the origin of novel structures breaking into the frame of the constituted world, transforming it *down to its sources*: the very quality of the originary experience? In other words, we submit that the constitutive routine cannot account for works of *human creativity*.

It is by contrast with the creative activity of man that the preceding denunciation of the hidden pre-suppositions of constitution comes to light. In fact, as I have shown elsewhere, the creative structurizing activity in its phenomenologically inspectable stage, is set off also by perception.[7] Even more the *creative impulse* which triggers the spontaneity of the creative process starts upon the foundation of the *already constituted world*. However, it starts by a dual attitude towards it: first, it assumes synthetically the existence of the constituted world and its present state; secondly, *it revolts against it*. The creative activity emerges *denying the validity to the constituted world* to be, in its given present state, an adequate interpreter of human reality. It is the revolt against the routine, the automatism and the inertly accepted rules of the constitutive system we passively flow with, and the involuntary submission to it which they call for; they are the target of the revolt of the *creative impulse* which tends to *transcend* this imposed, involuntary, anonymous trend of nature. To do it, it must break it first, then, after having within a self devised entity structurized a novel interpretation of reality, make it disrupt the regular channels of the constitutive world-process by making it intrude upon the constituted world and take foothold within it. Thus, the *creative process* stays in a twofold relation to the constitutive one: it situates itself *between the two different temporal phases of the constituted world by stemming from it and breaking off, then, breaking back into it in order to transform it.*

Indeed, the *creative perception*, as the decisive incipient point of the *creative process* reposes *first*, upon the already present constituted *life-world of the creative agent*. Preceded by unaccountable *subjective dealings* of the creator-to-be with the constituted reality, it actually triggers the creative activity by producing a new opening of sight within the fixated reality. Whether it is in an immanent perceptual vision in the case of poetry, science, fiction, or in the transcendent perception in the plastic arts, the inherited and passively constructed forms and affective contents of experience become repudiated. All our operational virtualities partly extracted from their chains in the constitutive system, become galvanized and set off in an effort towards reconstruction of the Real.

However, to show in which sense the 'passivity' of constitutive automatization is limited in its role and how the creative activity establishes itself in a contrasting *voluntary* and *free* agent, we must reach back to their common ground; its genesis as a specific function *orchestrating anew all the productive operative virtualities of man into a framework of the creative context* and revindicating the neglected role of the impulsive and affective realm of passions. (We speak here about 'creativity' in the precise and restrictive sense when the 'creative process' terminates in an aimed at, concrete and intersubjectively accessible 'product'.)

III. THE GENESIS OF THE CREATIVE FUNCTION: THE CREATIVE CONTEXT, ITS FRAMEWORK

Differentiation of the Two Functions with Respect to the Axiological Opposites: Voluntary, Involuntary

In my work, *Eros and Logos*, I have shown in the analysis of poetic creativity how the *creative impulse* surging from the rationally unattainable realm of man's dealing with his natural strivings, nostalgia and drives confronting the limited and imposed framework of reality, and tending to break through towards its more adequate version, releases all the presently knotted ties and opens the way for a new structurizing system to take over.

The question arises, however, what would be the ways and means of the *creative impulse*, that it could stimulate and bring about a new functional system? Furthermore, what would be the specific, radically different mode of this new functional agency, such that it could break away from the routine *life-world* constitution, free from the preestablished models, towards the invention of new patterns and types? Finally we must wonder whether phenomenological analysis may uncover these hidden springs of our virtualities and establish rationally the distinctive features of their dynamic flux that eludes the structures of reason.

In fact, in order to answer this last but crucial question in the affirmative we propose to outline the analytically established notion of the *'creative context'*. Sketching its role, origin and ways of operation we hope to outline the answer to the other questions as well. It will appear that the further distinction between the constitutive and creative functions lies along the radical divergence in their modalities opposing the *creative* 'voluntary' mode of operations to the constitutive 'involuntary'.

A. *The Outline of the Creative Process as the Fundamental Dynamism of the Creative Context*

The yet undetermined *creative impulse* finding the appropriate response in our functional virtualities, establishes itself searchingly in its own mode of operation; simultaneously it calls into active commitment a network of particular functional operations. In their consistent and purposeful orientation as they bring together the major operational arteries of man, these sub-sets of functional dynamics constitute the framework of creative operations: *the creative context*. The creative context plays the role comparable to a weaving loom, upon which the novel reality is going to be woven.

Starting, as I have already indicated, at the level of what is called 'originary experience' that is of perception, namely, *'creative perception'* it is at first there that the primordial ties between the affective qualities and forms, their productive agencies being cut from their constitutive pattern, enter into a new role for which they coalesce into a new pattern, a new operating scheme. The *creative context* taking from there on binds together, firstly, the constitutive operations of man, but going underneath the delineaments of logical structuration, it binds the elemental factors of passions, the impulsive, the emotive, the nostalgic strivings, the sensory and imaginative operational elements – this, together with their dynamic reservoirs, as well as the engendering forces which otherwise have been instrumental in the constitutive operations. The emerging creative agency determines itself in its distinctiveness having grasped hold of them at the crucial point of the very germinal origin of the *life-world*, where the form and lived quality of human reality is being decided, after having cut their ties with the constitutive system of structurization to which they have been blindly committed, binding them into a new type of orchestration. As the constitutive genesis aims at the structurization of an intentional – through the anticipation-fulfillment structure of its acts, already pregiven – object, so this new *orchestration* of man's functioning – through the mode of which the creative act, projects tentatively its delineament and simultaneously determines its new framework as well as its mode of operations – is oriented towards a novel, its *very own*, aim. But in opposition to the *constitution* of an intentional object, which is in advance pretraced within the mode of the constitutive process itself, this aim of the *creative* process, e.g., the work of art, of poetry, a scientific theory, etc., is not pre-traced by the intrinsic nature of the acts but the *object of a quest*. It is the entire creative process that carries on this quest, measuring up simultaneously with the revolutionary renovating tendency of the *creative impulse* and the already present human world which it is meant to transform. Concretely, it aims to discover its 'final form' and to incorporate it into an intersubjectively valid entity. To introduce it into the *life-world* means to break with its acquired and already sclerotized form for the sake of a novel interpretation of the Real. It is the task of the unfolding, ever searching and questioning, creative process. The *creative process* initiated in the process of *creative perception* comprises the following stages:

Beginning by the 'creative vision', which marks the representational aspect of the *creative impulse*, it advances through several stages of search after the most appropriate way of translating this *vision* into a newly devised 'original intentional object' and terminates in the 'phase of transition' that is the building stage in which this 'merely intentional' object is concretized into an intersubjective entity entering the already constituted human reality. Throughout its complete progress, the creative process – in radical contradistinction to the preestablished passive flow of constitutive genesis – is a questioning, a search, a moving on untrodden paths, a deliberation and choice among myriads of ever expanding possibilities; it is a search, not only after its construction and building elements but also after criteria of their choice and those of its own operating regulations.

The lineaments of this novel *orchestration* of human functions, which in its dynamism, progresses in a flexible state ready to be unbound and bound anew by different ties, is oriented toward a different mode of operations in a two-way traffic with the *creative project*, once indicating its transformations, then being transformed by its intrinsic demands to keep all the structurizing operations on a constant alert.

Then, this *creative system of functions* spread in the polarity of the *creative impulse* and the *creative orchestration* of man's operative virtualities and forces into the *creative context*, consolidates itself in its progress into a complete, self-oriented and self-propelling distinctive and autonomous, but above all, *self-regulating* (although dependent upon the constitutive function and the established world) *agency*. An agency which, in contradistinction to the *involuntary* progress of constitutive genesis is an agency of *will par excellence*, this agency, neither mystical nor mysterious entity within man, will emerge as none other than the consciousness in a new structurizing function: *creative function*.

Consciousness in the *creative function* is no more the submissive engineer fulfilling in an unquestioning way routine work, nor an administrator of the given; it is, as if a man 'awakened from a dogmatic slumber' discovers himself to be an inventor, or a *demiurgos* projecting the world he constructs. What are the final regulative principles of its orchestration which it follows in its projecting? Are they predetermined or 'freely' chosen? What would be the meaning of 'free choice' as distinct from the selection and deliberation, followed by the constitutive genesis which accounts for error and correction, and, according to Husserl guaranteed by the 'free play of imagination'? "What is their nature and source such that an ever new material for selection is present as well as appropriate new criteria for their choice?"

With these questions we anticipate two crucial points of every argument

underlying the attempt at the philosophical account for the human world
in terms of the human role in its origin: first, the question of the *a priori*,
second, that of the distribution of roles among human faculties, especially
the role of imagination. But before we enter into a philosophical controversy concerning it, we must indicate the functional extent of the
creative context within which alone, as we submit, the 'creative imagination' may emerge, so that we prepare our own ground for its treatment.

B. *Revindication of Passions and of the Elemental Nature in Man Within
the Creative Context*

Indeed, in its progress the *creative function* is a mechanism of discrimination, deliberation and selection. But so is the passive genesis. However,
in opposition to this latter which works with the material of rational
elements of structurization and their selective principles, the mechanism
of choice of the *creative function* is constantly oriented and reoriented
in its modalities by fluctuating inclinations, tendencies, expectations,
aspirations, hidden longings, aversions and sympathies; all of them
whether they are dispositional or acquired within the present world
(being matters of feelings, taste, belief), they seem to escape the authority
of our rational powers. The selective mechanism of the constitutive
function works chiefly at the level of rational structural elements, whereas
the *creative deliberation* seems to plunge with its roots into the deep well
of *passions*. They are affecting the deliberating function by their inner
workings within all the *elemental* forces and operational virtualities and
at all stages of our functioning. At the crucial creative stage, namely, that
of the incipient impulsive affectivity and sensibility which enter into an
interplay with the universe of ideas, this interplay stimulates and
galvanizes the *creative orchestration* into which it spreads; in return
entering into the generating scheme of its vast and unlimited range of
optional associations, it becomes the ground of the *creative renewal*.

It appears that, first, the mute upsurges of passions: of disquiet, pain,
fear, anguish, excitement and the dumb impulses of irritation, anger,
elation, the impulse to hit, to run, to embrace – which are hardly an experience, nothing reveals itself in them, nothing is given and brought to
light – are as the extreme edge of our psyche, registering mechanically the
processes of nature; at this level, the creative function lays its roots.

In the *constitutive process*, the mute operations of our functional endowment appear already as having reached the full extent of the structurizing and presentational proficiency in producing a perfectly intelligible perceptual object. As Plato already saw, there is an interval separating the one from the other; during this interval the instantaneous and dumb reaction travels the long road of the sensory-motor operations, of the proto-constitution of the primitive forms of spaciality and corporality projected in the perspectives of the past and the horizon of the possible futurities; all that with reference to the already established system of the constituted world. There is a necessary and spontaneous assimilation of each experience to this established whole. Thereby the lived quality of constituted experience goes through the filter of structurizing processes and is molded by them in such a way that, as Leibniz has so well expressed, 'each of them mirrors the whole world'.

On the contrary, the *creative impulse* rejecting the preestablished pattern and cutting of the ties among the functional agencies – the rules and points of reference of the organizing sovereign reason – it is upon the inarticulate movements of passions, impulses, moods and strivings that the emphasis falls. The *creative function* solicits their dynamism by projecting a double quest: *firstly*, it seeks after the interpretative points of reference for this natural striving operating within us, mute and yet most *significant* for the human being; *secondly*, it attempts to discover or invent the significance which this dynamic upsurge of *elemental* passions may have for the reinterpretation of reality it is originally asking for.

In the first movement, *Elemental Nature* lends itself to the meaning giving reason and is worked to raise from its anonymous impersonal status to become 'interiorized' in reflection; in the second, inverse movement, it is intellectual intuition, which interrogating, goes into Nature present in man, being molded through its whole range down to the most elementary operations.

It is in response to this double quest that a new *orchestration* arises: the operational virtualities of consciousness and of all our functions to their full extent, freed from their preestablished ties enter new ones establishing a new organizational pattern. Within its flexible frame, which we have called previously, 'the creative context', a new alliance is struck between the opaque, mute, 'bodily' functions – of which otherwise we receive only a faint echo through our impulses and affective glimpses – and the highest specialization of man's tendencies to organize his world: intellectual

intuitions, ideas, reasoning and speculative powers; a new alliance in the
infinite modalities of the transparent *Logos*.

The *creative context* is suspended upon the new alliance among all the
diverse functional dimensions of man.

IV. IMAGINATIO CREATRIX

Differentiation of the Two Functions with Respect to the Regulative
Principles: Transcendental 'A Priori', 'Creative Freedom'

The conception of human 'functional system' proposed as the answer to
the difficulties present in the classic controversy about the distribution of
roles among human 'faculties' in the origin of the human world.

A. *The Controversy Concerning the Role of Faculties: Imagination in*
Husserl and Kant

As we have already indicated at the outset, it seems that if both Husserl
and Kant failed in giving an adequate philosophical account of the origin
of the human world concerning man's role in it, it is chiefly because they
did not take as the point of departure a view of the world that was to be
accounted for in its nature, but narrowed their approach to one single
issue, namely that of the origin of science. Consequently, establishing
the system of consciousness from the bias of what made scientific knowledge possible, they have overlooked the specific distinctiveness of the
creative activities of man. Husserl, as we will attempt to sketch briefly,
does not seem to reach this issue. Kant, who takes it up in the *Critique*
of Judgment after the world constitution is in the main already established
in the first *Critique*, tries to fit in and falls short in accounting for the originality and novelty of artistic creation.

To bring to light the intricacies in their treatment of the matter, traced
from the initial assumptions as the reasons for their inadequacies, will
lay down a basic network of problems involved in the issue of *creative*
originality – into which we will then fit our own proposed solution.

The situation is indeed complex. If we place ourselves at the point of
view of the analysis of the *creative process* there is a striking resemblance
between the *creative process* in art and the discovery process in science:
they both share the essential element of 'invention'. Furthermore, the
creative process of a work of art and the process in which a scientific

theory is formed exhibit almost identical mechanisms. However, Kant
has already raised rightly the question of their distinctiveness insofar as
the *essence* of their products is concerned. He saw the *essence* of the work
of art as depending solely upon the working of the human *subject*, whereas the *essence* of the scientific discovery would depend upon the data of
nature. The scientific discovery is taken by both Kant and Husserl as
belonging to the higher operations of the constitution of the world, but
Kant raises the issue of the distinction of faculties upon which we would
have to call to explain the radical differences in the *scientific discovery*
and *artistic creativity*.

In order to explain how the *subjective* talent alone establishes the laws
for the *essence* of Art, whereas science aims at the discovery and explanation of the laws of nature within the framework of the constitutive scheme
established by pure reason, Kant calls upon the specific role of imagination (*Einbildungskraft*). Thereby, with the question of the origin of the
world and the role of human creativity in it, we enter into a controversy
about the distinctiveness of various human faculties and their specific
roles in it.

In fact, although mostly only lurking from the shadow, imagination
stays in the center of the problem of cognition. We find it, for instance,
in Descartes' analysis of the piece of wax, that seeking to discover the
reasons of its independent identity which persists through the changing
aspects of cognition, he has recourse to imagination. Indeed, he observes
an infinity of possible ways in which the piece of wax may vary in perception.

Only after deciding that imagination could not, on its own as a faculty,
bring forth to the mind this infinite series of possible variations of the wax
in perception, and that, consequently, the conception of the identical
piece of wax cannot proceed from the workings of imagination, that
Descartes feels compelled to acknowledge the reality of extension, or of
contingency.[8] We find here imagination brought in the case of sensuous
perception and its 'reality'; indeed, by and large, the conception of imagination appears bound with this of the perceptual experience and the question of empirical reality.

In Husserl this issue receives a particularly well-developed treatment.
However, introducing the notion of imagination at the same point as
Descartes, namely at the point of the *necessary conditions of the running*

off of the perceptual process in its *flexibility as well as in its various modes of presentification, remembrance,* etc. he sees its role at the opposite extreme to Descartes. In fact, imagination in its several forms assumed in Husserl's complete intentional apparatus, as representing the essential network of the complete functioning of man, is attributed an important role. Although in principle the conception of man's operations to be grasped within one network of conscious acts dismisses the classic distinction of faculties – the specific operational organization of fundamental intentionality seems to take care of all the aspects of constitution by its own mechanisms – and yet within this unified system, the role of imagination as the factor of modal modification of constitutive operations is quite distinctive.

In fact, within the transcendental analysis of classic phenomenology, imagination appears as a modality of the intentional acts, which having seemingly only a subsidiary role of complementing the *positional* consciousness, is in fact indispensable to make its operations possible. Its features, role, scope and prerogatives appear within the complete analytic framework as participating integrally in the way in which the constitutive system is grasped and its problems formulated, and can be understood only from within the complete system, down to its basic assumptions.

We may distinguish mainly two roles of imagination or 'phantasy' operating within the Husserlian constitutive system. First, phantasy functions as 'neutralization' of the positional acts, and belongs 'ideally' to every modus of *actual* consciousness, e.g., perception, belief, will, etc., as its *possible* counterpart; it corresponds to it exactly by its objective content, but modally presents its 'neutrality modification', that is void of any reference to its existential status or any claim to a distinctive *objective* position: the objective content which the images exhibit, refers according to Husserl's observation, in its essence, to the content of some act of *actual* consciousness.[9]

Secondly, fantasy as the reservoir of possibles corresponding to each actually being posited content of experience, allows for the 'free variation' operating within the process of every experience, starting by – as we have seen in Descartes – the positional act of perception allowing for its flexibility necessary to change, modification and correction of its course without losing continuity of the general constitutive progress of consciousness.[10]

Concerning its first role, imagination – in the strong sense of 'pure fantasy' – is seen as performing the role of expanding the horizon of reality by introducing the richness of infinite variety of the '*possible*' to the otherwise automatically restrictive constitutive genesis of the 'real'. Does this mean that it could, and is meant to, account also for originality and novelty breaking into the constitutive scheme? If Husserl himself does not see this role for imagination, what are his reasons? The question confronting the prerogatives and capacities of imagination can be formulated: "What is the distribution of roles among functions on Husserl's analysis, such that imagination had to appear deprived of the positional prerogative?" "Could imagination be the factor of originality and novelty, being deprived of the power to posit its objects?" and finally: "What conditions should imagination fulfill to be able to bring about self-given data?" These questions bring us to the scrutiny of its source.

Concerning the second role attributed by Husserl to imagination, namely, that to guarantee the flexibility of deliberation and expansive variety of progress in the constitutive processes, meaning 'freedom' from a rigid determination, we must ask the question about the nature of this 'freedom' and whether it would suffice to guarantee, not only the individual variation within the prescribed types, but also the original data of the radical *renovation of types*. With the issue of freedom, we will pass to Kant's cognitive network, the role he attributes to imagination, and the crucial metaphysical question of the *a priori*.

Ultimately we will attempt to show how the question of the selfgiveness of the data of imagination is answered by a new analysis of its functional role, aims and conditions.

As in Locke, in Husserlian analysis of consciousness, perception is recognized as the basis and starting point of all cognition. With reference to the perceptual temporal field within which the object of perception as the identical pole in which all the acts converge, is followed, Husserl distinguishes three essential modes of perceptual consciousness: (1) sensation ("*Empfindung*") as the means of the direct presentification (*Gegenwärtigung*) of the object in selfgiveness with retention following it immediately, (2) positional representation in remembrance (*Erinnerung*), and (3) representation in reproduction as imagination or 'pure fancy'. In fact, as we all remember, the notion of imagination or fantasy in general occurs in the Husserlian transcendental analysis within the framework of the analysis of the perceptual field in the stream of constitutive genesis; and specifically it is considered in relation to the modifications of the modes of presentification of the cognitive object through the different phases of temporal run off that make up the cycle of the perceptual genesis. Let us recall that, in principle, the object of perception in the Husserlian analysis of the constructive system of the *life-world* – the operations of which structures, organization and source phenomenology proposes to have established within the dynamic process of conscious constitution – is seen as a profile within a continuous time-conditioned series of acts that springs forth in the mode of the originary giveness of perception as its incipient moment, and with the perceived, as its initiating datum, passes into *reproductive* stages which terminate with the modifications of *pure fantasy*. Within this cycle of the perceptual spread, a radical distinction is drawn on the one extreme between the positional modality of selfgiveness of the object *in perception*, which then appears as the only source of selfgiveness and thereby the 'original experience', on the other extreme, the non-positional modality of imagination or 'pure fancy'. The modes of giveness of the perceptual object are, in point of fact, within the whole genetic cycle of its constitution, or as Husserl otherwise calls it, within the complete 'temporal field' which it covers, differentiated firstly, between those of direct presentification (*Gegenwärtigung*) of the object and its representation (*Vergegenwärtigung*).[11] While only the 'impressional consciousness' in the actual perceptual progress – which is continued in retention or 'primary remembrance' its 'comet's tail' – brings the temporal object in its actual presence, directly as self-given into the actual field of consciousness, the appearance of the object in consciousness does not terminate there – had this been the case, there would be no continuous, but only discreet world structure on the one hand; on the other, we might then have been deprived of the whole horizon of the intellect and be restricted to a purely animal vegetation – but continues in manifold reverberations; however no more in the mode of actual presence, but merely in various modalities of indirect representation and reproduction of the object originally given in perception. There is also a radical distinction between retention (primary remembrance), which actually continues the perceptual giveness, and representation (*Wiedererinnerung*) in the 'secondary remembrance'. 'The last one', writes Husserl, 'is the contrary of the originary giving act, no kind of the object may emerge from it'.[12] The reason for this distinction is that perception and retention, which continues it, are built upon sensation (*Empfindung*), whereas, to indicate the specific *hyletic* nature of the *reproductive acts* Husserl introduces the notion of 'phantasma'. Secondly, although situated within the modes of representation (*Vergegenwärtigung*), imagination in the strong sense, that is 'pure fancy', is still radically differentiated from remembrance in its triple form (*Erinnerung, Mitvergegenwärtigung* and *Erwartung*). In spite of the fact that all these *modi* are reproduced also within the form of fancy, 'pure fancy' such as presents a specific, basically *different modus of consciousness, which stands out as its complementary counterpart to originary actual mode of consciousness at large.* It is 'complementary' in the sense that, through its basically different status, it is in Husserl's analysis attributed a degree of *freedom* from the iron chains of the spontaneous unfolding of the originary temporal progress: thereby it is seen as guaranteeing for this progress, firstly, a spectrum of objective choice in structural selection, secondly, a flexibility in the structural operations.

Does it mean, however, that 'pure fancy' enjoys an independent status with respect to the constitutive spontaneity regulated by the ideas? Before this critical question could be treated, the status of imagination within the complete constitutive scheme has to be clarified.

In point of fact, as was already mentioned, imagination is not 'productive' of its objective content but 'reproductive' with reference to the perceptual process, which *alone* is productive. Fancy is, as Husserl insists, incapable of bringing about to givenness, an object or any of its features. "The essence of fantasy", writes Husserl, "is precisely not to give its object in its selfhood."[13] The other above mentioned modes of representation do not either present their object in its selfhood, *nevertheless* – and here the radical demarcation between the two basic, in this respect, modes of consciousness is to be drawn – they perform what Husserl calls, 'positional representation' (*setzende Vergegenwärtigung*) insofar as they reproduce the content of actual experience and *posit it as an integral part of the actual stream of consciousness* at the place which is reserved for it within the complete specific temporal field; the field within which the whole cycle of the concrete perceptual genesis of the object unfolds. On the contrary, pure fantasy reproduces the content of the actual originary acts but does not posit it in *any way* within *any* temporal field of the actual stream of consciousness; the reproduction of the object remains outside of the temporal actuality of the advancing originary genesis – which is not only the source of the *life-world* but its groundwork and foothold – it has no hold on this solid ground, it 'floats' (*es schwebt*).

If we consider that the 'selfhood' of the presentification within the field of consciousness on the *noematic* side and the 'positional' mode of operations on the *noetic* side in their alliance in the perceptual giveness – which are for Husserl the criteria of such a radical distinction as this between producing and reproducing consicousness – stem from the fascination which he seems to share with Locke by the overwhelming vividness of the immediate giveness in sensory, as well as immanent, perception – in the Lockean language of simple ideas – the division of conscious operations in 'productive' and 'reproductive' and the following 'reproduction' of reality in imagination appears as a downgrading of imagination and as a mere predecision on the issue of our concern. In point of fact, in order to bring to light the uniqueness of this type of experience, Locke distinguishes between the sensory giveness in general and imagination, and emphasizes that the type of experience which brings the simple ideas of the senses about may result exclusively from sensory perception; imagination is not in a position to provide such basic building blocks of cognition. Thereby, on the one hand, is denied to imagination the capacity to contribute any *new* material to the hence unfolding constitutive process. On the other hand, the role of originating the whole cognitive-constitutive system, attributed by Locke to sensory perception – as distinguishing it from all the conscious operations that follow it – is sharpened, first by his statement that ultimately the simple ideas of reflection (in Husserl's rendering 'immanent perception') derive from the observation of the cognitive operations of the sensory perception, and secondly, by Locke's blank statement that simple ideas can neither be destroyed nor produced by any means. In direct parallel, Husserl, although with considerably more nuance in analysis, also restricts the constitution to the one unique source: the *impressional consciousness* at the incipient stage of perception.

Husserl's analysis of the work of art falls perfectly in line with this general tendency in his analysis. In *Ideas I* the figures of the devil, knight, etc., represented in an etching by Dürer, are given as an example of the 'reproduction' of the actually given in perception.[14] That is, the objective

content of the work of art is in general seen by Husserl not as a 'presentification' but a 'representation', a 'reproduction' referring to the object of
actual perception for the original model. As work of 'pure fancy' it stays
in a different mode of consciousness than the object to which it refers;
instead of the positional consciousness – with either its presentation of
the object in bodily selfhood, as in actual perception, or its modified
representation posited by secondary remembrance – it has no positional
status with respect to its existence. The figures represented in the etching,
works of pure fancy, are neither 'being' nor 'non-being'. As works of
fancy they lie outside of the modality of being; they are what Husserl
calls 'neutralized modifications' of the actually given objects of experience. Pure fancy is seen as the modification of the given, neutralizing its
positional representation, that is, 'remembrance in the largest thinkable
sense'.[15] The works of fancy (neither 'being' nor 'non-being') 'float', and
yet they maintain themselves as conscious operations and products; they
lack – in the sensation (*Empfindung*), the very stuff of the temporal stream
of consciousness – the hyletic support of actual consciousness, but they
have it in 'phantasma'. The grounding in phantasma instead of sensation
allows imagination its relative independence from the actual stream of
genesis and frees its workings from the iron network of spontaneous
unfolding. We will investigate now whether it may open the way for
originality and novelty within constitutive constructivism, and what is
the meaning of its 'freedom'.

Indeed, upon this analysis seen as a neutralizing modification of positional consciousness, the work of fancy encompasses the whole realm of
conscious operations in their dual aspect: to each instance of experience,
as well as to every originally given individual, is attached a series of ideally
possible modifications in remembrance and of parallel neutralizing modifications of fancy. Therefore, Husserl may attribute to fancy a 'universal
significance'; as an ideal horizon of possible variations attached to each
actual originary giveness it is entrusted to play an important role in the
workings of positional consciousness: it introduces 'freedom'.

This 'freedom' is not limited to its role as allowing the flexibility in the
selection of particulars within a given ideal type. Being outside of the
spontaneous field of actual experience, fancy in its way of modifying the
given does not suffer the coercion of all the components of the given
temporal field – no actual act is really individual, all of them being in

their unfolding intertwined within a given temporal complex. 'Free' from the network of the actual field, fancy has room left for diverging from the given while transforming it in its reproductions. Furthermore, the running off of its productive process outside of the main stream, means not to be restrained by it; the process of reproduction may take different speeds, scanning selections, repetitions, different intensities of representation, of clarity, emphasis upon the different elements.... This 'free' run-off allows for incalculable variations in the divergence from the originary data it imitates.

In this understanding there is a considerable margin of qualitative transformability possible and Husserl might well believe to have, in this way, taken care of artistic creativity, especially since the artistic activity remains in his analysis misunderstood. However, it is obvious that such a role of imagination is relevant only to the problem of imitation in the issue of the relation between reality and art in *representational* art. The whole complex of questions concerning the *nature of art* is left out of Husserl's analysis. Does the possibility which the relative 'freedom' of fancy guarantees by the 'relative' margin of individual variation within the same type account for the *radical originality of types* which we see particularly strongly exhibited in *abstract art*?

Indeed, the whole spectrum of modification, with all its incalculable nuances possible, that fancy may expand in its transformations of the given object remains within the framework of the standard constitution as the imitation of the originally given; it is moving within the framework of its structure-type, which as we have established elsewhere, is, in the constitutive process accomplished with reference to the regulative role of ideas. *Ideas prescribe the reach of individual variations within the distinctivity of a type.* (Of course, there is also a possibility of a marginal modification within the content of a given idea, yet to assume of the essential transformability of ideas, thus of a continuous passing of one pattern of order into another, would ultimately imply a universal chaos, which is contrary to the experience of the world and man we have.) It might well be Husserl's – like Locke's – fascination with the enhancing quality of sensory perception, experience *par excellence* that made him identify selfgiveness as a criterion of productive and experiential validity and to restrict its spread to the selfgiveness of constitutive perception (transcendent and immanent) alone. Whereas, as we have shown elsewhere,[16] there might well be another typical instance of selfgiveness which is – in fact – an indispensable condition for the emergence of novel and original elements within the *life-world* – in another type of perception than that accomplished in the passive genesis. Oriented in his investigations of consciousness by his tracking blindly the genetic unfolding of the passive genesis, Husserl is bound to overlook it. The question "Is there only one source of selfgiveness?" emerges at this point and outlines the further progress of our investigation.

However, pursuing the source and nature of *origins*, Husserl himself is well aware of the limits of the repetitive constitutive system of consciousness. He sees the 'productive borderline' of consciousness in the 'originary impression' occurring as the component of the 'impressional consciousness', the basis of selfgiveness in actual experience. Whereas the productive unfolding of the spontaneous consciousness goes on from the impression into its serial modifications, the instances of impression already belonging to the productive system are themselves grounded in the 'originary impression' (*Urimpression*) that itself remains foreign to the constitutive consciousness; is not produced but found there as alien to it, 'radically new', 'the original product'; as such it is the ultimate originary source of the conscious productivity but also sets its radical limit.[17] Only the data of the 'originary impression' are radically 'new'. The *'Urimpression'* is then seen as the radical beginning which predelineates the total scope of the constitutive unfolding. Constitutive spontaneity may add to it nothing new.[18]

Having thus scanned with Husserl the major linking points of the constitutive system, we may state not only that, on this analysis, imagination bound to the scheme of the passive genesis and subject to its regulations could not account for the novelty and originality erupting into the *life-world*, but that the constitutive system itself, understood as rigidly predelineated by the incipient stage and repetitious, could not offer any other instance explaining it.

Husserl's recognition of the radical borderline between what the operational system of the transcendental consciousness receives, as radically alien to it and new, and what it itself produces – between 'impressional consciousness' with the notion of 'originary impression' and the productive flow of passive genesis taking from there on – is not without an analogy with Kant's distinction made between the 'manifold of experience' that expresses the state of consciousness being impressed upon (*affiziert*) by what it is not, and what it receives from without, and the complete aesthetic system of operations that works upon, forms it, but does not bring to it anything novel.

And yet Kant has reached deeper into the nature of the issue. He made a clear distinction between the type of 'originality' and 'novelty' of what man 'receives', with what he initiates his basic routine constitutive operations and 'originality' and 'novelty' as *breaking into* the routine functioning of man and which is of man's very own doing. As we have pointed out earlier in this argument, he has distinguished between the specific origins of *nature* and those of *Art*. Having thus diversified the issue, it is again in imagination that he sees the factor of 'freedom' from the pre-delineated routine functioning. Within Kant's scheme, imagination is attributed, in fact, the role of an *originary and decisive faculty*. May imagination, however, understood as an originary and independent faculty within the scheme of human functioning, fulfill the conditions indispensable for the role of promoting within the workings of this system, radical originality and novelty?

We submit that imagination (*Einbildungskraft*) is truly the source, the *dynamis* and the engine of the Kantian account for the origin of the human world. In point of fact, imagination performs in Kant's complete account both of the crucial roles. In cognition it is 'reproductive' and 'mediating' – although at a pre-perceptual level and not at the post-perceptual level as in Husserl – thus enabling the constitutive process and the routine world construction at its turning point in man's creative endeavor, but as a factor of the *elemental nature* of man it is the factor of 'radical freedom', freedom from all constraint.

Kant precedes Husserl not only in the notion of 'impressional consciousness' seen as the incipient stage of all cognition, but also, as we know, in the rigid automatism and repetitiveness of the transcendental system of cognition (it is obvious that in Kant's approach 'cognition' is as much 'constitutive' of the *life-world* – although not of the subject – as in Husserl's); moreover, what should be brought out, both of them, at first overwhelmed by the translucid rationality of the intellect, leave the opaque empiria in the dark background, then, having covered the empire of reason, are prompted to retrieve it and to revindicate its full rights. As such a radical overturn, we may consider the fact that imagination, which within the realm of cognition enjoys in Kant's scheme even less of 'freedom' than in Husserl's, in the realm of artistic creativity as *Ein-bildungskraft*, is attributed the 'radical' freedom of *elemental Nature* as a 'blind force of the soul'. Indeed, to account for the uniqueness and exemplariness of the work of art it is seen as an otherwise unbound power working through an agency not only distinct but separated from the cognitive faculties and only referring to them to check its work. Can imagination as an independent faculty and agent perform this role? Could the opaque realm of feeling and emotion evoked by the soul, represented by an undifferentiated agent of imagination, account for the uniqueness of a *novel* construct, which in order to be novel, has to count with the incalculable variation of *differentiated forms of constitution already present*? To investigate the situation, the inquiry into the role of imagination in cognition comes first.

In fact, when Kant summarizes his cognitive system in the introduction to the transcendental logic in the famous opening,

Unsere Erkenntnis entspringt aus zwei Grundquellen des Gemüts, deren die erste ist, die Vorstellungen zu empfangen | die Rezeptivität der Eindrücke, | die zweite das Vermögen, durch diese Vorstellungen einen Gegenstand zu erkennen | Spontaneität der Begriffe; | durch die erste wird uns ein Gegenstand gegeben, durch die zweite wird dieser im Verhältnis auf jene Vorstellungen | als blosse Bestimmung des Gemüts | gedacht.[19]

we are still missing the crucial link between the two: imagination – *Ein-bildungskraft*. Let us recall that the giveness of an object in Kantian constitution is chiefly dependent upon the synthetic organization of the various 'representations' (*Vorstellungen*) which are to be brought into the unity of *one* consciousness. Whereas this synthetic unity expressed in concepts is the function of the reason (*Verstand*), yet the possibility of the synthesis as such (*Synthesis überhaupt*) is attributed to a special 'blind', although indispensable function of the soul, without which we would have no cognition whatsoever,[20] in spite of the fact that we are only seldom aware of it. Thus a special, although apparently subsidiary, source or faculty is introduced between the sensibility with its pure forms of spatial and temporal extension and reason with its system of intelligible connections and categories. It is understood as a spontaneously acting agency upon which the operation of the complete cognitive-constitutive organism relies. Within the organizational network of this organism, emphasis is distributed between imagination and the identity of the ab-

stract I, present in all operations and bringing the dispersed and originally chaotic manifold into the synthesis of the *unity of apperception*. Indeed,
both levels of constitutive operations in Kant's conception, the giveness
and the intelligible synthesis, refer foremostly and directly to this necessary point. Necessary yet insufficient, since another type of synthesis is
indispensable. Kant is at every stage of his reconstructing of the cognitive system perfectly aware that he distills only a rational skeleton of
the complete operational system which has also a thick psychologico-
empirical dimension, and gives the neat priority to the rational network,
explaining that the empirical dimension alone would not suffice for the
giveness of objects of cognition, the order of the world and intelligibility
as such. However, although imagination shares in the way of exposition
this background position within the cognitive realm, will it in the way of
real constructive significance remain in this neglect?

Although, as Kant indicates, 'objects' could be given without being
thought, yet at the empirical level alone we would hardly have cognition,
for which Kant assumes the necessity of an intelligible form. Identifying
the type of connections which enter into the intrinsic structure of the
constituted objects of cognition with those of logical judgment, he de-
duces, following Aristotle, from the logical function, a series of universal
concepts, 'categories', to preside over the synthesizing activity of the intellect in the final organization of the manifold of sensibility. Kant concedes
that the categories as 'pure forms of thought' (*blosse Gedankenformen*)
cannot apply directly in their universality to the sensuous manifold
(*Gegenstände der Sinne*).[21] The 'figurative synthesis' (*Synthesis speziosa,
figürlich*), which organizes according to his view, the unity of apperception, needs a faculty to mediate between the sensuality and intellect. But
it is more than a bridge between the transcendental aesthetics and transcendental analytic: its role is not to be a structural link but an *operational function* allowing for the continuity in the structurizing progress.
That is, the role of imagination stretches from the 'impressional' level of
the received manifold of senses due to the capacity of being affected,
through the 'positing' of the received in forms of time and space (sensuous forms), towards their establishment into the final objective forms
of giveness with reference to the categories.

This dynamic role of imagination underneath the network of structurizing activity of sensibility and intellect – which conducts their opera-

tions – starts already in the most concrete empirical dimension and runs through the transcendental one. To be precise: the first that is *empirically* given, as Kant sees it, is 'appearance' (*Erscheinung*), which if 'conscious' is called 'perception' (*Wahrnehmung*). But 'appearance' is a compound of a manifold of elements and the 'content' of 'perception' is chaotic and disconnected. (What Kant, viewing the perceptual process 'all at once', considers as the 'manifold of elements', corresponds to what Husserl, viewing perception as the genetic temporal spread, sees as 'individual glimpses' (*Abschattungen*), each of them contributing to the constitution of the object of perception, but none containing it on its own.) Left at this empirical level, experience would not amount to the constitution of an organized *life-world*. Towards the possible organization of the manifold within one and the same appearance and then the interconnection of disconnected empirical perceptual glimpses, which otherwise cannot coalesce towards the fullfledged perception 'positing' an object, we need a synthesizing organization of this manifold. This is the the first function attributed to imagination; it is meant to bring the manifold of perceptual glimpses into one 'picture' (*Bild*). But since the individual elements of the manifold appear without exhibiting any coalescing principle among themselves – a haphazard and heterogeneous variety – a basis for their synthesis has to be established. That is, imagination has, first of all, to bring about such a common denominator in order that her own activity, that to establish their unity of a 'picture', may be accomplished. According to Kant, it happens in a twofold way. Firstly, imagination 're-produces' for her own purpose each of the heteroclitic elements, so that it reflects them all together in one modality, namely that of 'apprehension'; secondly, by 'association' it operates the organization of these empirical representations.[22]

And yet, the associative synthesis of the empirical alone could not suffice for the structural construction of the perceptual object; it introduces merely at random, accidental connections, whereas a universally valid combination of elements is required, such that the universal objectivity of the consistent world-order could come about. At this point the 'unity of apperception' of all possible acts within the ego-pole serves Kant as such a ground of a most universal 'homogeneity' (*Affinität*) of all the disperate empirical elements: they are brought to the common denominator as the acts of the same I and can be discriminated among themselves. Without the reference of all acts to the 'pure I' as a qualitatively undifferentiated, empty and abstract pole of reference in the apperception, ever-perduring within the stream of appearances – no unity of consciousness as such would be possible. These two principles of unity, the *ego*-pole and imagination conjoined, form the axis of the conscious apparatus: the one bringing about the unity of consciousness in all its heteroclitic operations and acts, the other – being already the concrete vehicle of constitution – establishing the compossibility of the basic constitutive material and the synthesis of the singular fragments as imports into the structure of the complete objective unit.

Yet the question arises, by which principles can imagination, that in itself is 'blind', *that is, deprived of a discriminatory apparatus*, be a factor of the associative discrimination? With respect to what principles are the dissociative features left out and the associative links brought together? If the *dynamis* of imagination were a totally 'free' flux – that is unrestricted by directions, prototypes, regulative framework of reference – even the empirical synthesis would be nothing else but a product of chance and maybe we would remain at the level of Bergson's view of the spontaneous, that is whimsical, unfolding of Nature's creative élan. Yet Bergson, with respect to the human world, introduces into this boundless whimsical flux the differentiating activity of the human intellect. And so Kant cannot remain at the level of appearances, assuming that without an *a priori* of reason, the cognitive-constitutive process at the level of appearances – which in reality is nothing but an abstracted phase of a cognitive cycle incapable of standing on its own – could stop. In order that appearances are brought up to the level of being posited as consititutive of an object – that is tantamount in Kant's terminology, to the level of 'cognition' – they have to be made 'intelligible'.

Their structural differentiation and composition prepared already at the level of the synthetic unity of apprehensions is again operated by imagination, not as a 'free' agent, but with reference to the categories. That is, the constitution of a unified object of perception out of fragmentary, chaotic and diverging instances of the appearances, is accomplished in the work of imagination by 'reproducing' them into the homogeneous shape of 'apprehensions', and then, by discriminating them and synthesizing according to the *a priori* principles and structural regulations of categories. How does imagination apply universal categories to singular apprehensions of appearances? Or, in other terms, how does the singular element enter into a composition, the pattern of which Kant – unlike Husserl who sees there an autonomous 'essence' – seems to identify with the concept. Moreover, on this analysis, there is a need to orient the selective process of constitution of the object appropriately in order that the *a priori* rules of the intellect (*Verstand*) become applicable. Let us recall that Kant makes appeal to a special faculty, transcendental *'Urteilskraft'* in order to "distinguish whether something falls under a given rule (*casus datae legis*) or not". [23] The power of judgment (*Urteilskraft*) is meant to make the rules of categories applicable (*subsumieren*) to the concrete. To this effect, Kant sees the necessity of a qualitative identity of the content in the objective 'representation' of the concept, since the concepts are quite heterogeneous from the empirical (sensuous in general) apprehensions of appearances. The *tertium quid* that is on the one side of the same kind (*gleichartig*) with the categories and, on the other side with the appearance – this 'representation' (*Vorstellung*) which is pure from the empirical and yet, both intellectual and sensuous (*sinnlich*) – called 'the transcendental *schema*[24] is the product of imagination. As, at the empirical level, imagination performs the role of producing the synthesis of the 'impressional' manifold into a unified 'picture', so the *schemata* of the sensuous (*sinnlich*) concepts – as the pure synthesis according to the unity rules of the concepts in general, which is set by the category and means the possibility to constitute objects at all – are the work of imagination at the *transcendental level*.[25] The role of imagination supplying the *schemata* for the work of the intellect is not only the primordial condition of the whole cognitive-constitutive system as Kant sees it, but Kant himself recognizing it is compelled to introduce firstly, the notion of the soul left so far in the background, secondly, to refer to a realm of 'Nature' which lies below the level of the cognitive-constitutive correlate: 'the island' of the transcendental apparatus of cognition – 'the uncharted sea' of things in themselves, below the transcendental system. He states that imagination is the art hidden in the depths of the human soul; its real workings would remain difficult for us ever to wring from nature.[26]

Recapitulating our critical inquiry: imagination in Kant's perspective is at the cognitive-constitutive level, 'productive' but in a restrictive sense. It 'produces' the intermediary link for the structural operations; however it does not introduce anything *new* to the given. Although it is the primary agent of the giveness, its role consists merely in allowing what is received to become formed according to the *a priori* regulations of the mind. Seen by Kant as the solution to the same problem of perception as that for which it is primarily used by Husserl, namely, that *of the genetic passage from the dispersed concrete singular 'impressional' level of the manifold of 'apprehensions' towards establishment of their coalescence within a coherent series under the auspices of a universal pattern that they may together establish within one and identical objective structure,*[27] it has an essentially bigger part to play but it does not have any more productive 'freedom' than the 'free play' of *variation* in the Husserlian scheme of analysis. Its 'freedom' at the origin, as an undetermined and undifferentiated 'blind force', is in its productive operations caught between the received nature of the 'appearances' and by the categories.

Nevertheless, and it is worth while to point out and keep in mind that firstly, this operating *dynamis* crucial for the work of the transcendental apparatus might well be harnessed by and led by it, yet it stems from and dwells in the otherwise neglected dimensions of the soul as meaning both: the well of empiria and its dynamic condition. Secondly, that through the notion of imagination Kant is compelled to break through the 'secondary' concept of 'nature', as constituted within the transcendental system, to the notion of 'Nature' as the primeval ground and *which seems to be represented in man, as the dimension of the elemental empiria of the soul.*[28]

It is to imagination as this *elemental* power of *Nature* in the human soul that Kant recurs again to account for novelty and originality in works of art, which he acknowledges in their full significance as unique and yet integral elements of the human world. However, the acknowledgement of the extraordinary status of the work of art within the *life-world* precludes within Kant's approach, that the cognitive-constitutive apparatus could bring it about, inversely, according to Kant artistic achievement does not yield cognition. Consequently artistic creation, to be accounted for, calls for a further division of faculties in man's functioning as its specific agencies.

In fact, Kant is striving to find such a specific way, divorced from the routine constitution of the standard world, that could account for creativity. Having sketched already the issues involved within the general outline of his constitutive system, its background will serve us to recall the main features of his views on creativity. In fact, in opposition to the object-positing function of cognition, Kant differentiates still among all the virtualities of the soul (*Seele, Gemüt*): the *capacity of desire* (*Begehrungsvermögen*) motivating the will and finally, the *feeling of pleasure and displeasure* (*Lust* and *Unlust*) – the last one *positing no object* but expressing the subjective condition of man. While the capacity of cognition is served by the faculty of intellect and that of desire by will, the feeling of pleasure and displeasure is served by judgment.[29] In point of fact, intellect and reason apply their representations to the objects, the faculty of judgment does not bring about concepts of objects but is oriented solely towards the *subject himself*. Furthermore, while intellect and reason contain an *objective* relation between the representations and corresponding objects, the feelings of pleasure and displeasure are merely receptive of the state of the determination of the subject.[30] The specific domain of the faculty of judgment lies with its strictly purposeful and subjective orientation which neither constitutes nor reflects an object, but indicates – as it is in the *aesthetic reflective* judgment – the relation of a representation to the feeling of pleasure or displeasure.[31] With this a foundation for the approach to the creative activity is laid down.

Firstly, for Kant, as we know, human creativity is purposefully oriented by its reference to the faculty of judgment, in its particular kind, the judgment of taste; that is, in order to break with the routine and iron logic of transcendental constitution, we have to refer to a set-up of faculties that is extraneous to it, 'extra-routine', extraordinary. In its orientation as well as in its source, artistic creativity is brought back to the specifically *subjective resources of the human individual*. The universal 'schematism' of the cognitive apparatus is superseded by the uniqueness of the endowment of an exceptional individual: the *Genius*. Considering exclusively the works of great art as works of creativity to explain their *uniqueness, originality, exemplariness*, Kant refers *directly* to the role of imagination. Of course, he establishes several points of rapport between the judgment of taste and the cognitive faculty, in order to account for the *objective* side of the work of art as part of the *intersubjective* world. The main ones of them seem to be: firstly, the reference of the judgment of taste, which is the criterion of appreciation and discrimination within the process of art, to concepts of objects, but exclusively with respect to the feeling of pleasure and displeasure. Secondly, in spite of the uniqueness and irrepeatability of both, of the work of creativity and that of the creating subject, there is granted by Kant a universality of *subjective conditions* of judgment – which accounts for the universal validity of pleasure and displeasure we take in the representations of an object. Both of these points would mean some sets of universal *objective* directions for the discrimination within the creative progress with respect to the balance between subjective and objective conditions a work of art has to fulfill in order to be 'original' within the world but not altogether strange to it. Thirdly, Kant attributes to the judgment of taste *a priori* roots in the consciousness of the formal purposefulness present in the play of the cognitive powers.[32] Most interesting of all, he considers even the possibility of purely subjective judgments of objects at the pre-constitutive level – at which the cognitive faculties are not yet orchestrated into the transcendental apparatus but stay in a 'free play' – appearing not as cognition, but which relate to representations without concepts as the point of connection between the respective working systems of cognition and creativity.[33] This seemingly extensive relation to the cognitive system may lure us into believing in a concrete involvement of the cognitive apparatus into the process of creation, as seen by Kant. Indeed, we may at first, following Gadamer, assume that although 'free from constraints' of the cognitive *schematism*, creativity has all the powers of man available to work with at the pre-constitutive level.[34] However, then the crucial question occurs: "Where does it take its rules and regulations from?"

The conception of the Genius is meant by Kant to be the answer to this question. Indeed, each constructive act presupposes rules which would guarantee the possibility of its product. The very conception of Fine Arts as depending upon judgment might well lead toward its freedom from the automatism of the *cognitive routine*. But did Kant's subjective approach lead in the direction of discovering for creativity a different operational system to establish itself?

In this respect there is an ambiguity in Kant's conception of the genius as the answer to both the question of the source of originality and that of the rules in creation. Kant tells us that 'Genius' means a 'talent' (*Naturgabe*) that gives to art its rules. Genius is directed by 'taste'. Ultimately what does the talent of the Genius consist of if not of a special 'creative' role of imagination? Imagination is the factor of originality on its own; instead of entering into the *schematism* of the pure reason, it surges within the subject 'free' from the constraints of the constitutive system and its mechanisms, as the powerful streamlet of the *Elemental Nature* having the intellect (*Verstand*) as its sole partner and as its sole regulative instance to bring the forces of the soul into effervescence: wakening them to a new life and releasing their spontaneities, imagination becomes '*belebendes Prinzip im Gemüt*'.[35] Thus the making of the genius consists of these two forces supposedly not restricting each other but in a 'free concordance':

in der freien Übereinstimmung der Einbildungskraft zur Gesetzlichkeit des Verstandes eine solche Proposition und Stimmung dieser Vermögen voraussetze, als keine Befolgung von Regeln, es sei der Wissenschaft oder mechanischen Nachahmung, bewirken, sondern bloss die Natur des Subjeckts hervorbringen kann.[36]

We may then, first, with Gadamer, see the creative role of imagination that at last came into its own, in its power to awake all the potentialities of the creating subject – which is certainly the case and a prerequisite of the creative process. Nevertheless, the extremely intricate deliberation and discrimination progress which is necessary in the creative endeavor to establish the criteria for novelty with respect to the already given, remains unexplained by imagination so understood. Both imagination and taste, as the sole promoters of the creative work cannot, as strictly subjectively oriented – that is, divorced from the inner workings of the cognitive-constitutive structuration – account for the necessary constructive mechanisms. The 'radical' freedom of imagination through its origin and the lack of relative constraints on the part of the cognitive apparatus means also a *privation of means to exercise the function of deliberation and selection*. A set of functions performing these operations is indispensable for establishing a balance between the old and the new, the unprecedented and the altogether strange. It seems that to be 'creatively free' a certain framework within which these opposites take their meaning is implied and that a 'radical' freedom is inoperative, irrelevant. Moreover, with Kant's emphasis upon the exemplariness and novelty of the work of human creation, we have a strongest possible affirmation of the *positional* status of its content, without which art as product of imagination would remain, as in Husserl, a mere reflection of cognition. Within the framework of our own analysis, in order to challenge the

constitutive set-up by its *positional* originality, the creative work has to
be established in the *mode of self-giveness*. How could imagination as a
constructively undifferentiated agency operate the selfgiven positionality
of its product? To be 'free' means already a reference with respect to
'what' and it seems, as we have already attempted to show in our own
analysis, that in reality the *creative process works within the framework
of the constituted world and it is with respect to this world that the balance
between the old and the new has to be estimated, in each work of creation
anew*. Thus could imagination as an objectively independent faculty
operate it without availing itself of the mechanisms of the constitutive
function? It seems that it could emerge, operative, in 'creative freedom'
only from a *special integration of all the functions, that she would galvanize
to trigger from the operations of their novel configuration an original invention*. Finally, Kant claims that originality and novelty would stem
from Nature 'itself', of which imagination and taste are only the promul-
gating agents, appearing so 'free from any constraint as if it were the
product of Nature alone'. How could, however, the *elemental Nature*,
meaning merely incalculable virtualities and dynamic powers, account
for originality, which as we have tried to specify is the outcome of
differentiation?

Although coming to their full right, both the elemental realm of the
soul and the freedom of imagination as understood by Kant, seem to fall
short of the capacity to fulfill the task he attributes to them.

B. *'Imaginatio Creatrix' and the Functional Orchestration Within the
Creative Context: The Regulative Choice in the 'Creative' Versus the
'A Priori' of Ideas in the 'Constitutive' Function*

As we have attempted to show, the conception of imagination as an
independent faculty in Kant could not fulfill the role which Kant has
recognized as its own, that is, to account for the break into the standard
progress of the constituted world by the work of human creativity,
whereas Husserl does not seem to have even risen to the occasion. Nevertheless, we expect to have by their succinct appreciation and criticism
sketched the network of issues with reference to which we might now
outline our proposed new approach already prepared by the conception of the *creative orchestration of human functions* within the *creative
context*.

Thus, unlike Husserl's 'free play of imagination' which is subservient to the constitutive system following its structurizing principle of *a priori* ideas, or Sartre's conception of imagination as completely independent specific faculty following its own whims, neither definable nor explained, we find *imaginatio creatix* as the decisive factor within the *creative context*. If it is in a position to assure novelty in the creative process, it is not as participating in the system of passive genesis, and transcendental constitution in its full extent, but as emerging from within a *specific orchestration of functions* which brings together the mechanisms and forces of the constitutive apparatus with those of the complex realm of passions. Husserl and Kant rejected this latter into the background as the realm of the soul, as irrational, inaccessible to the sovereign reason, and once rejected could not reintegrate it into its proper role.

There, at the point of interplay between the ideas and the feeling, intellectual intuitions and affective responses, from their point of fusion along unforseeable lines emerge new, barely outlined or alluded to, qualities of feeling, profiles of forms, which in turn, intermingling in a vast range of ever variating elements in the process of deliberation that goes on and on without stopping, generate in profusion ever new, ephemeral and soon vanishing – but not without having generated others – qualities of feelings and emotions in the whole range of their modalities and fragments, profiles of forms, shapes, interrelations; both quality and form so intimately fused into an experiential, barely sketched unit that hardly a sharp distinction is possible. Their *generative* power going in all directions through associative references to the already established – although sclerotized, emptied of the fresh pulp reality – becomes firstly the reservoir, the creative function in its quest and deliberation draws upon.[37] Secondly, they do not appear in the usual mode of constitutive experience, as *factual* giveness, encircled rationally into a definite objective *datum*, but chaotic and incomplete, elusive and fluctuating, they surge with a *suggestive* power as proposing themselves straight to such or other structurizing problem. Thirdly, their appearance may be barely suggestive but is *effusive* and *dynamic*, that is it awakes the creative quest to venture into indefinite novel channels to seek even more than they concretely may offer. Thereby this dynamic game between the realm of impuls and that of the rational and the affective intuition *distills a force*, both posing problems to our quest and invigorating the will to seek an answer. A novel alliance upon new intertwinings, emerges in the form of a new relational pattern among the types of operations, new dimension of qualitative moulds for sensations, emotions, feelings... an infinitely advancing selfgenerating system... with a new suggestive, allusive, evocative force... invigorating will, a prompting spirit... *imaginatio creatrix*.

Imagination appears as mediating between two producing levels, the one of the generative forces of passions and the other, scrutinizing and selecting power of reason at all their strata. It may thus seem that our analysis gives support to Kant's view about the mediating role of imagination. However, he saw it in cognition, especially perception, whereas we find it in strikingly different function: *creative function*. The role which Kant attributes to imagination in artistic creation, is as we have shown, different from our views. To perform the role we have just described it must participate in the operations of both. Its very own and distinctively unique mode of operation consists in the *suggestive, evocative* and *invigorating mode of presenting to the deliberating agency of the creative function* the previously distinguished elements thus uplifting them from the flow of experience – otherwise they would vanish with the progressing stream – to the level of self-presentation. These acts of selecting *generate new allusive qualities and forms*. Furthermore, being brought into a spectrum of the already present reservoir of the possible, they intergenerate and with each incoming series, the range of diversified possibilities of choice expands into ever renewing ones.

Thus imagination can emerge in this distinctive creative way only within the *creative context* on the basis of its complete set-up. It operates within the *creative orchestration* as the chief mediator at all levels bridging towards their cooperation the virtualities of reason and those of passions. It advances without deciding about anything – it merely serves the deliberating agency of the *creative function*. However, in its mediating role, imagination draws conclusions from the progress of the creative process, prompting it incessantly by new suggestions which this progress itself indicates. Neither a routine agency of constitution as in Husserl, nor an elemental faculty, *Einbildungskraft*, through which Nature would prescribe laws to the inert genius, as in Kant, not an independent transcendental faculty aloof from reality and from the constitutive genesis, as in Satre, but a result and integral factor of *a novel orchestration of man's functioning*.

Indeed, if the *creative function* may perform its role of introducing novelty and originality into the constituted world, – which otherwise would flow its routine way towards its prompt extinction since after having reached all the permutations and combinations in the unfolding of the first outlined types it could not renew them and thus would be condemned to use all its resources and come to an end – it is due, *firstly*, to its giving absolute preeminence to *will* and choice, *secondly*, to the role of imagination in freeing it from the preestablished, constitutive *a priori* structural of regulations.

We thus reach a question of *capital metaphysical scope:* "What are the ultimate structurizing principles the *creative function* follows in its operations in the selection of the construction elements and their moulding to be concretized into an object?"

As we have shown in our previous work, the constitutive genesis is in its process referring for its continuity as well as for a point of reference for identifying and structurizing its intentional object, to the ideas. We have also argued that the ideas do not make an integral part of the constitutive system, but must be *transcendent* to it as *a priori* principles: a framework of reference for the structurizing function of constitution.[38] Could we, in fact, without having such unchangeable, unavoidable, universal regulative principles for establishment of the organization promoting life, maintain the continuity of life and a homogeneous universe? However, by the same stroke, this introduction of the automatized routine of generating forms would preclude any *deviation from the type* that each idea is prescribing as the range for the structurizing diversification and individualization.

The extreme opposite, as we have seen, the *creative function* to remedy and complement is not an agent of a passive and an inborn routine but of will and deliberation *par excellence.* Its aim is not to serve the purposes of life; to the contrary, it emerges as an act of defiance against all that does it unquestioningly and automatically. We have attempted to show how the new orchestration of the creative function, as Kant has already seen, mobilizing all the modes of operations and faculties at all functional levels has freed them from the preestablished routine as well as from its regulations. Furthermore, we have seen how from the new scheme of interplay between elemental functions and reason spontaneously generate in profusion, novel unforseeable and unpredictable structurizing ele-

passive.

ments; finally, how the *creative imagination* mediating between the self-germinating qualities and forms, and the selective agency of the creative function, proposes ever new structural elements, suggesting their possible ramifications, sequences and associations. In this way the *creative imagination* accomplishes still an additional task: within the set up of the searching and selective deliberation it proposes novel avenues and, adjusting to the decisions already made, indicates their far reaching consequences offering simultaneously alternate solutions. In this way the structurizing agency in its decisions is neither bound by any preestablished *a priori* delineating structural regulations or principles, nor is it left to the whims of the deliberating agent. The decision of the agent is, in fact, itself an outcome of a long and impenetrably complex process going along the whole progress of invention, moulding and structuration; it is guided by the possibilities generated from the interplay of passions and reason as well as by its own progress. Ultimately it refers to the abysmal wealth of *Logos* and *Eros* which, as we know, cannot accept any constraint to their freedom and their choices cannot be rationally accounted for.

Concluding, let us suggest, that the great question of Leibniz *whether our world is the only one possible*, which in classic and current phenomenology is emphatically answered in the affirmative, might have, with the distinction of the *plurifunctionality of man, the constitutive function being just one of its configurations and the creative function another*, received a basis for a new perspective.

NOTES

[1] The basically logical nature of intentionality is firmly stated by Husserl: *"Die Wahrnehmung und ihre parallelen Bewusstseinsweisen der Anschauung sind aber die ersten Grundgestalten des Bewusstseins, die für den Aufbau des spezifisch logischen Bewusstseins in Frage kommen... wir sind dabei schon Logiker, ohne es zu wissen" Analysen zur Passiven Synthesis,* Appendix II, p. 319, footnote, *Husserliana,* Vol. XI, Nijhoff, The Hague, 1966.
[2] Cf. by the present author: *Eros et Logos, esquisse de phénoménologie de l'intériorité créatrice,* Nauwelearts, Louvain 1972.
[3] Cf. by the present author: 'Ideas as the Constitutive *a priori*', *Kant Studien,* 1959.
[4] *Ibid.*
[5] *Analysen zur passiven Synthesis,* p. 319, footnote.
[6] *Ibid.*
[7] Cf. *Eros et Logos.*
[8] *"...je me concois capable de recevoir une infinité de semblables changements et je ne saurais néanmoins parcourir cette infinité par mon imagination et par conséquent cette con-*

ception que j'ai de la cire ne s'accomplit par la faculté d'imagination." 2nd Meditation, 12, PUF, Paris, 1966, p. 47.

[9] Cf. Edmund Husserl: *Untersuchungen zur Phänomenologie des Inneren Zeitbewusstseins, Husserliana*, M. Nijhoff, The Hague, p. 232.

[10] From the many texts devoted to 'free variation'. Cf. *Ideas I*, pp. 223–224.

[11] Cf. *Phänomenologie des Inneren Zeitbewusstseins*, pp. 10–38.

[12] *Ibid.*, p. 43.

[13] *Ibid.*, p. 51.

[14] *Ideen I*, p. 226.

[15] *"Näher ausgeführt ist das Phantasieren überhaupt die Neutralitätsmodification der 'setzenden' Vergegenwärtigung, also der Erinnerung im denkbar weitesten Sinne"*, *Ideen I*, p. 224. Cf. also *Analysen zur Passiven Synthesis*, p. 322–330.

[16] Cf. by the present author: 'Constitutive and Creative Perception', *Proceedings of the International Congress of Aesthetics*, Amsterdam 1964.

[17] *Phänomenologie des Inneren Zeitbewusstseins*, Appendix I, p. 108.

[18] *"Die Eigentümlichkeit dieser Bewusstseinsspontaneität aber is dass sie nur Urerzeugtes zum Wachstum, zur Entfaltung bringt, aber nichts 'Neues' schaft"*, *ibid.*, p. 100.

[19] *Kritik der reinen Vernunft*, A50, B74/R. Smidt (ed.), Felix Meiner, Hamburg 1956.

[20] *Ibid.*, B103, p. 116.

[21] *Ibid.*, Para. 24. *Von der Anwendung der Kategorien auf Gegenstände der Sinne überhaupt*, 1646, 10, B/198.

[22] *Ibid.*, 178a.

[23] *Ibid.*, *Elementarlehre II, Teil I. Abt. II. Buch.* p. 193/A132/20.

[24] *Ibid.*, pp. 197–199.

[25] *Ibid.*, p. 200/A142/20.

[26] *Ibid.*, p. 200, 10/B181.

[27] *Analysen zur Passiven Synthesis*, Vorlesungen, pp. 275–276.

[28] Such a reaching in depths is for Kant possible although it was not for Husserl because he – as has been pointed out by A. Gurwitsch and M. Merleau-Ponty – has abandoned the 'hypothesis of constancy' between the 'impressional consciousness' and a possible 'external' agent supposed to produce the impression.

[29] Husserl himself seems to see these points of analogy. Cf. *Analysen zur Passiven Synthesis*, p. 126.

[30] *Kritik der Urteilskraft, Introduktion.*

[31] *Ibid.*, p. 189.

[32] *Ibid.*, p. 205.

[33] *Ibid.*, p. 292.

[34] *Ibid.*, p. 286.

[35] *"Die Kunst des Genies besteht darin, das freie Spiel der Erkenntniskräfte mittelbar zu machen."* H. G. Gadamer, *Wahrheit und Methode*, I. C. B. Mohr, Tübingen, 1960, p. 50.

[36] *Ibid.*, p. 389.

[37] *Ibid.*, p. 393.

[38] Cf. *Why is There Something Rather than Nothing, Prolegomena to the Phenomenology of Cosmic Creation*, Van Gorcum, Assen, 1964, and the prequoted 'Ideas as the Constitutive *a priori*'.

PART I

THE *A PRIORI*

WELCOMING REMARKS

(Erwin Straus on September 4, 1972)

Ladies and Gentlemen,

When Professor Tymieniecka put the honor, burden, and pleasure of presiding upon my shoulders, she did not tell me what was on her mind. But, I suspect that the privilege was accorded to me because I am the only one here who has known Husserl personally, since he was one of my professors in Göttingen. It was in 1913 and 1914.

The burden and honor of presiding would be very hard for me if everyone on this program were not so well known, so that no one really needs any introduction.

A 'brio' is usually played at the beginning of the second act of *Fidelio* in Europe. Here, I would like to place it at the beginning of this meeting as a symbol of hope. There will be no discussion following the inaugural lecture. I think this is appropriate because our ideas, like the *'logoi spermatikoi'*, to use a term coined by the Stoics, should take time to grow and fertilize in each one of us.

Tymieniecka (ed.), Analecta Husserliana, Vol. III, 45. *All Rights Reserved.*
Copyright © 1974 by D. Reidel Publishing Company, Dordrecht-Holland.

J. N. MOHANTY

'LIFE-WORLD' AND 'A PRIORI'
IN HUSSERL'S LATER THOUGHT

I

1. It is generally agreed upon that about the year 1925,[1] more definitely about 1929, there came about a remarkable and profound change in Husserl's thought – a change which may be *indicated*, though not adequately characterized, by the fact that he began to make more and more use of the term *life-world*. Two such changes, each one of which may be maintained as having been equally radical, had characterized the development of his thought in its earlier stages: one was the turn from the psychologism of *Philosophy of Arithmetic* (1891) to the essentialism of *Prolegomena to Pure Logic* (1900), and the other, slowly making its appearance in the second volume of *Logical Investigations* (1901) established itself definitively in *Ideas I* (1913) as a turn from essentialism to transcendental idealism centering on the key concept of a constituting transcendental subjectivity. The concept of the *life-world* now, about the late twenties, seems to lay claim to replace or at least profoundly modify the concept of transcendental subjectivity as the key to Husserl's later thought. We thus may regard the three concepts: essence, transcendental subjectivity (TS) and *life-world* (LW) as being the key concepts of the three major phases of Husserl's thought. I do not believe that any of the three terms mentioned above was ever a total departure from its preceding phase. It further seems to me that there is a basic continuity in the development of Husserl's thought, a continuity which is sustained not merely by a common methodological concern but also by certain basic philosophical commitments. This continuity is at every stage enriched by new orientations and almost continuous self-examination. It is against the background of this belief that I propose to examine how his later concept of the LW stands related to the earlier concepts of essence and TS.

2. This general problem may be further analyzed as follows: Phenom-

Tymieniecka (ed.), Analecta Husserliana, Vol. III, 46–65. All Rights Reserved.
Copyright © 1974 by D. Reidel Publishing Company, Dordrecht-Holland.

enology began as the descriptive science of *a priori* essences (stage 1). It then turned to transcendental subjectivity in which both *essences* as well as *facts* are constituted (stage 2). Finally, it turns to the LW as the original basis from which all essences, material as well as formal, arise by a process of idealization, which is the *Sinnesfundament* of all higher order constructions. At stage 1, whose main conceptions are summarized by Husserl in the opening section of *Ideas*, Vol. I, the dominating distinction is that between 'facts' and 'essences.' At stage 2, the basic distinction is that between the constituting consciousness (TS) and the constituted objectivities. At stage 3, we have, in the same role, the distinction between the *Sinnesfundament* and the idealized constructs. From this over-all account, the specific questions that emerge are:

(A) How does the move from stage 1 to stage 2 affect the prior concept of *a priori*, and also the resulting concept of TS? More specifically, what would be the concept of the *a priori* without this move? And, what would been the concept of TS had it not been reached through such a move?

(B) How does the move from stage 2 to stage 3 affect the prior concepts of *a priori* and TS? Furthermore, how does the fact that the concept of the LW is reached through the stages 1 and 2 affect the concept of it?

In other words, how are the concepts of *a priori*, TS and LW related to each other; how do they modify, limit and influence each other? The two sets of questions formulated above should not be construed as if they concerned Husserl's philosophical development. We wish to make use of Husserl's philosophical development to throw light on the inter-relation between these concepts.

3.   Since the primary concern of this paper is to suggest some answer to the questions under group (B), I may rather quickly indicate a few points relevant to the questions belonging to group (A). First, it seems indisputably to be the case that the concern with the *a priori* persists throughout Husserl's thought, that there is no phase of his thinking, leaving aside for the present *Philosophy of Arithmetic*, in which he did not set it as a task for his philosophy to lay bare the *a priori* structure of some domain or other. To begin with, phenomenology was to be a description of *a priori* material essences; then it concerned itself with the act-*noema* correlation, which is an *a priori* structure of TS; and finally, there is, in *Crisis*, the tasks of describing (i) the *a priori* structure of the

LW and (ii) the 'universal correlation *a priori*' between the structures of
the LW and the structures of the TS. Thus, the concern with the *a priori*
persists; but we have yet to examine what the conception of the *a priori*
is, and if that too remains unchanged. In the second place, even in *Formal
and Transcendental Logic* (1929) where the concepts of the universal life
of consciousness (p. 98) and the 'Logos of the aesthetic world' as that on
which exact significations are built up, as also the notion of 'genetic constitution', are already explicitly formulated, Husserl writes:

*Nichts hat die klare Einsicht in den Sinn, in die eigentliche Problematik und Methode der
echten Transzendentalphilosophie so sehr gehemmt, als dieser Antiplatonismus....* (p. 229)

Which shows that, for Husserl, a correct understanding of the ideality
of the *a priori* (material as well as formal) is necessary for an adequate
conception of TS. The point I am trying to make may be reformulated
thus: if the TS had been arrived at only through the search for the constitution of a world of individual objects and facts, of the empiricist's
world, drained of all *a priori* structures and ideal significations, then the
resulting conception of TS would have been immensely poorer. In fact,
since according to Husserl every identity, even the identity of a physical
object, is constituted by the passive synthesis of ideal *noemata*, the proper
notion of constitution and therefore of the constituting TS would be
lacking – were there no a priorities either amongst the constituted or in
the structure of the constituting subjectivity. The consequence, I am
afraid, would have been that the conception of TS would have been
indistinguishable from the Berkeleyan mind which *contains* the ideas. It
is only with the recognition of ideal – *a priori* objectivities (formal and
material), it is only after the thesis that the world does exhibit *a priori*
structures – that the genuinely phenomenological problem of constitution can be posed and in principle solved, and that an adequate concept
of TS as constituting such a world arrived at. Therefore, but for the
move to stage 2 *via* stage 1, the conception of TS would not have been
what it in fact is in Husserl's philosophy. Finally, the turn from stage 1
to stage 2 also profoundly affects the status of the concept of the *a priori*.
At stage 1, the *a priori* essence seemed to provide the reason for what
individual facts are. Individual facts presuppose essences. Empirical
sciences are said to presuppose *eidetic* sciences. With the realm of the
essences, it seemed, one was reaching the rock bottom of rational en-

quiry – that layer of being which, while presupposing none else, rendered all empirical beings intelligible. With the turn to stage 2, this assurance is revealed as having been only provisional, and even the essences no less than individual facts are now brought under the *epoche*. Both belong to the constituted world, and therefore in need of phenomenological clarification. The *a priori* as such is no more presuppositionless, the self-evident, the self-clarifying, the self-constituting. The essences may be mundane, constituted, and so in need of constitution analysis. With the *a priori* objective essences dislodged from their dignity, the *a priori* structures of the TS are accorded that status. The concept of the *a priori* thus is rendered suspect: it covers both the constituted and the constituting, the ontic and the ontological, the mundane and the transcendental.

II

4.   Since *Logical Investigations*, Husserl's basic concept of the *a priori* seems to have undergone little change, although the application of this concept and his evaluation of its role certainly changed considerably. The components of this concept which seem to have remained unaltered are the following:

(i)   *A priori* validity is established not by inductive generalization but by *apodictic* evidence (*Logical Investigations*, Findlay edn., I, p. 99) based upon *eidetic* intuition.

(ii)   The mark of the *a priori* is unconditioned generality and strict essential necessity. (*First Philosophy* I, pp. 402–3; *Phenomenological Psychology*, pp. 70–71; *Experience and Judgment*, Section 97c).

(iii) The *a priori* is *not derived from* the structure of human consciousness. Any such derivation would entail a sort of relativism (*Logical Investigations*, Findlay edn. I, p. 145) and anthropologism (*First Philosophy* I, p. 403), which would make the *a priori* contingent (*Crisis*).

(iv)   *A priori* knowledge is based on *a priori* structure: in the widest sense, the *a priori* = *Eidos*, the morphological essence. *Formal and Transcendental Logic*, p. 219, fn. 1).

(v)   Distinction is made between formal *a priori* and material *a* priori. (*Logical Investigations*, II, p. 456; *Formal and transcendental Logic*, §55.

There is the corresponding distinction between analytic *a priori* and synthetic *a priori* truths. In *Formal and Transcendental Logic*, material a priorities are called contingent, since they possess a '*sachhaltiges Kern*' which goes beyond the mere form, and the formal *a priori* alone is called 'pure' (Section 6). Also cf. *First Philosophy* I, p. 224 ff. fn.)

a

(vi)   Empirical facts owe their rationality to their *a priori* structure (*Ideas* I, *Cartesian Meditations*, Section 64), but all *a priori* is not self-justifying or self-founding.

(vii)   Consequently, *Formal and Transcendental Logic* (Section 98) distinguishes between the 'ontic *a priori*' which is the *eidos* and the 'constitutive *a priori*' (of 'possible experiences, possible modes of appearing').[2] Only the constitutive *a priori* by its very conception is able to be self-founding. These two, the ontic *a priori* and the constitutive *a priori* are inseparable.[3]

(viii)   Thus the term *a priori* has two meanings – *a priori* knowledge of essences and essential structures; but also 'all concepts which, as categories, have the significance of principles and on which essential laws are founded.'[4]

(ix)   The *a priori*, in all cases, prescribes a rule which determines the domain of possibility, the scope of possible experience. Of an *eidos*, one may distinguish between its *eidetic* extension and empirical extension. By itself an *eidos* or an *eidetic* truth is free from all existential presuppositions, from all *Setzung* (*Ideas* III, 28). Reaching the *a priori* requires a '*Befreiung vom Faktum*' (*Phenomenological Psychology*, 71).

(x)   To transcendental subjectivity there belongs, by virtue of its essence, a universal and inexhaustible system of *a priori*, which may even be called an '*eingeborenes a priori*' (*Cartesian Meditations*, 181, 182; 28–30, 38).

Crisis introduces one novel distinction, not to be found in any of the earlier works: this is the distinction between objective-logical *a priori* and the *a priori* of the *life-world* (Section 36). The distinction in any case does not coincide with that between the formal *a priori* and material *a priori*. The material *eidos*, the essence, is no doubt an objective *a priori* as much as the formal *a priori* is. The *a priori* of the *life-world* is now said to be 'subjective–relative' (*Crisis*, p. 140), 'pre-logical' (*ibid.*, p. 141), it is presupposed by the objective sciences (*ibid.*, p. 139). But again, the *a priori* of the LW, we are reminded (*ibid.*, p. 139), is not itself relative. "We can attend to it in its generality and, with sufficient care, fix it once and for all in a way equally accessible to all." [5] Part of the task of this paper is to evaluate this concept of the LW *a priori* and its relation to both the objective-logical *a priori* and the subjective-constituting *a priori*. But for this purpose, we need, first of all, to consider briefly the concept of LW itself.

III

5.   Two different concepts unite in the concept of the LW: the concept of 'life' and the concept of 'world'. Of these two, Husserl's interest in the concept of the world is earlier and sustained; his relation to the concept of life has been full of vicissitudes – partly because of his changing attitudes towards Dilthey and his *Lebensphilosophie*.

5.1.   'World': In *Logical Investigations*, the world is said to be "merely the unified objective totality corresponding to, and inseparable from, the ideal system of all factual truth." [6] In *Ideas* I, the concept of 'world' is discussed in the context of the 'naturalistic standpoint': the 'world' is taken to indicate the total field of possible research from the theoretical position called 'natural' standpoint (Section 1). In Sections 27–28, we are given a description of the world of this standpoint in terms of endless ordering in space and time with an indefinite zone of indeterminacy. The world is said to consist of things and animals with positive and value qualities and practical significations. All the other worlds, e.g. the arithmetical world, are there from some standpoint or other, but the natural world is constantly there for me. The natural standpoint then is more basic than the other standpoints (like the arithmetical standpoint): it still remains 'present' even when I have adopted new standpoints. Husserl does not want to say that the natural world is the most comprehensive sphere within which the arithmetical world finds its place. On the contrary he denies this, and maintains that the two worlds are present together, though disconnected: their connection lies in their common relation to the ego who can freely direct the glance to the one or the other. Furthermore, although Husserl calls this world the world of natural standpoint, he does not regard it as a theory, it is called however a *thesis* – which again is not an act of predicative, existential judgment although it may be expressed in the form of such a judgment. Not being an act proper, it would be misleading even to characterize it as a belief (which is an act). The 'potential and unexpressed thesis' is prior to the explicit judgment (Section 31). This thesis, Husserl goes on to tell us, has its source in sense-perception. It may not be thus wrong to say that the world of natural standpoint is the world of sense-perception whose central category is the concept of the *thing*.

An appendix to *Ideas* II (dating from the early 20's [7]) identifies the 'natural world' with the LW. This possibly is one of the earliest passages in which the concept of LW is found to replace the earlier 'natural world'. The basic relation of the LW is said to be not causality, but motivation: the subject can be motivated only through that which it experiences as possessing 'value'. These 'value' characters appear as motivated as much as motivating. Things are not mere bodies but are 'valuable'. In the LW, the other is directly perceived through empathy. The mode of givenness is 'subjective.' The object-in-itself of the natural sciences is said to be constituted by a formal-methodological rule of unity (compare Kant, *Kr.d.r.V.* A 105/6), a formally ordered intentional unity of unending appearances 'for all men'.[8] However, every subject is said to have an actual and not a mere formal essence: it is something for itself.[9] The natural sciences originate from this LW of persons which itself is left untouched by them. The sciences follow a theoretical interest which leaves the LW actuality to start with, and return back to it in the form of technology and 'application' of the sciences to life. The theoretical interest itself however belongs to the subjective sphere.[10] The task which Husserl sets here is to describe this subjectivity, its life, its workings, its achievements, its constructs as constructs, and to 'explain' them in the sense which follows from such description.[11]

The replacement of the natural world by the LW is further strengthened by the perception of (i) its historicity and (ii) its intersubjectivity. *First Philosophy* II (1923–24), for example, says that we belong to the "totality of endless *Lebenszusammenhang* of one's own and of the intersubjective historical life."[12] Only *Cartesian Meditations*, in its rather desperate attempt to exhibit the constitution of the alter ego, returns to what it calls 'my sphere of ownness' which it regards as 'the original sphere',[13] the 'primordial world' in which the objective, intersubjective world is said to be constituted. Paul Ricoeur thinks that the 'primordial sphere of ownness' of the Fifth Mediation is nearly what the *Crisis* would call the LW, but he rightly corrects this identification in view of the intersubjectivity of the LW.[14] *Formal and Transcendental Logic* (1929) emphasizes the unity of the *life* of consciousness determined by a universal constitutive *a priori* of intersubjective intentionality (Section 98). The new concept of *life* gradually comes to the forefront.

P/39, 173

5.2.   Husserl's early antipathy to Dilthey's *Lebensphilosophie* derived from his suspicion that the latter may lead to *a sort of* psychological relativism, a denial of the ideality of meanings and thought-structures, an immanence philosophy which seemed to be oblivious of the phenomenon of intentionality. However, once *Logical Investigations* had established the ideality and transcendence of meaning and thought-structures, particularly after Dilthey had welcomed the *Investigations* as giving new and invaluable insights into the achievements of mental life, it was possible for Husserl to make use of the concept of *life* without his earlier compunctions.[15] The *noetic-noematic* correlations, the fact that *noematic* identity and ideality are achievements of *noetic* acts, the idea that empty intentions receive fulfillment from living concrete experience, the further discoveries of the constitution of acts and objects alike in the temporal flux of subjectivity, the way acts of consciousness imply and motivate each other 'intentionally', the phenomena of intersubjectivity – all these seemed to support a new, enriched concept of *life* of subjectivity, stronger than that of Diltheyan empiricism. Even the purifications enforced by the reductions and the transformation of empirical subjectivity to the transcendental did not deter Husserl from making use of the concept of *life*: now we are told of transcendental life of the ego. But this concept of life – it should be remembered – is intentional, intersubjective and accomplishing of ideal objectivities. It consists not merely in the positioning, objectifying acts of consciousness, but also in the non-positional, anonymous, pre-objective, 'operative' intentionalities constituting the sense of being 'already given' that belongs to the perceptual world as perceived or what *Formal and Transcendental Logic* calls the 'aesthetic world'.

5.3.   The manuscripts of 1929–30 compiled by Landgrebe as *Experience and Judgment* speak of the world as the field of unity of passive *doxa* which is prior to, and the universal ground of all cognitive confirmation,[16] as the "horizon of all possible judgmental substrates" (Section 9), i.e. of individuals which are possible values of the empty place 'something'. 'Experience' is defined as the evidence of individual objects (Section 6), not merely in the originary mode of self-givenness but also in its further modalizations (e.g., in the 'as if' givenness in phantasy). The LW, then, is characterized as the 'world of experience': "*Der Ruckgang auf die Welt der Erfahrung ist Ruckgang auf die 'Lebenswelt'*" (Section 10). This is, no doubt, a world of individuals, but it is also permeated by 'logical accomplishments'[17] – not merely in the sense that the determinations of modern science belong to the LW of the modern adult, but also in the more important senses: (i) that every object presents itself within a horizon of typical familiarity (within which the further distinctions of 'determinable' and 'determinate', of the 'known' and the 'unknown' find their places,[18] and (ii) that this pre-given, pre-logical world, nevertheless contains proto-logical structures out of which the structures of higher order logic arise through a process of idealization (e.g., 'negative judgment' arises out of the pre-logical frustration of expectation (Section 21a)). However, the original LW is not an exact world; the space, time and causality, for example, which characterize it are not exact.[19] Exactness, as also the concept of in-itself is the accomplishment of the method of idealization, as *Crisis* puts it.

5.4.   It is, however, *Crisis* which securely established the concept of the LW on the philosophical map, and made it a central interest of phenomenology. Did it signify a radical break in Husserl's thinking? Even if the change be not as radical as many think it is, what significance are we to attach to it insofar as Husserl's other basic philosophical concepts – especially those of the *a priori* and TS – are concerned? Again, let us take a closer look at this concept as it is to be found in *Crisis*.

  That the *Crisis* concept of the LW is vitiated by equivocations has been admirably pointed out.[20] In fact, we have not one homogenous concept of the LW but a whole spectrum of concepts which tend to shade off into one another. Husserl's primary motive, however, is clear: he wants to return to the world of pre-scientific, pre-logical and pre-predicative experience.

It is the perceived world (p. 49), "the original ground of all theoretical and practical activity" (p. 49), " the constant ground of validity" (p. 122), "the source of self-evidence" (p. 127), and "the source of verification" (p. 126), and "constantly exists for us" (p. 113). But it is also "the world of all known and unknown realities" (p. 50), in which everything has "a bodily character" (p. 106) and in which we ourselves live "in accord with our bodily (*leiblich*), personal ways of being" (p. 50). It is also a realm of subjective phenomena (Section 29), a realm of anonymous functioning (p. 112); it is again "an accomplishment", "a universal mental acquisition", "the construct of a universal, ultimately functioning subjectivity" (p. 113). In it there are no idealities (p. 50), geometrical space or mathematical time, no ideal mathematical points, straight lines, planes, no mathematical exactness (p. 139); yet it "remains unchanged as what it is, in its own essential structure and its concrete causal style" (p. 51). It is in principle intuitable, while the objective-scientific world is not (p. 127), but the intuition it is capable of is "subjective-relative" (p. 125); it is given "prior to all ends" (p. 138 fn), and yet its truths are "as secure as is necessary for the practical projects of life that determine their sense" (p. 125). The LW is not an entity, "the plural makes no sense when applied to it" (p. 143), it is the field, the horizon which is constantly and necessarily pregiven (p. 142); yet, the mode of actuality is predominant in it (pp. 145–6). It is the "world for us all" (p. 209); yet, "each of us has his life-world, meant as the world for all" (p. 254). It is 'pre-scientific' but it also includes the sciences as cultural facts of the world, so that objective science as an accomplishment of scientific community belongs to the LW without altering its concreteness (p. 130).

One way of resolving the complexities in the above description is to distinguish between several different worlds as follows:

A. The world of scientific objectivity
B. The world of perceptual objectivity
C. The many special worlds determined by specific pre-scientific interests
D. The Life-world in the strict sense.

Brief comments on each of these concepts may be in order:

World *A*: The world of scientific objectivity is constituted by scientific – theoretical interest, by the interest of theoretical *praxis*. I suppose its constitution presupposes the constitution of logical and mathematical idealities which are then 'applied' to sensible things, sensible qualities and sensible shapes in order to 'mathematicize' them directly or indirectly.

World B: There is a sense in which we perceive objects which are taken to have determinate shapes and sizes, determinate qualities which are measurable, which retain a measure of invariance, etc., etc. This world of perceptual objectivity may not be the 'original' world of perception, it may have been constituted out of a more original stratum of phenomenal field. This perceptual objectivity is in a rather inexact sense the 'same' for all perceivers, it is not subjective-relative, it consists of publicly observable objects, and so on.

World C: Prior to scientific interest, there are no doubt various pre-scientific interests arising out of the various professional and work motivations. These interests constitute special worlds: the world of the scholar, the world of the businessman, the world of the carpenter, etc. These worlds no doubt intersect and overlap insofar as men living in these special worlds nevertheless live in a common world also and in some sense do perceive the same objects. A real estate man, a carpenter, an architect and an antiquarian of course may be perceiving the same house, but the house as perceived by each of them is 'incorporated into' each one's specific world and thus presents a different facet to each. One could perhaps say that the house in itself (which is perceived by all of them in common) is a construct, or is constituted out of the different *noemata* (defined in terms of these different worlds). One could even say that world B is constituted out of world C, and world A out of world B. World B is a sort of hybrid and unstable middle region!

What then is world D, what we have called provisionally "*the life-world*?" There seem to be three possibilities:

(i) The *life-world* is the perceptual world but conceived not as a world of determinate objects, i.e. not as world B, but as what is perceived indeterminately, relatively to a subject. This would be the pre-objective world of perception of Merleau-Ponty (without committing oneself to, or ascribing to Husserl, Merleau-Ponty's unique concept of the body-subject). This may be juxtaposed between C and D. But let us call it LW_1.

(ii) The *life-world* may be regarded as the totality of all other worlds. However, this may be ruled out [21] on the ground that the idea of totality makes no sense when applied to the various worlds. The various worlds are not juxtaposed side by side, they are not given in attitudes which can be cultivated and maintained simultaneoulsy; if I used Wittgenstein's language, I could say they do not belong to the same logical space. The idea of the *all* of these worlds is not a legitimate idea.

(iii) Finally, we may regard the *life-world* as the horizon within which all other worlds are constituted and given, it is not itself another world beside them. It is, as the horizon within which they are given, also the condition of their possibility. Let us call it LW$_2$. It is with reference to it that we are to understand Husserl's oft repeated thesis that the LW is not a structure constituted by any interest, scientific or pre-scientific. In the life of interest, the LW is not given as LW. And yet in all such inner-worldly attitudes one relates to the LW not as theme but as horizon. Husserl's distinctive attitude towards the LW develops out of this situation, for he wants to make it thematic. This requires the disinterested attitude of reflection, the extension of *epoche* to it, and rediscovering it within the all-constituting structure of TS. This certainly distinguishes his concern with the LW from that of the existential phenomenologists.[22] Furthermore, it is only of LW that we can meaningfully say, as Husserl often does, that it is always *pre-given*.[23] The notion of 'pre-givenness' is a most strange notion indeed. I suspect that once we meditate on this notion, we may even be led to the conclusion that the LW as the horizon of all possible worlds is also a condition of their possibility and is thus itself a most comprehensive *a priori*.

We now begin to realize that Husserl's use of '*life-world*' in *Crisis* and elsewhere ambiguously, and sometimes misleadingly, covers what I have called the worlds *B* and *C* and also the two senses of *life-world* which have been distinguished. We need not stop here to enquire into the reasons for the ambiguity. But it may be mentioned that part of the reason for this lies in the context in which the LW concept is introduced in *Crisis*. This context requires him to go behind the scientific objectivity, but behind scientific objectivity there is a whole spectrum of worlds all of which were lumped together under the title LW.

IV

We are now in a position to return to the questions formulated earlier. We had there asked: how are the concepts of *a priori*, TS and LW related to each other, how do they modify, limit and influence each other? Three of Husserl's theses in *Crisis* provide us with the keys to answer this question. These are:

(1)   The objective-logical *a priori* is the result of idealization of the *life-world* (pp. 140–41).

(2)   The *life-world* has its own *a priori* structure (p. 139).

(3)   The *life-world a priori* is a 'stratum' within the universal *a priori* of the transcendental subjectivity (p. 174).

Husserl gives us, in *Crisis*, an account of how the ideal objectivities of geometry, for example, are constituted. He calls this process 'idealization'. He tells us, amongst other things, how starting from given sensible shapes (e.g. circular sensible shapes) one imagines other possible sensible shapes belonging to the same type. One then arranges these shapes in a series – e.g. in the order of being more or less circular – the entire series pointing to the limiting form: the perfect circle which is the geometrical form. The entire series is not, and cannot be, gone through in imagination. One proceeds *as if* it has been.

Several features of this account need to be emphasized. First, one wonders if the account Husserl gives does not involve a circularity of a sort. May it not be that the very ordering of sensible shapes in a series (in one sort of series rather than in another) presupposes the setting up of a limiting concept towards which the series tends as its limit? In that case, is it not the case that the process of idealization, as portrayed by Husserl, presupposes a recognition (*a priori*?) of the geometrical form under consideration? How else is the series to be formed? One possible answer to this question is that the series is formed not on the basis of *a priori* recognition of a geometrical form but on an intuitive apprehension of a phenomenal affinity of a certain sort, i.e. of a phenomenal circularity, of a sensible pattern, of what Merleau-Ponty called 'physiognomy'. It seems that in Husserl's view, the LW experience is characterized by acquaintance with such vague patterns at a purely intuitive level, by what he calls vague typicalities and habitualities, by its 'style'. In order that this answer may really be able to take care of the objection of circularity, it should be shown that these *life-world* patterns themselves are not results of a lower level of idealization.

But in order that LW experience may be said to be characterized by such patterns, we need to get clear about how such a conception of LW would differ from two other more commonly held conceptions of pre-conceptual immediate experience. One of these is the Humean model according to which the pre-conceptual experience consists in atomic, unrelated, discrete impressions. The other, the James-Bradley model, looks upon pre-conceptual experience as an undifferentiated whole, or flux, of sentient experience with no internal or external distinctions. In the former case, thought relates, connects and synthesizes what is, to begin with, unrelated. In the latter case, it both analyzes the whole and synthesizes the so analyzed parts. Now one criterion of adequacy of a model of pre-conceptual experience is that it should be powerful enough to be able to account for the higher order idealizations to develop *out of it*. Neither the Humean nor the James-Bradley model satisfies this requirement. The idealities, in these models, have their origin from some other source – the transcendental subjectivity, for example. If, however, the idealities have to develop out of pre-conceptual LW experience itself, the latter must have, within its structure, something analogous to them. It is not enough to say that the idealities are products of idealization. That the human activity of idealization could accomplish something like them requires that LW experience itself was not a mere homogeneous, undifferentiated flux, or a series of atomic impressions, but was characterized by vague typicalities which anticipated what idealizing activities were to accomplish. The talk of 'genesis' then requires for its appropriateness no more than exhibiting forms in LW experience which are *analogues* of, or correlates of, the idealities whose genesis we are seeking. A phenomenological account of genesis can do no more than such exhibition of correlates, analogues and intermediate links.

At this stage, one may also raise the question: is this account of genesis compatible with Husserl's earlier theory of the ideal objectivity of the logical and the mathematical entities and essences? Does the process of idealization simply discover the idealities which were there, or are they generated by the process? Husserl in fact does speak of 'accomplishment'. He also says that objectivity is the result of method.[24]

By way of replying to this, it should be emphasized that Husserl's talk of 'genesis' is to be *so understood* that his account of the genesis of idealities is not incompatible with the thesis of the ideal objectivity of the logical and mathematical entities and of the essences. This is how we can

avoid bringing upon *Crisis* the charge of either psychologism or anthropologism. Of course, geometrical forms and geometrical truths are there
prior to their discovery by the geometrician. It simply *makes no sense* to
say that before a certain point in time geometrical truths did not exist,
that they *began* to exist as a result of someone's geometrical investigation.
The reason why this makes no sense is that with regard to such truths, or
even with regard to geometrical entities, certain uses of temporal predicates are just inapplicable. It belongs to the very sense of such entities
that their being cannot be temporalized. Now if this be the case, the task
of genetic phenomenology is to find out the 'genesis' of such a unity of
sense. Therefore no genetic phenomenology could lead to a denial of the
unity of sense whose genesis is being explicated. The unity of sense is the
starting point of a genetic enquiry; it provides the guiding clue. A genetic
phenomenology cannot therefore be incompatible, by the very nature of
its project, with that whose genesis it seeks to unfold. It is this *respect* for
the given unity of sense that distinguishes a phenomenological genetic
account from all other varieties of theories of 'origin'.

I will not here enter into the very interesting question, what precisely
is the difference between the method of *eidetic* variation which was earlier
used by Husserl to extract the essences from particular cases and the
present account of the method of idealization through which idealities are
generated out of the LW and whether these two methods, despite their
differences, do not share a common structure. It is also worthwhile to
recall here that the process of idealization conquers, as Husserl tells us
(p. 346), the infinity of the experiential world by another sort of infinity,
"the infinity of reiteration" (p. 348 fn.), but the latter infinity is not gone
through but anticipated, thereby overcoming the limits of our finite
human capacity. I will not also consider the difficulties of interpreting
this, but will only note that for Husserl all constitution involves an infinity, the possibility of reiteration and the idea of '*immer weiter*'. This latter
infinity is in an important sense more 'manageable' than the infinity of
sensible appearances.

After having distinguished earlier between LW_1 and LW_2, I cannot but
ask the question: which of these forms the basis on which the idealization
process is supposed to work? Which LW may be said to have those vague
typicalities and patterns which are transformed into exact essences?

In reply to this question, I may begin by considering the second and the third theses of Husserl's *Crisis*: the theses namely that the *life-world* has its own *a priori* structure and that the *life-world a priori* is a 'stratum' within the universal *a priori* of transcendental subjectivity.

Now, to say that the *life-world* has its structures and to say that it has *a priori* structures are not the same things. It is one thing surely to hold, as Husserl unquestionably did hold, that our most primitive experience of the world is characterized by typicalities, even certain relatively invariant 'essential' types which "furnish us in advance with all possible scientific topics" (p. 226), types which function as "general separations and groupings" (p. 227). Furthermore, he also held the view, elaborated in *Experience and Judgment*, that there is a sort of proto-logical structure of such experience, in the sense that most of the forms of logic are *anticipated* in certain features available for inspection even at the level of the most primitive experience of the world. However, these structures may not be called the *a priori* of the LW. In fact, they may be empirical in the fullest sense. At the same time, we cannot deny that Husserl also held the view that the LW is characterized by its own *a priori*. I would like to suggest that LW_1 is fully empirical, its forms and types are all empirical types. However, experience of LW_1 always presupposes the 'pre-givenness' of LW_2 as the universal horizon within which any world at all could be experienced. Being 'pre-given', and so prior to any specific world – even prior to the naturalistic world – LW_2 is wholly *a priori*. In fact, it consists in a set of *a priori* structures which predetermine the general pattern of any experience. One could say that no world is conceivable unless it conforms to this structure. To the *a priori* structure of LW_2 no doubt belong, according to Husserl, spatio-temporality, causality, historicity, orientation around a human center, openness understood as horizontal character, and the character of passive associative synthesis. None of these however is to be understood in the strictly idealized sense of exact and fully determinate forms. In spite of their vagueness, they nevertheless represent a general structure of any world-experience *qua* world-experience. The objective sciences *presuppose* these *life-world a priori*, and the formal sciences develop them (p. 139) into determinate forms (e.g. the spaces of different geometries).

There is no doubt that Husserl's own exposition suffers from not distinguishing between LW_1 and LW_2, and therefore between the 'set of fixed types' which characterize the former (and form the basis for the genesis, through idealization, of the material essences of *eidetic* regional ontologies) and the *a priori*, invariant, structure of the latter. These *a priori*, invariant structures of LW_2 are determinables of which the different worlds – actual and possible – are different determinations.

Phenomenologists tend to blur over the distinction between 'essences' and 'categories'. Of essences, it may meaningfully be said that they are – either in a strictly ontological or at least in a quasi-ontological sense. The same cannot be said of categories. The essences – formal or material – are idealizations of LW_1 typicalities. The *a priori* of LW_2, on the other hand, are categories: they are the *a priori* conditions of the possibility of any entity, or world, whatsoever. It is of the former that one may ask the following sort of questions: "How is the geometrical form 'circle' related to the circular physiognomy, the visual *gestalt*?" "How is the system of colour concepts related to the felt distinctions of colour functions or powers?" I am formulating these questions in the words of Merleau-Ponty, for he of all phenomenologists faces up to this situation. Consider now the following sentences from his *Phenomenology of Perception*:

> The first perception of colours, properly speaking, then, is a change of the structure of consciousness, the establishment of a new dimension of experience, the setting forth of an *a priori*.[25]

Once a sensationalistic and atomistic concept of experience has been set aside, the *a priori* is not any longer to be conceived of as form imposed on empirical data *ab extra*; it is rather the full fruition of that 'rational intention' which is inherent in the phenomenal field.

> The *a priori* is the fact understood, made explicit, and followed through into all the consequences of its latent logic: the *a posteriori* is the isolated and implicit fact.[26]

Note the phrase 'its latent logic'. The process of idealization which transforms the empirical typicalities into essences may be regarded as making explicit such a latent logic in empirical phenomena. But these *a priori* essences are not conditions of the possibility of the empirical structures. (What Husserl may be taken to have maintained is that *eidetic* structures are presupposed by empirical sciences, not by experience itself.) While saying this, I am aware of the limits of the appropriateness of the talk of 'explicit' and 'implicit', for this model tends to blur that phenomenological discontinuity which exists between empirical types and the corresponding objective-logical essences.

Aron Gurwitsch has recently argued[27] that one of Husserl's major concerns, in connection with the LW problem, is the problem of access. The medium of access is consciousness. To make this mode of access, with its *a priori* forms, thematic, it was necessary for Husserl to bring the LW under the *epoche*. The medium of access is thereby "stripped of the sense of mundanity", and becomes transcendental subjectivity. LW_1 undoubtedly is brought under the *epoche*. But if by LW_2 we have meant the universal pre-given horizon within which any world experience at all is to be possible, then is it not the case that this LW_2 precisely is the transcendental subjectivity? Such seems to me to be the position of many interpreters of Husserl.[28] However, it seems certain that Husserl surely never identified the two and rather held, on very good grounds, that after the *epoche* the LW *a priori* shows itself to be a 'stratum' within the universal *a priori* of transcendental subjectivity.[29] I say "on good grounds", for the life of transcendental subjectivity is not exhausted in that anonymous operative intentionality which constitutes LW_1, and therefore it is not *merely* the horizon within which LW_1 experience comes to be. TS is also that in which the horizon constitutes itself as horizon, i.e. the reflective consciousness which recognizes the anonymous intentionality *as having been* its own unreflective functioning. The *a priori* of that horizon is then reflectively correlated to the *a priori* of TS. Thereby, a perfect analogy is made evident between: (i) the relation of LW_1 typicalities to the idealized essences, and (ii) the relation between the *a priori* structure of LW_2 and the *a priori* structure of TS. In neither case does the correlation amount to an abrogation of the distinction of levels.

To sum up: I had asked: "How does the move from stage 2 to stage 3 affect the prior concepts of *a priori* and TS? Furthermore, how does the fact that the concept of the LW is reached through the stages 1 and 2 affect its nature, function and status?" My answers are as follows:

(1) The *a priori* essences, now – in the context of the concept of the LW – are not simply discovered by a methodology such as *eidetic* variation but are constituted through a process of idealization. They are now shown to be dialectically related to their empirical prototypes. The essences are not conditions of the possibility of empirical types. They rather fully explicate the implicit rational intention pre-figured in the latter. The essences become quasi-ontological and more earth-bound.

(2) The TS is revealed in its most primitive role as anonymous operative

intentionality which prescribes the framework for any world experience
at all, but is not exhausted in that role, for it is *also* the reflective awareness
of its own past anonymity.

(3) The concept of LW, by being reached through eidetic phenomenology and as the *Sinnesfundament* of exact and empirical *sciences*, is
kept secure from possible romantic misconstructions of its nature. To
serve the purpose for which it is introduced, it has to be structured and
permeated by an *a priori*. At the same time, the problematic of LW is taken
up into the larger context of transcendental phenomenology and the
problem of access, which was ever Husserl's concern, is reviewed at this
new and primordial dimension.

Thus, *Crisis* does *not* mark a totally new beginning in Husserlian
meditations, but extends the philosophical commitments already made
to a new dimension of experience.

NOTES

[1] Cf., e.g., Iso Kern, *Husserl und Kant*, Martinus Nijhoff, The Hague, 1964, p. 255.
[2] E. Husserl, *Formale und transzendentale Logik*, Max Niemeyer, Halle, 1929, pp. 219–220.
[3] Cf. D. M. Levin, *Reason and Evidence in Husserl's Phenomenology*, Northwestern University Press, Evanston, 1970, p. 161; S. Bachelard, *A Study of Husserl's Formal and Transcendental Logic* (Eng. trans. by Lester E. Embree), Northwestern University Press,
Evanston, 1968, pp. 181–184.
[4] E. Husserl, *Die Idee der Phänomenologie* (ed. by Walter Biemel), *Husserliana* II, Martinus
Nijhoff, The Hague, 1950, pp. 51–52.
[5] E. Husserl, *The Crisis of European Sciences and Transcendental Phenomenology* (Eng.
trans. by David Carr), Northwestern University Press, Evanston, 1970, p. 139. All references
in the body of the paper to Husserl's *Crisis* are to this English edition.
[6] E. Husserl, *Logical Investigations* (Eng. trans. by J. N. Findlay), Vol. I, Humanities Press,
New York, 1970, p. 143.
[7] Cf. E. Husserl, *Ideas* II (ed. by Marly Biemel), *Husserliana* IV, 1952.
[8] *Ibid.*, pp. 376–77.
[9] *Ibid.*, p. 377.
[10] *Ibid.*, p. 375.
[11] *Ibid.*, p. 376.
[12] E. Husserl, *Erste Philosophie* (ed. by Rudolf Boehm), Vol. II, *Husserliana* VIII,
Martinus Nijhoff, The Hague, 1959, p. 153.
[13] E. Husserl, *Cartesianische Meditationen* (ed. by S. Strasser), *Husserliana* I, Martinus
Nijhoff, The Hague, 1959, p. 135.
[14] P. Ricoeur, *Husserl, An Analysis of his Phenomenology*, Northwestern University Press,
Evanston, 1967, pp. 139, 173.
[15] E. Husserl, *Phänomenologische Psychologie* (ed. by Walter Biemel), *Husserliana* IX,
Martinus Nijhoff, The Hague, 1962, pp. 33–35.

[16] E. Husserl, *Erfahrung und Urteil* (ed. by Ludwig Landgrebe), Claasen, Hamburg, 1954, p. 24.

[17] *Ibid.*, p. 39.

[18] *Ibid.*, p. 33.

[19] *Ibid.*, p. 41.

[20] D. Carr, 'Husserl's Problematic Concept of the Life-World', *American Philosophical Quarterly*, October 1970, 331–339.

[21] W. Marx, *Vernunft und Welt, Zwischen Tradition und anderem Anfang. Phenomenologica*, no. 36, Martinus Nijhoff, The Hague, 1970, esp. pp. 63–77.

[22] Cf. A. Gurwitsch, 'Problem of Life-World', in M. Natanson (ed.), *Phenomenology and Social Reality: Essays in Memory of A. Schutz*, Martinus Nijhoff, The Hague, 1970, pp. 35–61.

[23] Husserl, *Crisis*, p. 142.

[24] *Ibid*, pp. 345, 348.

[25] Merleau-Ponty, *Phenomenology of Perception* (Eng. trans. by Colin Smith), Routledge and Kegan Paul, London, 1962, p. 30.

[26] *Ibid.*, p. 221.

[27] Gurwitsch, 'Problem of Life-World', pp. 35–61.

[28] Cf. L. Landgrebe, 'Husserl's Departure from Cartesianism', in R. O. Evelton (ed.), *The Phenomenology of Husserl*, Quadrangle Books, Chicago, 1970. According to Landgrebe, "Transcendental Subjectivity is nothing else than the inseparable unity of world experience and its intentional correlate, the intended world experienced within it...". (p. 286).

[29] Husserl, *Crisis*, p. 174.

RICHARD T. MURPHY

THE TRANSCENDENTAL 'A PRIORI' IN HUSSERL AND KANT

It is our purpose to elucidate the transcendental character of the *a priori*: its essential relation to the possibility of experience and its origin in transcendental subjectivity. It is to Kant's theory of the *a priori* that we look for a point of departure. However, in order to reveal fully the transcendental import of Kant's teaching, appeal will be made to Husserl's modified and expanded theory of the Kantian *a priori*. Central to Husserl's teaching, and of utmost importance to our purpose, is his phenomenological analysis of transcendental subjectivity. Only such an analysis, we are convinced, offers the hope of uncovering the transcendental function of the *a priori* on the primal level of the pre-predicative, pre-reflective experience underlying ethical and aesthetical reflection.

Kant employs the term 'transcendental' to designate not that knowledge which is concerned with objects but rather "such knowledge as concerns the *a priori* possibility of knowledge, or its *a priori* employment." [1] The essential core of this transcendental knowledge is a critique which investigates "the understanding... in respect of its *a priori* knowledge." [2] According to Kant, knowing is a dynamic act in which the manifold representations given in intuition are unified by the understanding under the concept of an object and thereby attain a determinate relation to said object. For such a synthesis unity of consciousness is an original and *a priori* requirement. Therefore, that upon which the employment of the understanding rests is "the original synthetic unity of consciousness." [3]

The necessary manner in which the manifold given in intuition belong to one and the same consciousness and, consequently, make possible self-consciousness and its original unity constitutes an *a priori* knowledge of all objects of possible experience. For this necessary connection and unity of the intuited manifold in consciousness are required for representing any object whatever of possible experience and "precede all experience, and are required for the very possibility of it in its formal aspect." [4] Hence we may conclude that the *a priori* synthetic unity of

Tymieniecka (ed.), Analecta Husserliana, Vol. III, 66–79. All Rights Reserved.
Copyright © 1974 by D. Reidel Publishing Company, Dordrecht-Holland.

original apperception is transcendental in the sense of comprising a formal principle of that unity required *a priori* for the possibility of experience. Kant makes a decisive step, oftentimes not recognized or not accepted, when he argues that the transcendental unity of apperception, precisely because it is synthetic, presupposes an act of synthesis. This synthesis must be *a priori* since the manifold given in intuition "must possess that unity which constitutes the concept of an object." [5] This act of synthesis is the operation of the understanding insofar as it is but the faculty of apperception. For the spontaneous act of synthesis belongs to the understanding because it is "the faculty of combining *a priori*, and of bringing the manifold of given representations under the unity of apperception." [6]

Kant has stated explicitly that the *a priori* unity required for the concept of an object has its ground in that "pure original unchangeable consciousness I shall name transcendental apperception." [7] But if this *a priori* synthetic unity is effected by the understanding as the faculty of apperception, may we not infer legitimately that Kant is holding that transcendental apperception as the *a priori* principle or ground of that act of synthesis whereby such synthetic unity is effected? If this be granted, then transcendental apperception or consciousness is the fundamental ground of experience in a twofold manner. On the one hand, the transcendental unity of apperception is the *formal* principle *a priori* of that necessary synthetic unity of the intuited manifold on which the possibility of experience depends. On the other, transcendental apperception is the *constitutive* principle *a priori* whose absolutely essential function is exercised through the pure understanding in an *a priori* act of synthesis.

Now, according to Kant, the manifold to be synthesized by the understanding can only be given in sensible or passive intuition. The act whereby the manifold of sensible or passive intuition is brought under the formal *a priori* unity of transcendental apperception Kant entitles "the transcendental synthesis of *imagination*". [8] Admittedly, Kant's doctrine of transcendental imagination is obscure and even contradictory when the first and second editions of the *Critique of Pure Reason* are consulted. With Husserl we look to the second edition for the mature and more authentic teaching on the role of the pure understanding. Accordingly, we claim that the *apriori* synthesis of the intuited manifold under the original unity of apperception is primarily the work of the pure understanding. If this is correct, then the pure understanding, despite its pure or *a priori* concepts or principles of synthetic unity, "is yet able to determine the sensibility."[9] If the concepts or the categories of the understanding originate *a priori* in the pure understanding, then it is difficult to see how such concepts or categories can determine the intuited manifold in light of the radical dichotomy Kant himself erects between understanding and sensibility.

This difficulty accounts not only for his conflicting explanations of transcendental imagination but also for the unsatisfactory analysis in general of the *noetic* aspect of transcendental apperception or consciousness. To offset Kant's shortcomings and to clarify the *a priori* in its transcendental function of synthesis we turn to Husserl. The point of convergence between these thinkers is located in the area of the understanding. For Kant, the synthesis of the given manifold of intuition under the transcendental unity of apperception is effected by the 'subsumption' of that manifold under the *a priori* concepts of the pure understanding. Such 'subsumption' is the work of the understanding as the faculty of apperception. For Husserl the synthesis is effected by the 'intentional constitution' of the manifold according to *a priori* laws by transcendental subjectivity. Such 'constitution' is the work of the understanding.[10]

In the second edition of the *Critique of Pure Reason* Kant states that apperception, "as the source of all *combination*, applies to the manifold of *intuitions in general*, and in the guise of *categories*, prior to all sensible intuition, to *objects in general*."[11] The combination or synthesis of the manifold of an intuition in general is thought in the pure concepts or categories of the understanding as the faculty of apperception. These categories are "mere *forms of thought*" so that the synthesis is "purely intellectual."[12] But human consciousness is characterized by a certain *a priori* form of sensible intuition, namely time. Time itself is a pure or *a priori* intuition and contains a manifold. This manifold of *a priori* sensible intuition must be combined in accordance with the synthetic unity of apperception. This synthesis or combination, "which is possible and necessary *a priori*, may be entitled *figurative* synthesis...."[13] Since all objects of possible experience must be given under the *a priori* form of sensible intuition, they must necessarily be subject to this synthesis and its *a priori* unity. This synthesis is the function of the understanding since exercised according to the pure concepts or categories. Hence it is "an action of the understanding on the sensibility; and is its first application – and thereby the ground of all its other applications – to the objects of our possible intuition." [14]

The Kantian distinction between the 'intellectual' and the 'figurative' synthesis of the understanding suggested to Husserl an analagous differentiation in the 'constituting' function of the understanding. 'Intentional constitution' is operative on both the reflective level of the sciences and the pre-reflective level of the *life-world*. This latter level is originating since preceding and making possible scientific reflection. What Husserl has signalized as of utmost significance is that the *life-world* follows *a priori* laws so that it "has the same structure as that of objective science ...." [15] This indicates to Husserl that reason or the understanding is no less operative in the 'intentional constitution' of the *life-world* than in that of the world of the objective sciences. Kant has emphasized that the understanding is the faculty of transcendental apperception or subjectivity. To explicate the origin of the *a priori* structure of the *life-world* Husserl will claim that the *life-world* is the product of the *a priori* 'constituting' intentionality of transcendental subjectivity. [16] In other words, the 'being-sense' (*Seinsinn*) of the *life-world* is 'constituted' (may we say 'constructed'?) *a priori* in transcendental consciousness according to universal and necessary categories or laws of synthetic unity.

If we are to clarify the transcendental significance of the *a priori*, Kant and Husserl have set us a twofold task. We must explicate the structural law-abidingness of the world of possible experience (the *life-world*) by appeal to *a priori* laws of 'subsumption' or 'constitution' by the pure understanding. Such an explication demands a genetic explanation in terms of the transcendental subject. We attempt to fulfill this task by appealing once again to both these transcendental philosophers.

Kant is to be credited with the pivotal distinction between 'Transcendental' and 'General' or 'Formal' Logic. The latter "abstracts from all content... and deals with nothing but the mere form of thought." [17] This Formal Logic deals with the *a priori* laws of thought apart from any relation to possible objects. The validity of these laws is grounded on the principle of contradiction. [18] We may designate the propositions of this logic and the categories or forms of thought explicated therein "analytic *a priori*." [19] Transcendental Logic, on the other hand, "concerns itself with the laws of understanding and reason solely in so far as they relate *a priori* to objects." [20] The validity of these laws or concepts (categories) depends on this, that, although *a priori*, "they can yet relate *a priori* to objects of experience," [21] They can do so because "they must be recognized as the *a priori* conditions of the possibility of experience...." [22] Since these laws or categories refer to experience as its condition of possibility, we may designate the propositions of Transcendental Logic and the categories or forms of thought of an object enunciated therein 'synthetic *a priori*'. For such propositions and the laws or categories enunciated therein cannot be validated by the principle of contradiction. [23]

But Kant has stated that the categories are laws or principles of the connection and unity of the manifold given in intuition and hence "are required for the very possibility of it (experience) in its formal aspects." [24] The categories, then, are merely formal; and the *a priori* – even if synthetic and, accordingly, possessing a transcendental significance as the necessary condition of possibility of experience – is solely formal. Yet, it is to be recalled that Kant distinguished two levels on which the categories are operative, that of intellectual synthesis and that of figurative synthesis. "Both are *transcendental*, not merely as taking place *a priori*, but also as conditioning the possibility of other *a priori* knowledge." [25] On the level of intellectual synthesis the categories apply to an intuition in general and, consequently, to objects in general whatever be the mode of intuition. On the level of figurative synthesis, the categories apply to a determinate form of our *a priori* sensible intuition under which all objects of possible experience must stand. Hence on this level the categories apply "to objects which can be given us in intuition." [26]

It seems to us that on the level of intellectual synthesis the categories are the formal laws or principles of unity through which any object whatever acquires its 'being-sense' as object of knowledge. These formal principles constitute *a priori* the object in its being-as-an-object and thereby comprise the logical conditions of possibility of the object as an object. This intellectual synthesis and its *a priori* categories we would assign to what Husserl calls 'formal ontology'. As Husserl has pointed out, analytic logic as formal theory of science possesses an *a priori* universality and is directed to objects of knowledge. "Its *a priori* truths state what holds good for *any objects whatever*, any object-provinces whatever, with formal universality, ... as objects of judgments...." [27] Husserl characterizes such truth as "analytic *a priori*". [28]

But, as we have seen, Kant holds that the pure or *a priori* categories of the understanding apply to what is given under our determinate mode of *a priori* sensible intuition. Likewise, we have seen that Husserl claims that the understanding and its categories function on the pre-reflective level of the *life-world*. Can we not speak, then, of a 'material *a priori*'? By that we mean an *a priori* that predelineates the form of an object of possible experience "thought of concretely as materially determined and determinable...." [29] Husserl has designated the material *a priori* as a "synthetic *a priori*". [30] It is true, however, that Husserl's notion of the material *a priori* is quite Platonic. The material *a priori* comprises those essences which constitute the material object in its *eidetic* singularity. But his theory can bear a more Kantian interpretation. The material or synthetic *a priori* comprises those *a priori* principles or forms that constitute (construct?) the world of possible experience. [31] Since they determine the object in a more particular way, that is, in its 'being-as-an-object' given in a determinate mode of sensible intuition, these principles or forms are operative on what Kant calls the level of 'figurative synthesis' and determine sensibility. [32]

Let us define the categories comprised within the 'material' or 'synthetic' *a priori* 'material' categories and those comprised within the 'formal' or 'analytic' *a priori* formal categories. The question immediately arises: Are these categories identical or not? Kant has said that they are identical. [33] This is the reason that the chapter on 'Schematism' assumes pivotal importance within the Kantian problematic. Again, for Husserl the formal *a priori* involves formalization and is more proper to the level of scientific reflection. [34] Yet, he insists on the essential likeness of the world of scientific reflection and the *life-world* 'constitued' in pre-reflective experience. This likeness presupposes that the same *a priori* norms are operative on both levels. [35] Accordingly, we argue that the *a priori* is transcendental in a twofold manner: it is constitutive both formally and materially of the object of possible experience.

If the transcendental *apriori* is 'constitutive' both formally and materially, then we must abandon the radical dichotomy Kant places between 'matter' and 'form'. There is no pre-given 'matter' to be subsumed somehow under a purely formal *a priori*. Rather, the distinction between 'matter' and 'form' may be employed to differentiate between a material and a formal *a priori* level of 'intentional constitution'. [36] On the formal level where the synthesis – that is, the work of 'subsumption' or 'constitution' – is more properly intellectual and involves the categories in their formal mode, this work of 'constitution' may be assigned to the pure understanding in the more strict sense. On the material level where the synthesis is more closely allied to the determinate mode of our sensible intuition, the work of 'constitution' may be assigned to the 'pure' or 'transcendental' imagination. If we accept the second edition of the *Critique of Pure Reason* as Kant's mature and more authentically transcendental teaching, then the imagination is not a separate faculty but the pure understanding in its *a priori* role of determining sensibility.[37] How, then, can Kant assign imagination to sensibility as he does even in the second edition of the *Critique of Pure Reason*?[38] We feel compelled to abandon Kant's explicit distinction between sensibility and understanding. Following Husserl's lead we affirm that sensibility (passivity) and understanding (spontaneity) refer to distinct but essentially interrelated and interdependent levels of *a priori* 'intentional constitution'. The formal level on which the object is 'constituted' in its very being-as-an-object belongs to the understanding proper. The formalization whereby such a 'constitution' is effected according to formal *a priori* categories is characteristic of scientific reflection. The material level on which the object is constituted precisely as a possible object of *a priori* sensible intuition belongs to the understanding in the more ampliative sense (Kant's 'transcendental imagination'). This level characterizes pre-reflective experience or primal perception where the *life-world* is intentionally 'constituted' *a priori*.

If we are reading Kant and Husserl aright, all the preceding considerations of the transcendental significance of the *a priori* (its essential role as condition of possibility of experience) lead to the realization that the *a priori* is rooted in the 'constitutive' function of the transcendental ego and owes to transcendental subjectivity its transcendental significance. It is incumbent on us to show that transcendental apperception is both the formal and material *a priori* principle 'constitutive' of the world of possible experience. In order to accomplish this task our analysis must move from the level of structural interpretation to that of genetic interpretation. To make clear this transition it seems necessary to retrace our steps somewhat.

For Kant the objective world of valid experience is the world of nature investigated in the natural sciences. This world is characterized essentially by its necessary conformity to *a priori* laws or principles of structural unity. Without such structural law-abidingness, objectively valid experience and science would be impossible. Since these *a priori* laws or principles are the categories, Kant locates the ground for the *a priori* structural unity of the natural world in pure understanding, the faculty of apperception. Accordingly, the structural interpretation gives way to the more basic genetic interpretation which is founded on the constitutive function of transcendental apperception."[39]

Husserl's insight into the twofold function of the understanding substantiates Kant's structural interpretation of the *a priori* synthetic unity of the world of possible experience. The identity of the laws of 'intentional constitution' on both the pre-scientific and the scientific levels guarantees the validity of the laws of science. The Kantian assumptions – the irreformable validity of Aristotelian logic and the necessary validity of Euclidean geometry and Newtonian physics – need no longer be invoked.[40]

In Kant's system the structural is predominantly *noematic* in the sense of being directed primarily to the objects of possible experience. The subjective or *noetic* aspect is secondary and even in Kant's eyes nonessential.[41] On the other hand, Husserl's phenomenological description seeks primarily to clarify the subjective or *noetic* aspect, that is, the act itself of synthesis whereby the understanding 'constitutes' the *a priori* structural unity of the world. This emphasis by Husserl on the subjective or *noetic* aspect marks the transition from a static or structural phenomenology to a genetic phenomenology. Husserl repeatedly credited Kant with this turn from the ontological, objective – *a priori* to the *a priori* of subjectivity itself. The latter *a priori*, as Kant has seen, is originary and originating.[42] No less than Husserl's, Kant's teaching supports a genetic interpretation.

The genetic interpretation stresses the importance of Kant's teaching that apperception is the origin of all synthesis or combination and that the function of synthesizing belongs to the understanding as the faculty of apperception. However, this synthesizing function is exercised by the understanding only insofar as it determines in conformity to its own *a priori* categories a pre-given manifold of sense intuition.[43] Such determination seems inexplicable in light of the radical dichotomy Kant places between understanding and sensibility. The *a priori* must not be viewed as the mere form of logical unity thought in the 'transcendental object $= x$ to be applied somehow by the understanding to an 'extraneous matter' given in sense intuition.[44] Rather, so it seems to us, the *a priori* is constitutive of the object precisely as object of experience irrespective of any distinction between 'matter' (the given content) and 'form' (its structural unity). Kant himself has engaged the understanding and its *a priori* synthesizing function deeply into the level of sense intuition. Not only are space and time *a priori* forms of intuition but also are pure or *a priori* intuitions the unity of whose manifold presupposes a synthesis by the understanding that is itself *a priori*.[45] Can we not say with some plausibility that the *a priori* synthesizing function of the understanding is not something alien to sense intuition? It is not operative merely on the predicative or reflective level of conceptual thought, but also on the pre-predicative or pre-reflective level of primal perception. Indeed, we believe that this outlook embodies the thrust of Kant's chapter on 'Schematism' in the *Critique of Pure Reason*. To explicitate the implications of this outlook we turn to Husserl.

Transcendental Logic, whether it be Kant's or Husserl's, is concerned with understanding and its *a priori* concepts or categories. What is original to Husserl's Logic is its turn to the transcendental dimension of the *a priori* forms of cognizing subjectivity itself.[46] By this turn to the subjective or *noetic*, Logic becomes genetic. Central to this genetic logic is the realization that "All the subjective structures have an *a priori pertaining to their functions*."[47] Because the *a priori* is grounded in transcendental subjectivity, its functions extend to both the formal and material levels. These are but two different constitutive levels on which the transcendental subject is operative.[48] The formal level is more proper to reflective, predicative spontaneity whereas the material is more proper to pre-reflective, pre-predicative experience. But both levels are intertwined with one another and interdependent in transcendental subjectivity and its *a priori* structures.[49]

This intertwinement and interdependence seem to indicate that the phenomenal world of objective experience is 'constituted' according to *a priori* laws founded in transcendental subjectivity. The reductive description of transcendental consciousness will attain its objective if and when it brings to intuitive clarity the *a priori* synthesizing function of constitution already operative on the pre-predicative level underlying not only science but logic itself. This pre-predicative level is that of primal perception. Here the full force of Husserl's interpretation of the twofold function of the pure understanding outlined by Kant in the *Critique of Pure Reason* becomes decisive. The understanding functions on both the more properly formal level of reflective 'constitution' as well as on the 'material' level of pre-reflective 'constitution' since it is the faculty of transcendental apperception or consciousness. This twofold function of the understanding overcomes the Kantian dichotomy of concept and intuition and manifests the *a priori* grounding role of transcendental subjectivity.

To summarize our preceding analyses we may assign two quite distinct but interrelated tasks to transcendental philosophy. The first task is a Kantian type of reflection which seeks to uncover the transcendental significance of the *a priori* through an eidetic description of the *a priori* structural unity of experience and, accordingly, of the objects of possible experience. Such a description, if successful, will culminate in an eidetic or *a priori* intuition which isolates the *a priori* (both the formal and the material) as the formal, i.e., the formative or constitutive principle of the synthetic unity of possible experience and, accordingly, of the objects of possible experience. This structural *a priori* will be explicitated in what Kant would call 'synthetic *a priori*' judgments. This structural analysis is both *noetic* and *noematic* since all *noematic* aspects of this structural *a priori* is grounded in the *noetic* functioning of the understanding as the faculty of transcendental apperception or consciousness.

But this turn from the *noematic* to the *noetic* requires a genetic analysis which penetrates ever more deeply into the 'constituting' function of transcendental subjectivity. This genetic analysis will reveal in a more basic sense the transcendental significance of the *a priori* by revealing the constitutive function of the *a priori* on the primordial level of pre-reflective perception. This can be accomplished if and only if we take the 'transcendental turn' or 'reversal' demanded by Kant and Husserl. This 'turn' makes intelligible the position that the subject, taken in its transcendental character of original unity of apperception, performs the *a priori* syntheses by which the world attains that necessary unity in virtue of which it becomes a world for the experiencing or cognizing subject. This world attains its 'being-sense' through such syntheses. Furthermore, if, as we have attempted to show, this *a priori* is operative on the pre-reflective level of the world lived by the experiencing subject, the *life-world* then transcendental reflection must terminate in an eidetic reflection on the experiencing subject. In brief, transcendental philosophy becomes an egology, a philosophy of the 'ego' in its essential function of *a priori* synthesis or 'constitution'.

What we are trying to say, to speak in Husserlian terms, is that the phenomenal world attains its 'being-sense' as world for the experiencing subject in and through transcendental synthesis, that is, through 'subsumption' or 'constitution' by the understanding as the faculty of apperception or transcendental consciousness. If the *noematic* or, as Kant would say, the 'objective', is grounded *a priori* in the *noetic* structure of 'constituting' subjectivity, then an eidetic description of the concrete subject in its transcendental function of 'subsumption' or 'constitution' is in order. Kant is, to say the least, quite diffident in regard to such a description, but Husserl undertakes no less than an experience of the transcendental or 'constitutive' *a priori* which is at the same time an intuition of the 'ego' in its essential role.

If the transcendental *a priori* is grounded in the 'constituting' intentionality of the transcendental subject, the transcendental subject, in turn, is concretized in the factual ego. Here the threat of a transcendental solipsism and empiricism intervenes, a threat which Husserl himself saw.[50] If the *a priori* is to be genuinely transcendental, then 'intentional constitution' which grounds the 'being-sense' of the world must be itself *a priori*. Yet, this 'constitution' is effected by the concrete or factual ego. For the factual ego is the identical pole of those multiple acts of *a priori* 'constitution' in which the world attains its 'being-sense'. In so 'constituting' the world the factual ego 'constitutes' itself as "the identical ego which lives actively or passively in the subjective processes of consciousness and through them relates to object poles."[51] Husserl makes a decisive step in elucidating this auto-constitution of the ego which preserves the priority of the transcendental *a priori*. For the subjective processes have an eidetic form "which contains an infinity of forms, of *a priori* types of actualities and potentialities of life."[52] Hence the factual ego 'constitutes' itself as a possible ego in conformity to the "universal essential lawfulness of temporal egological coexistence and succession."[53] The auto-constitution of the factual ego form which originates the 'constitution' of the world follows the *a priori* law of temporal compossibility.

The temporal 'constitution' of self by the factual ego is, accordingly, both transcendental and eidetic. As a result the transcendental problematic is basically genetic in character. It concerns the 'primal institution' (*Urstiftung*) If the 'sense' and 'being-status' of the world for the experiencing subject and, correlatively, the 'sense' and 'being-status' of the subject's being-in-the-world. This 'primal institution' is a 'passive synthesis' effected on the anonymous strata of pre-reflective, pre-predicative perception. For all 'active constitution' presupposes a pre-given residue. But this irrational precipitate is "only possible as integrated into the system of *a priori* forms which belong to it as egological... the system of the consrete *a priori*." [54] This 'concrete *a priori*' is the factual ego in the totality of its actual and potential life of consciousness. Consequently, "the phenomenological explication of this... ego – the problem of its constitution for itself – must include in general all problems of constitution." [55] In other words, in making thematic the pre-reflective, pre-predicative sphere phenomenological description reveals that the transcendental 'constitution' of the world is regulated by *a priori* laws eidetic to the factual or concrete ego. From this it follows that transcendental 'constitution' is free from the contingency and particularity of the merely subjective or solipsistic. The *a priori* character of the transcendental is verified in the transcendental significance of the *a priori*. The *a priori* is transcendental since 'constitutive'; the 'constitutive' is transcendental since *a priori*. The apriority of 'constitution' rules out solipsism and introduces intersubjectivity. Ultimately, as Husserl has pointed out, the problematic of transcendental subjectivity (and, accordingly, of the transcendental *a priori*) is that of transcendental intersubjectivity.

Transcendental intersubjectivity involves the problematic of a twofold *a priori* 'constitution': the 'constitution' of the Other as Other and the 'constitution' of an intersubjectively valid world. Husserl's attempts to solve this problematic have focussed on perception and have met with notorious ill-success. It seems to us that Kant's assignment of primacy to practical reason suggests an apparently more feasible approach. To 'constitute' the Other as Other is to 'constitute' the Other as 'person'. Such 'constitution' involves moral or ethical values. To 'constitute' an intersubjectively valid world is primarily to 'constitute' a 'human' world. Such 'constitution' involves primarily aesthetical values. Phenomenological description of these values may help to resolve the problematic of intersubjectivity. But such a description must be transcendental in the authentic sense of being truly genetic. The origin of these ethical and aesthetical values must be sought in the 'constitutive *a priori*' function of the transcendental ego concretized in the *eidos* of the factual ego. This reductive description will be both *noetic* and *noematic*. This reduction remains a major task yet to be completed by phenomenologists.

NOTES

[1] Immanuel Kant, *Critique of Pure Reason* (transl. by N. K. Smith), St. Martin's Press, New York, 1965, p. 96. Although we shall utilize Smith's translation, our subsequent references to the *Critique* will employ the original pagination of the German edition – the first edition being referred to as *A*, the second edition as *B*. In this instance the original pagination is A56=B80.
[2] Kant, *Critique of Pure Reason*, A13=B26.
[3] Kant, *Critique of Pure Reason*, B137.
[4] Kant, *Critique of Pure Reason*, A130.
[5] Kant, *Critique of Pure Reason*, A104–05.
[6] Kant, *Critique of Pure Reason*, B134–35.
[7] Kant, *Critique of Pure Reason*, A107.
[8] Kant, *Critique of Pure Reason*, B152.
[9] Kant, *Critique of Pure Reason*, B153.
[10] Cf. Edmund Husserl, *Erste Philosophie (1923/24)*, Vol. I: *Kritische Ideengeschichte* (ed. by R. Boehm) *Ha*. Vol. VII, Nijhoff, The Hague, 1956, pp. 398–99.
[11] Kant, *Critique of Pure Reason*, B154.
[12] Kant, *Critique of Pure Reason*, B150.
[13] Kant, *Critique of Pure Reason*, B151.
[14] Kant, *Critique of Pure Reason*, B152.
[15] Edmund Husserl, *Die Krisis der europäischen Wissenschaften und die transzendentale Phänomenologie* (ed. by W. Biemel), *Ha*. Vol. VI, Nijhoff, The Hague, 1954, p. 142. Hereafter this will be cited simply as *Krisis*.
[16] Husserl, *Krisis*, pp. 70, 121.
[17] Kant, *Critique of Pure Reason*, A54=B78.
[18] Cf. Kant, *Critique of Pure Reason*, A60=B83–84.
[19] Cf. Immanuel Kant, *Prolegomena To Any Future Metaphysics* (transl. and ed. by L. W. Beck), Bobbs-Merrill Company, New York, 1950, pp. 14–15.
[20] Kant, *Critique of Pure Reason*, A57=B82.
[21] Kant, *Critique of Pure Reason*, A56=B81.
[22] Kant, *Critique of Pure Reason*, A94=B126.
[23] Cf. Kant, *Critique of Pure Reason*, A6=B11–A10=B14.
[24] Kant, *Critique of Pure Reason*, A130.
[25] Kant, *Critique of Pure Reason*, B151.
[26] Kant, *Critique of Pure Reason*, B150.
[27] Edmund Husserl, *Formal and Transcendental Logic* (transl. by D. Cairns), Nijhoff, The Hague, 1969, p. 120 (106–07). The numbers within brackets refer to the original pagination.
[28] Cf. Husserl, *Formal and Transcendental Logic*, p. 150 (134).
[29] Husserl, *Formal and Transcendental Logic*, p. 150 (134).

[30] Husserl first made this identification in the first chapter of the first section of his *Ideen zu einer reinen Phänomenologie und phänomenologischen Philosophie*, Vol. I: *Allgemeine Einführung in die reine Phänomenologie*, Nijhoff, The Hague, 1950. This volume, edited by W. Biemel, will hereafter be cited simply as *Ideen* I.

[31] Cf. Husserl, *Ideen* I, p. 367.

[32] Cf. Kant, *Critique of Pure Reason*, B152.

[33] Cf. Kant, *Critique of Pure Reason*, A79 = B105, B150−51.

[34] Cf. Husserl, *Formal and Transcendental Logic*, pp. 119–20 (106–07).

[35] Cf. Husserl, *Krisis*, pp. 142, 154–55.

[36] Robert Sokolowski has outlined clearly in his *The Formation of Husserl's Concept of Constitution*, Nijhoff, The Hague, 1964, the historical transformation which Husserl's concepts of 'matter' and 'form' have undergone.

[37] Cf. Kant, *Critique of Pure Reason*, B153.

[38] Cf. Kant, *Critique of Pure Reason*, B151.

[39] Cf. Kant, *Critique of Pure Reason*, A125–28.

[40] Cf. Husserl, *Krisis*, p. 142.

[41] Cf. Kant, *Critique of Pure Reason*, 'preface', Axvii.

[42] Cf. Iso Kern, *Husserl und Kant*, Nijhoff, The Hague, 1964, pp. 192–94.

[43] Cf. Kant, *Critique of Pure Reason*, B153–54.

[44] Cf. Kant, *Critique of Pure Reason*, A108–10.

[45] Cf. Kant, *Critique of Pure Reason*, B160–61 (footnote 'a').

[46] Cf. Husserl, *Formal and Transcendental Logic*, pp. 33–39 (29–35).

[47] Cf. Husserl, *Formal and Transcendental Logic*, p. 191 (169).

[48] Cf. Husserl, *Formal and Transcendental Logic*, pp. 271–73 (240–41).

[49] Cf. Edmund Husserl, *Erfahrung und Urteil* (ed. by L. Landgrebe), Claassen, Hamburg, 1954, pp. 239–40.

[50] Cf. Edmund Husserl, *Cartesianische Meditationen und Pariser Vortrage* (ed. by S. Strasser), *Ha.* Vol. I, Nijhoff, The Hague, 1950, p. 104. This will be cited as *Cartes. Med.*

[51] Husserl, *Cartes. Med.*, p. 100.

[52] Husserl, *Cartes. Med.*, p. 108.

[53] *Ibid.*

[54] Husserl, *Cartes. Med.*, p. 114.

[55] Husserl, *Cartes. Med.*, p. 102.

R. D. SWEENEY

THE AFFECTIVE 'A PRIORI'

The question of an affective *a priori* is a pervasive yet generally secondary theme in the history of phenomenology. There are few phenomenologists who refer to an *a priori* at all who do not also, in some way, include the area of affectivity under that rubric; but very few of these give it any kind of thematic treatment, and then only in schematic form. In short, our theme is a kind of step-child in phenomenology. Moreover, it is a step-child whose very name is a matter of some confusion. Searching in the standard lexicons for definitions of affectivity one concludes that they move in an "uncomfortably tight circle" from "emotion to feeling to affect and back to emotion".[1] It should suffice here to say that we mean by the affective the domain or range of experience associated with the terms feeling or emotion. But, of course, efforts to refine the notion more carefully will be noted as we proceed with our investigation. The meaning of *a priori* is probably susceptible to more precision, depending on the context; but to keep it hospitable enough for the variety of approaches found in phenomenology it will be used simply as the equivalent of the essential or eidetic taken in both 'formal' and 'material' or 'non-formal' senses, particularly this latter as extending to what Husserl refers to as the 'contingent *a priori*'.[2] Such an *a priori*, for Husserl, is "not an *a priori* of pure reason" but it is still "freed from everything empirically factual".[3] As such it includes the standard connotations of universality and necessity – at least within its region of application – and the character of being presupposed by observational or inductive awareness within that region.

A further complication in the development of our theme arises if, with Dufrenne, we consider the specifically ontological meaning of the *a priori* and find it to signify the basic accord of man and world, the mutual implication of the subjective and objective. For then we also find that it is impossible to express that basic accord in its immediacy, since any such attempt could entail a separation of the two aspects – of what are essentially inseparable.[4] We are reminded thus that we must elect to deal primarily with one side (aspect) or the other – at least as a matter of

Tymieniecka (ed.), Analecta Husserliana, Vol. III, 80–97. All Rights Reserved.
Copyright © 1974 by D. Reidel Publishing Company, Dordrecht-Holland.

emphasis – and so our focus will be on the subject-dide (*noetic*) since this is the aspect usually intended by discussions of the affective, and since, moreover, the object-correlate (*noema*) in this context is more often than not dealt with in non-affective terms as, for example, in terms of 'value'. Even within this focus, however, it is clear that on occasion inclusion of the *noematic* is indispensable. Similarly, other neighboring issues, like 'self' and 'will', will normally be prescinded from, but of course cannot always be.

Within these restrictions, then, we can attempt to discern in the phenomelogical literature on affectivity a pattern of development that is not so much chronological as logical, i.e., dialectical, beginning with Husserl. Husserl actually deals with affectivity thematically only in one place in his published writings, namely, a brief discussion in *Logical Investigations* where he endorses Brentano's thesis that all feelings are intentional and explains that there are two intentions in any feeling: "the underlying founding intention gives us the *presented* object, the founded intention the felt object. The former is separable from the latter, the latter inseparable from the former." [5] He further explains that this thesis holds even for the pleasurable or painful aspects of the experience, provided we make it clear that we are applying the term feeling to the intentional act proper, and not to the accompanying physical sensation. For example, joy "holds in its unity an idea of the happy event and an act-character of linking that relates to it", along with a 'sensation' of pleasurable excitement attaching to the idea and experience both in the subject and on the object – "the event seems as if bathed in a rosy glow".[6] But we have a tendency to identify the feeling with the sensation because of the latter's stronger and more lasting character and the consequent ease in isolating and naming it.[7]

Elsewhere, Husserl adverts to affectivity only in passing, as, for example, in *Ideas I*, Section 84: "...perceiving is the perceiving of something, may be a thing; judging the judging of a certain matter... then, acting concerns action, loving the beloved, joy the object of joy".[8] But it should be noted also that in the next section (85) affective qualities are included among the 'sensile' material or *hylé* (equivalent to sensations above) that is 'animated' by the intentional act or *morphé*, thus effecting a greater integration of the two levels – intentional and sensational – that are relatively separated in *Logical Investigations*.

In any case, from such passing references we can discern a 'vectorial' [9] approach to the affective *a priori* in Husserl coupled with a cognitive emphasis, i.e., following the cognitive model exemplified in acts of perception, intuition, etc., wherein specific contents are grasped with evidence. Judging from Roth's descriptions, the same themes are discernible in the unpublished manuscripts where, e.g., Husserl speaks of value as "being grasped originally and immediately in the glance of feelings".[10] Or again, a parallel with cognitive acts is suggested by Roth's description of Husserl's distinction between active, affective, evaluating acts, and passive sense feelings.[11]

With Max Scheler, the affective *a priori* becomes central and thematic, although subordinated to his value theory. His aim was to establish a theory of the intuition of value, and his strategy was to isolate such intuitive acts in the area of feelings. Specifically, these are 'feelings of...' (*Fühlen von*) – functions that are neither representational, nor conceptual, nor predicative (the intellect and *a fortiori* the sensorium is 'value-blind', according to Scheler, but affective intuitions of particular values, regularly coupled with allied acts of 'preference' and 'subordination' which reveal the hierarchy of values.[12] These 'feelings of' are best seen in contrast to *non-intentional*, non-revelatory, feeling-states (*Gefühlzustände*) which are static contents or phenomena, and which are only indirectly and causally connected with their 'objects' through the mediation of representations, associations, or symptoms. Anger is an extreme form of a 'feeling-state': we get angry 'about' something but only by dwelling on values or disvalues already perceived.[13] But there are other examples of 'feeling-states', viz., 'response-reactions', which are more closely connected with the 'feelings of' because they share the orientation of the latter and are 'called for', so that, indeed, we are saddened if the response is not forthcoming.[14] They are thus referred to as 'correlated', even though they are not, as such, intentional; that is, they do not "mean an object and in their effectuation make an object appear".[15] It might be noted here that Scheler gives no precise examples of 'feelings of' except for the aesthetic instance of 'feeling the beauty of snow-covered mountain'; in all other instances, he speaks rather of the correlated 'response-reactions' which, unlike the 'feelings of', can be objectivated by inner perception and hence, observed and described. This leads one to suspect that, in the absence of phenomenological description, the 'feelings of' are more postulated than exhibited.

Although secondary, response-reaction forms of feeling-states are included in a discussion concerned with the vertical '*depth*' of feelings entitled the "Stratification of the Emotional Life". The aim here is to show why any attempt to 'manage' bliss (a feeling higher than happiness) – which is what Scheler considers to be the basic meaning of eudaemonism – is self-defeating and leads inevitably to hedonism. To accomplish this, he distinguishes four levels of feelings: sensible, vital, mental, and spiritual (the level of bliss and despair). These levels are given *a priori* in our experience but are detectable also by several criteria, e.g., localization, duration, and the degree of controllability by the will (sensible feelings are controllable, bliss is not). They are further characterized by differing modes of self-reference, of closeness to the 'I': sensory feeling has *no* relation to the person, and is only indirectly related to the 'I'; mental feelings are qualities of the 'I' (ego) and spiritual feeling of the person. But of special interest here, is the criterion of intentionality which, in its modalities, distinguishes the higher three levels among themselves, and in general from the lowest level, e.g., joy is 'about' something, pain is not.[16] Thus, since most feelings are included here, Scheler is extending the role of intentionality to non-cognitive feelings, thereby affirming – somewhat inconsistently, it seems – that there is a form of intentionality involved with responses after all.

But the cognitive model is still dominant in Scheler, as can be seen in the two other major variations on his theory of affectivity (*ordo amoris*) – the intentional acts of sympathy and love. Central in establishing a sense of community, sympathy involves a feeling for the other and his feelings, and thus has response elements. Primarily, however, it is a mode of *knowing* the other as other.[17] Similarly, in the case of love, Scheler goes to considerable length to make it clear that love is not to be confused with any intellectual or judgmental function, nor with a simple intuition of value, as in the 'feeling of'; yet, at the end of his conceptual acrobatics on love, he settles on 'discovery' of the values of the other as central to love, albeit in combination with a 'dynamic upsurge', the net effect of which is to cause the values of the other to fly up "like sparks from a fire".[18] That is, the lover is oriented not simply to the present values of the beloved, but to her potential values and to promoting them. But while it is 'spontaneous', Scheler shies away from any suggestion of love as 'creative': and so, especially since it is 'pioneer and guide' for all other affective acts, love would seem, on balance, to indicate the dominance of the cognitive model.

With von Hildebrand, although he is frequently identified with Scheler, the emphasis seems to move from the cognitive to a response model. To be sure, in his early writings, von Hildebrand put considerable stress on the affective perception of values (*Werterkenntnis*) and its distortion in 'value-blindness'.[19] But within affective perception, he distinguished between value-seeing and value-feeling; while the former is marked by a certain distance or detachment, the latter involves a 'creative contact', a 'being touched', in which the value is 'lived'.[20] Going further in this direction, he developed the theme of the 'attitude' (*Stellungnahme*): whereas the value-perception has its 'content' on the object-side, the 'attitude' has it on the subject-side.[21] This same distinction is amplified in von Hildebrand's later works in the notion of affective response. To distinguish between affective response and cognitive acts, von Hildebrand utilizes the notion of intentionality but stipulates a bi-directionality within it (what we might also call a 'polarity'): "In the cognitive act, the intention goes, so to speak, from the object to ourselves: the object reveals itself to our mind, it speaks and we listen. In response as such, the intention goes from us to the object," as, e.g., when „we are full of joy about something."[22] Furthermore, in cognitive acts we are basically *receptive* toward the object, although not passive, whereas responses are spontaneous.[23] But von Hildebrand is also concerned with differentiating types of response within the category of affective response, and so he again uses the bi-directional model, this time to distinguish between the value-response and the 'response to the subjectively satisfying'. In the value-response, as found, e.g., in our admiration for a heroic deed, there is a self-transcending, 'centrifugal' character. In the response to the subjectively satisfying, on the other hand, e.g. our enjoying a cigarette, there is an 'immanence', a self-confinement, a 'relating of the object to ourselves' – in short, a 'centripetal' character.[24] In a more formal analysis, von Hildebrand specifies that in the value-response the determining ground (*principium*) is the value on the object-side, and the response is the 'determined' (*principiatum*). In the response to the subjectively-satisfying, on the other hand, the *principium* is the response itself (i.e., the responding subject as dominated by self-concern) while the object is the *principiatum*. As von Hildebrand explains it, the delight of a Tannhäuser in sensual pleasure *bestows* the importance on the object, while the delight in witnessing an act of forgiveness flows from the awareness of the importance-in-itself of the object.[25] Thus there is at the heart of intentional affectivity, in von Hildebrand's understanding, a kind of 'appropriateness' or 'due relation': the value-response is an 'answer' to the 'call' issuing from the value: it places the stamp of the 'heart', of the total person (in contrast to the 'thin' and partial responses from other 'centers' of the person found in volitional and intellectual responses) on the value or value-event.[26] This permits a sharp differentiation, then, of the value-response from the response to the subjectively-satisfying, since the latter is markedly inappropriate, whimsical, arbitrary. Indeed, on these grounds von Hildebrand claims to be able to distinguish between intrinsic qualities of response within a general type, e.g., 'holy joy' and 'malicious joy'. The first is appropriate to a value-laden good, the second, e.g., in a rejoicing over another's misfortune, is in-appropriate.[27] But this raises the immediate question of whether, if appropriateness is central, the response to the subjectively satisfying is, strictly speaking, intentional at all; it would seem, rather, to be a matter of a pseudo-intentionality, a simulation of the intentionality or genuine responses that borrows its surface plausibility from its imitative quality.

We can see then that von Hildebrand, while maintaining the vectorial structure, has, in effect, displaced the cognitive model and its emotional-intellectual paralellism – we might even say its crypto-intellectualism – and replaced it with a response model. This model is structured in terms of a bi-directional interpretation of intentionality, on which is placed the main burden of differentiation and stratification of affectivity.

Further variations on these models exist in philosophers more prominent for their extensive work on other questions. We could say, for example, that Heidegger is representative of the cognitive interpretation of intentionality, but only with major qualifications. Most obvious is the fact that the term intentionality hardly occurs in his writings, and then only in reference to other writers. It can be maintained, however, with Henry and others, that 'transcendence' is an alternative concept, since it has similar connotations of 'going-beyond', 'other-directedness', etc.[28] But 'transcendence' is clearly different from intentionality too, because of its existential thrust, its centrality to the disclosure of world, nothingness, and Being.[29] In any case there would seem to be a revelatory quality to

'State-of-Mind' and to the 'mood', and in this sense Heidegger might be said to echo Scheler's cognitive model even while disavowing any close adherence.[30] A mood, Heidegger says, "makes manifest how one is," that is, it "discloses *Dasein* in its thrownness..." and accompanies all other forms of disclosure, all understanding, as a tonality.[31] As Schmitt explains it, moods are a form of 'non-theoretical knowing', indeed, the precondition for all knowing. They are cognitive, but do not yield 'objective knowledge'.[32] All moods are ontological for Heidegger, but dread is the ontological mood *par excellence* by which the world as world and its 'nothingness' are disclosed.[33] Moreover, all moods are temporal, for they are grounded primarily in the past. But at the same time, temporality provides a criterion for authenticity since, e.g., authentic dread "springs from the *future* of resoluteness, while fear springs from the lost Present...."[34]

If Heidegger can be considered a distant echo of the cognitive model, we find in Sartre, within the vectorial paradigm, a certain similarity to the non-cognitive or response model. The word 'response' could be quite misleading, however, if we applied it literally to Sartre, since the significance of an emotion for him is claimed to lie in the context of action, not reaction. It is an attempt to transform an obstructive, frustrating situation that resists change, by the 'constitution of a magic world'. The woman who wants to avoid embarrassment but cannot do so through direct action, will faint, and this epitomizes all emotional behavior.[35] Emotion is thus a change in the intentional direction of consciousness, in its embodied relationship to the world, which is not, strictly speaking, an action but a pseudo-action – a substitute for action. Sartre's approach seems more plausible when applied to more violent, negative emotions, than to milder, positive emotions. But Sartre is undaunted by difficult cases that may not at first seem to fit his theory. In joy, for example, there is no obvious avoidance and no particular future orientation (as there is in fear, for example, according to Sartre) for joy tries "to realize the possession of the desired object as an instantaneous totality". But since this possession is impossible, the joyful person turns away from the object by means of an 'incantation' – the dance of joy.[36] It would almost seem, then, that for Sartre, rather than appropriateness, a kind of inappropriateness is a key mark of the intentional structure of affectivity, reflecting his underlying assumption that all emotions, except for anguish, are basically inauthentic, i.e., a form of *mauvaise foi*.

With Michel Henry we find a counterposition – virtually an antithesis – to all the approaches to the affective *a priori* which emphasize intentionality: "Believing that they could grasp its affectivity and its most important character in intentionality, modern philosophy has simply missed its essence and lost it completely".[37] Scheler is singled out as an egregious case, because in his distinction between 'affective perception' (feelings of) and affective states, the latter are seen as pure affectivity and yet as non-revelatory, blind, and contingent – as simple empirical or ontic contents that can have only extrinsic meaning by reason of their capacity to support emotional functions and behavior. But for Henry this amounts to a kind of dismemberment of the essence of affectivity between states, on the one hand, and meanings based on the intentionality of non-affective, perceptual acts, on the other.[38] Henry allows a partial qualification of these strictures with respect to sensible feelings in Scheler's hierarchy, because here the being of the feeling (hence its meaning) – since it contains no intentional dependence – consists in its being constituted.[39]

Similarly, according to Henry, Heidegger's locating of the revelatory power of affectivity in transcendence has the net effect of losing the distinctive meaning of affectivity in the ontological comprehension of Being. Initially, Henry argues, affectivity was simply presupposed by Heidegger as juxtaposed to every ecstatic transcendence. But then it was identified with transcendence in such a way that it was submerged and lost.[40] Also, for Henry, Heidegger's revelation of self to self through the temporality of transcendent affectivity is actually possible only if there *is* a self in the first place, and this is made possible – revealed – through the immanence of affectivity.[41] Feeling, as such and in general, according to Henry, is nothing but a blind and brute fact, except for this self-revelatory power, this 'self–tasting'. Henry would therefore de-emphasize affective intentionality, seeing it only as a contingent, occasional concomitant of the basic function of affectivity which is to be revelatory of the self and its being, that is, its ontological meaning in the mode of immanence.[42]

In Paul Ricoeur we find an effort to delineate affectivity by coordinating what we have outlined as its two major aspects: the intentional, 'vectorial', or 'object-related' aspect, on the one hand, and the interiorizing, self-affecting aspect, on the other. There are several levels to this effort, beginning with the *eidetics* of *Freedom and Nature*, which examines the emotions as 'moving feelings' in the context of bodily spontaneity, itself part of the broader context of the reciprocity of the voluntary and the involuntary, the main theme of the book. Here the emotions are seen, not as motives for the will but as organs – as means rather than ends – as an involuntary that sustains the voluntary by preceding and limiting it, by reinforcing and inhibiting it.[43] In all forms of the emotional – central or fringe variety – there is meaning because of a 'synthetic' or '*Gestalt*' quality, but more importantly, because of an intentionality which 'incorporates emotion to thought in its broad sense' so that the 'shock of thought' (the lightning judgment or evaluation involved in all emotion) and the 'bodily revolt' combine to register an intentional object by taking it into 'organic density'. In turn, this density is 'transcended' in a new appearance of the world of action (Sartre's 'magic consciousness', it might seem – but only in a very modified sense).[44]

A similar analysis relates tendencies (needs, motives) to feelings (*sentiments*) taken in their broadest sense. A tendency is both the 'objective direction' of a mode of behavior constituting our tie with beings and aspects of the world, *and* the aim of a feeling. Or better, the feeling is "nothing but this direction of conduct insofar as it is felt".[45] This explains how the same experience can designate an aspect of the thing, and thereby express intimacy with the self. Thus, in contrast with representation, which opposes us to objects, 'feeling' attests to our 'affinity', our 'coaptation', our 'elective harmony with realities of which we bear an affective effigy'. Ricoeur calls this, indeed, a 'connatural union' (after the scholastics).[46]

As in Husserl, then, the affective intention is built on, founded by, the objectivating intention: "it is on things elaborated by the work of objectivation that the feeling projects its affective correlates, its sensed qualities: the lovable and the hateful, the desirable and the abominable'.[47] These correlates or qualities, however, are not objects facing the subject – at most they are 'quasi-objects' – but 'the intentional expression of an undivided tie with the world', and so they appear at the same time as a coloration of the self, i.e., self-affection. There is a reciprocity – even a kind of chiasm – here as Ricoeur describes it: "This landscape is cheerful and I am gay: but the feeling is my belonging to this landscape which is in turn the sign and cypher of my intimacy'.[48] But this sort of analysis is difficult to handle in our ordinary language, Ricoeur points out, because that language is constructed on the dimension of objectivity where subject and object are distinct and opposed. Hence feeling can only be described paradoxically as a strange mixture of intention and affection.[49]

To speak of 'interiorization', 'intimacy', etc., is not, according to Ricoeur, to make feelings 'subjective', it is not to detach them or 'reduce them to a centripetal movement'; feeling is also "centrifugal inasmuch as it manifests the aim of feelings. And it is only insofar as it manifests this aim that it manifests myself as affected'.[50] It is on this basis that feelings can serve an important *regulatory* purpose: "by exhibiting the aim of the tendency on the predicates of things, feelings can furnish to action the objective signals by which it can regulate itself'.[51] In addition, this intentionality of affectivity serves a *clarificatory* function; it is only the existence of objective correlates that enables me to sort out my affective experience: "the correlates serve as a guide in the labyrinth of feelings'.[52] But these correlates are only of the order of the good and the bad, the suitable and the unsuitable, taken in a general sense. Specific values are really only quasi-objects, 'false substantives', since they are not a matter of intuition (affective or otherwise); rather, in addition to motivation, they presuppose two special, constructive acts, viz., 'reduction to essence (the *a priori* of the good here and now)' and a preferential intuition of the *order* of values.[53]

So far, then, we can see that intentionality is active in Ricoeur's approach but that it is construed neither in the cognitive nor the response mode. The cognitive aspect – or better, prerequisite – of affectivity would seem to be the function of motivation, needs and tendencies, with the exception of the 'lightning judgment'. However, this 'judgment of novelty' is not, strictly speaking, emotional, but the role of thought brought to bear by the voluntary – incarnate thought that becomes 'physically imposed'.[54] To be sure, Ricoeur speaks of tendencies as being 're-sensed' in feelings, but this would seem to apply only to their self-affecting character, and not precisely to their intentional structure. Let us say, then, that Ricoeur accepts an element of the cognitive in affectivity, but that this is by reason of its supportive role with respect to motivation, not because affectivity is cognitive *per se*. Similarly with the response model, as we have described it above: it would seem to apply here, but only partially. There is a general sense in which our feelings are an 'answer' to the world. And they do have their content on the subject-side. But it would seem that, for Ricoeur, this content is more a quality of interiorization and integration, than any specific reflection of the object. Or rather, it would be more accurate to say that the subject-object dichotomy does not really apply, except as a tool of analysis; it is precisely what affectivity overcomes. It is true that the affective act has a dual reference, for Ricoeur – to the object and to the self (outward and 'me-ward'); but the thrust of this relation is to pull the two terms together. In von Hildebrand's approach, on the other hand, the effect of the bi-directionality of affective responses is to reinforce the subject-object dichotomy. Von Hildebrand's analysis of 'appropriateness' would seem to underscore this point, since it seems to imply a specific distancing of subject and object in the light of which a certain correctness can be established. Rather than 'appropriateness', Ricoeur speaks of a truth or falsity of moods like anguish, but in the sense of 'authenticity' or 'depth': modern anguish is inauthentic if it should remain on the 'vital' level and avoid the depths of historical and 'metaphysical' anguish.[55]

In a new exploration in *Fallible Man*, intended to be more empirical and dynamic, Ricoeur compares 'vital feelings' (*epithumia* in Plato) which have their unifying theme in the desire for pleasure, with 'spiritual feelings' (*eros* in Plato) which are oriented by the striving for happiness. While vital feelings are primarily a matter of adaptation to social roles, in the region of spiritual feelings we see Man not simply as the problem-solver but as the one who poses questions and problems, and thus we can detect an 'opening in the closing of Care', namely, the perspective engendered by reason understood in the Kantian sense of 'demand for totality of completeness', which means here, 'feeling as an openness to happiness'.[56] Reason is then 'the intention according to which "I can continue my existence"' that is personalized by feeling – *eros* in Plato's sense of 'that through which we are in being'. In turn, *eros* branches into a multiplicity of 'spiritual feelings' which are 'no longer adaptable to any finite satisfaction' and which are oriented both toward the 'we' and toward the 'supra-personal – the totalization of persons in the Kingdom of Ends'.[57] Sacrifice illustrates this self-transcendence of the 'heart' – the organ and symbol of these feelings – by combining friendship and loyalty.[58]

Moods fit in here as 'atmospheric feelings', i.e., feelings that are closest to the 'formless state' or material substratum (*hyle*) from which all feelings emerge. Joy and anguish are, as it were, the upperside and underside, respectively, of the ontological mood or 'feeling of belonging to being'.[59] Thus there is here the Heideggerian theme of mood coupled with a response motif not too distant perhaps from von Hildebrand's value-response but closer to the later Heidegger and the thinking that is 'thanking'.[60] And yet there is also here a partial similarity with Henry's philosophy of immanence and its thesis that 'affectivity reveals existence'; with one fundamental difference, however – for Ricoeur, no 'moment of immanence' can have meaning by itself but only by reference to a 'moment of transcendence'. That is, intentionality is intrinsic, not extrinsic, to affectivity.

Most central to the dynamics of affectivity, however, is not the vital and rational dynamisms themselves but the intermixing and conflict between them. As Ricoeur explains it, in achieving a unification of the world through the interiorizing of intentionality, affectivity also creates a new division – of the self from the self: "it stretches the self between two affective projects, that of the organic life which reaches its term in the instantaneous perfection of pleasure, and that of the spiritual life which aspires to totality, to the perfection of happiness'.[61] This division, and the consequent tension and struggle, are grounded in the *thumos* or heart, which Ricoeur borrows from Plato and reworks as 'intermediary', the 'mixture of the vital and the spiritual', of sensibility and reason. Access to the meaning of *thumos*, Ricoeur finds in the 'passions' understood in a Kantian manner as illustrated in the triad of 'possession', 'power', and 'worth' (esteem) – *Habsucht, Herrsucht*, and *Ehrsucht* – the three fundamental demands of the self.[62] Ricoeur calls these passions '*a priori* feelings' because of their transcendental structure as regulated by intentional objects, that is, regions of objectivity represented by economics, politics, and culture. Despite standard connotation, passions are not deviations as such; what we understand in them at first are the 'primordial modalities of human desire which are constitutive with respect to man's humanity'. Only afterwards, by comparison with the primordial quests, do we see them as 'departure, deviation, downfall'. The initial understanding is accomplished by an imaginative variation which (as in Husserl) 'manifests the essence by breaking the prestige of the fact', that is, by imagining another state of affairs, a 'kingdom of innocence'.[63]

For example, the reflection on 'having' (*Habsucht*) begins with a description of the 'economic object' but then moves back to the corresponding feeling. What is discerned then is that, in contrast to the simple need which is only an 'oriented lack', the desire for the economic object is relative to the object's *'availability for me'*. Through this feeling I experience both my control over the having and my dependence with regard to what is other than myself.[64] In addition the whole area of relations with other persons is involved in *Habsucht*: "Mine and yours, by mutually excluding each other, differentiate I and You through their spheres of belonging" – establishing thus a relation of 'appropriation' where the interiorization reaches even to my mind and its thoughts.[65] Since we cannot imagine Man without 'having' (an essential limit, hence *a priori*) we see that there can be no innocence of non-possession – only more just and more communal modes of possession – and then, by comparison, we recognize perversions of having in greed, avarice, etc.[66]

Now such perversions, deviations, etc., are possible because the *thumos* invests the primordial feelings with the urgency of the drive for pleasure and the infinitude of the striving for happiness. Thus each is readily assimilated to the other through the mediation of the *thumos*. In particular, desire can lose, through passion, its character of orientation to 'provisional repose' – momentary completion – and become instead a 'desire of desire', endless in its pursuits.[67] Hence the fanaticism of the passions. Or else the infinitude of the striving for happiness degenerates into the sheer indefiniteness of restlessness. But the two dynamisms cannot be integrated, and so the *thumos* is seen as the synthesis of the vital and the spiritual, but an incomplete, restless synthesis. Thus the *thumos* is central to 'disproportion' and hence the main source of human fragility. But at the same time it is also the framework in which the self is developed, as can be seen especially in the area of feelings of worth (esteem): if humanity is what I esteem in another and in myself then "my self... is received from the opinion of others". Affective appreciation is the highest point to which self-appreciation can be raised.[68]

If the passions, as Ricoeur understands them, represent the fully human, concrete, lived level of affective experience, then their meaning is crucial to a particularized *a priori*. But reaching this meaning requires

– once again – an understanding of their objective correlates, and here these are understood to be immersed in history and culture. Can there be, then, an historical *a priori*? Husserl raised this question in the *Crisis* and seems to have answered in the negative.[69] Ricoeur has not thematized this issue but he does, as we saw, refer to passions as '*a priori* feelings'. In another context he speaks of *a prioris* in the plural and contends that "such an *a priori* cannot be detached from the history or civilization that presided at its birth; honor preserves its feudal aura, tolerance its eighteenth century overtones...".[70] This is a more 'historialized' version of the *a priori* than Scheler's, for example, and in general, I suggest, a modification and extension of the traditional *a priori*. Nevertheless, it still applies here, that is, some mode of *a priori* insight is available in the life-world, similar, perhaps, to the culturally modified expressive quality that Merleau-Ponty calls the 'emotional essence'. But to reach this insight requires interpretation, since the sedimented, objectified correlates of the passions are ambiguous, plurivocal – that is, they are symbols in Ricoeur's sense with the characteristic duality (or multiplicity) of levels of meaning. But symbols, for Ricoeur, must be interpreted on the basis of a hermeneutics – here, thus, a hermeneutics of affective symbols. Ricoeur has not actually developed such a hermeneutics specifically, but only as an aspect of broader questions. For example, his *Symbolism of Evil* can be read in part as a description and interpretation of the progressive interiorization of the primary feelings of evil, particularly guilt.[71]

But in *Freud and Philosophy*, Ricoeur has developed a brief but compact statement of some points pertinent to affective hermeneutics. Here again, Ricoeur's concern is with how the passions embody themselves in regions of the human world – they 'internalize a series of object-relations and are guided thereby'. This is true even in the sphere of worth which is primarily the area of the constitution of subjects, of intersubjectivity and mutual recognition. But this 'mutual constitution through opinion' is guided by distinct figures or cultural objects – not by things but by the works and monuments of law, art, and literature. Here the various cultural objects embody man's possibility, give his images the density of 'thingness' and thus mediate the 'dignity of man', that is, they are the traces of the process of the recognition of the self in another self. But this area is also the "level at which man can become alienated from himself, degrade himself, make a fool of himself, destroy himself".[72] As

symbols characterized by the 'dialectic of overdetermination' – the combination of disguise and disclosure and the tension between them – these
objects express both *libido* and creativity – both desire and striving for
happiness.[73] In other words, all these spheres (of having, power, and
worth) incorporate the dynamics of the *libido*, but as transformed in its
direction, i.e., sublimated.

An illustration that Ricoeur gives us is the interpretation of one rather
complex cultural object, a privileged example derived from the remote
past yet quite current by reason of its updating in Freudian theory, viz.,
Oedipus Rex. While Freud's interpretation found it to be exclusively a
drama of incest and parricide that awakens recognition and consequent
horror in us, for Ricoeur it can also be interpreted as a tragedy of truth
and growth in self-consciousness, i.e., in the collapse of Oedipus' presumption that the guilty person could be himself. But the two readings
(interpretations) and their correlates combine in the unity of the symbol:
the second-order tragedy belongs to the first because the mutilation that
is the tragedy of sex is also the darkening of truth.[74] In a similar way, all
cultural objects are *noematic* correlates of sublimation or sublimated
affectivity, thus incorporating a 'profound unity of disguise and disclosure' – of the *oneiric* and the poetic.[75]

In the light of our starting point, it is noteworthy here that Ricoeur
finds a parallel between his approach and Husserl's in the fact that for
Husserl, intentionality animates the sensuous matter or *hylé*, which includes affects and feelings; likewise, for Ricoeur, the Freudian interpretation gives us a *hyletic* of affects by enabling us "to set forth the genealogy of the main human affects and to establish the table of their derivatives". It thus verifies "Kant's insight that there is only one 'faculty
of desiring'; in Freudian terms, our love of money is the same love we
had as infants for our feces".[76] That is, the Freudian approach can assist
us in reaching a developmental, genetic, but non-reductive unification
of affectivity. However, while it can speak of 'transformations of instinct', the Freudian approach "cannot account for the innovation or
advancement of meaning that is inherent in this transformation".[77] That
can only be accounted for by an intentional analysis that combines
Freudian archaeology and its 'inert fate', with Hegelian teleology of
consciousness understood as guided by progressive figures of self in the
dialectic of mutual recognition.[78] Such an analysis would illuminate the

dimensions of self which stretch backwards and forwards in time, but are not accessible to immediate experience. In this way, instead of being a free-floating function, the interiorizing function of affectivity can be grounded in the dynamics of desire and growth. And it shows, in addition, that the intentional dimension of affectivity need not be sacrificed to the immanent, but that they can be reconciled when the dynamics of the self are seen in a relation of mutual implication with the dynamics of the life-world and its plurivocal value-symbols – each term of the correlation growing apace.

These observations concluding what is basically a survey and therefore necessarily sketchy and abstract, are not intended to assert the availability of a clear and complete synthesis, but simply to indicate that there is greater compatibility between various perspectives on the affective *a priori* than might initially seem to be the case. They are also intended to suggest the importance of further studies in this neglected area.

NOTES

[1] G. Schrader, 'The Structure of Emotion,' in J. Edie (ed.), *An Invitation to Phenomenology*, Quadrangle, Chicago, 1965, p. 252.
[2] E. Husserl, *Formal and Transcendental Logic* (transl. by D. Cairns), Nijhoff, The Hague, 1969, p. 29.
[3] *Ibid.*, p. 30.
[4] M. Dufrenne, 'The *Apriori* and the Philosophy of Nature,' *Philosophy Today* **XIV** (1970) 206–207.
[5] E. Husserl, *Logical Investigations* (transl. by J. N. Findlay), Humanities Press, New York, 1970, p. 570.
[6] *Ibid.*, p. 574.
[7] *Ibid.*, pp. 574–575.
[8] E. Husserl, *Ideas* (trans. by W. Gibson), Allen and Unwin, New York, 1931, p. 243.
[9] This term is used by C. Schrag, *Experience and Being*, Northwestern U. Press, Evanston, 1969, pp. 83–84, to explain the origin of meaning in intentional consciousness, where it resides in "the vectorial pathways and connecting tissues within the experienced life-world".
[10] A. Roth, *Edmund Husserls Ethische Untersuchungen*, Nijhoff, The Hague, 1960, p. 109.
[11] *Ibid.*, p. 167.
[12] M. Scheler, *Der Formalismus in der Ethik und die materiale Wertethik*, GW, Vol. II, Francke, Bern, 1954, pp. 269–272. It might be worth noting here that for Scheler, in contrast to Husserl, emotive cognition takes precedence over non-emotive: "the apprehension of values through them ('feelings of') is the bais of our subsequent apprehension of objects."
[13] *Ibid.*, p. 272.
[14] *Ibid.*
[15] *Ibid.*, p. 273.

[16] *Ibid.*, pp. 341–356. English Translation: 'Towards a Stratification of the Emotional Life', *Existential Phenomenology* (ed. by N. Lawrence and D. O'Connor), Prentice-Hall, Englewood, 1967, pp. 19–30.

[17] M. Scheler, *The Nature and Forms of Sympathy* (transl. by P. Heath), Yale U. Press, New Haven, 1954, pp. 14–26.

[18] *Ibid.*, p. 157.

[19] D. von Hildebrand, 'Die Idee der Sittlichen Handlung', *Jahrbuch für Philosophie und phänomenologische Forschung* **III**, Niemeyer, Halle, 1016, p. 211.

[20] D. von Hildebrand, 'Sittlichkeit und ethische Werterkenntnis', *Jahrbuch für Philosophie und phänomenologische Forschung* **V**, Munchen 1922, pp. 469–470.

[21] D. von Hildebrand, 'Die Idee der Sittlichen Handlung', p. 142.

[23] D. von Hildebrand, *Cristian Ethics*, McKay, New York, 1952, p. 196.

[23] *Ibid.*

[24] *Ibid.*, pp. 37–38.

[25] *Ibid.*, p. 37.

[26] *Ibid.*, pp. 244–256.

[27] *Ibid.*, pp. 320–321.

[28] M. Henry, *L'Essence de la Manifestation*, II, PUF, Paris, 1963, p. 738.

[29] *Ibid.*, pp. 747–748. M. Heidegger, *Being and Time*, SCM, London, 1962, pp. 188, 245, 414–415.

[30] Heidegger, *op cit.*, pp. 73–74; p. 178.

[31] *Ibid.*, pp. 172–179.

[32] R. Schmitt, *Martin Heidegger on Being Human*, Random House, New York, 1969, pp. 171–175.

[33] Heidegger, *op. cit.*, pp. 172–179.

[34] *Ibid.*, pp. 415–417.

[35] J.-P. Sartre, *Sketch for a Theory of the Emotions* (transl. by P. Mairet), Methuen, London, 1962, pp. 63–66.

[36] *Ibid.*, pp. 70–73.

[37] Henry, *op. cit.*, p. 608.

[38] *Ibid.*, p. 733.

[39] *Ibid.*, pp. 758–770.

[40] *Ibid.*, pp. 737–739.

[41] *Ibid.*, pp. 753–754.

[42] *Ibid.*, pp. 718–722.

[43] P. Ricoeur, *Freedom and Nature* (transl. by E. Kohak), Northwestern U. Press, Evanston, 1966, p. 252.

[44] The point that Ricoeur concedes to Sartre is that there is a moment of transformation of the world in magic consciousness. But this is part of the anticipatory moment of desire and not its energy; it does apply to the passionate aspect of the emotions, in particular, but tends to substitute 'spontaneity of consciousness' for bodily initiative.

[45] P. Ricoeur, 'Le Sentiment', *Edmund Husserl*, 1859–1959, Nijhoff, The Hague, 1959, p. 262.

[46] *Ibid.*, p. 263.

[47] *Ibid.*, p. 264.

[48] *Ibid.*

[49] *Ibid.*, p. 263.

[50] P. Ricoeur, *Fallible Man* (trans. by C. Kelbley), Regnery, Chicago, 1965, p. 132.

[51] *Ibid.*, p. 135.

[52] *Ibid.*, p. 137.

[53] *Ibid.*

[54] P. Ricoeur, *Freedom and Nature*, p. 256.

[55] P. Ricoeur, *History and Truth*, Northwestern U. Press, Evanston, 1965, pp. 287–305.

[56] P. Ricoeur, *Fallible Man*, pp. 153–155.

[57] *Ibid.*, pp. 157–158.

[58] *Ibid.*, pp. 160–161.

[59] *Ibid.*

[60] M. Heidegger, 'Postscript to 'What is Metaphysics?'', (transl. by Crick and Hull), in *Existence and Being* (ed. by Brock), Regnery, Chicago, 1949, p. 358.

[61] Ricoeur, *Fallible Man*, p. 201.

[62] *Ibid.*, p. 169.

[63] *Ibid.*, p. 170.

[64] *Ibid.*, pp. 173–174.

[65] *Ibid.*

[66] *Ibid.*, p. 175.

[67] *Ibid.*, p. 176.

[68] *Ibid.*, pp. 184–185.

[69] E. Husserl, *The Crisis of European Sciences and Transcendental Phenomenology* (transl. by D. Carr), Northwestern U. Press, Evanston, 1970, pp. 140–141.

[70] P. Ricoeur, *Freedom and Nature*, pp. 76–77.

[71] P. Ricoeur, *Symbolism of Evil* (transl. by E. Buchanan), Harper and Row, N.Y., 1967, *passim.*

[72] P. Ricoeur, *Freud and Philosophy* (transl. by D. Savage), Yale U. Press, New Haven, 1970, pp. 506–510.

[73] *Ibid.*, p. 514.

[74] *Ibid.*, pp. 514–519.

[75] *Ibid.*, p. 519.

[76] *Ibid.*, p. 512.

[77] *Ibid.*

[78] *Ibid.*, pp. 523–524.

DISCUSSION

Special Contribution to the Debate

HENNING L. MEYN

The Life-World and the 'A Priori' – Opposites or Complementaries?*

I shall concentrate on a central point in the phenomenological enterprise, namely the function of the *life-world* (LW) concept in constitutional analyses. The merit of Professor Mohanty's distinction between
LW_1 and LW_2 is that it commits us ultimately to reject any kind of pure
foundations interpretation of Husserl's work. It should be clear that
neither LW_1, nor LW_2 can provide the justificatory ground of incorri-
gible statements from which, say, particular scientific statements are
derived (or to which they can be reduced). But if an appeal to the *life-
world* does not serve this function, what purpose does it serve? Professor
Mohanty's answer is that the "vague typicalities" of LW_2 "are the *a
priori* conditions of the possibility of an entity, or world, whatsoever."
It is to this rather Kantian answer that I shall direct my following remarks.

Let me begin by calling your attention to the fact that the kinds of
transcendental arguments employed by Kant, and in their revival by
Strawson and Shoemaker, have recently come under attack (from Stroud
and Gram, for example).[1] Kant argued (A737) that pure reason could
establish certain absolutely secure principles from concepts of the understanding in their relation to possible experience, principles – like the
causal principle – which "[make] possible the very experience which is
[their] own ground of proof, and that in this experience [they] must always [themselves] be presupposed." Professor Mohanty, you recall,
stated that "experience of LW_1 always presupposes the 'pre-givenness'
of LW_2 as the universal horizon within which any world at all could be
experienced." However, before considering the recent attack on this
position we would perhaps do better asking what purpose Husserl's
appeal to the *life-world* might serve, in order to see what other alternative
interpretations are left open.

(1) It might serve the purpose of uncovering in pre-theoretical experience the Archimedean point from which we can justify or explicate our
cultural and theoretical structures, horizons or frameworks (I shall use

Tymieniecka (ed.), Analecta Husserliana, Vol. III, 98–115. All Rights Reserved.
Copyright © 1974 by D. Reidel Publishing Company, Dordrecht-Holland.

these terms interchangeably). This interpretation we might call the foundations interpretation. Though this might seem to be consistent with Husserl's sometime Cartesian stance and his talk about perception as the paradigm of evidence, it clearly fails to do justice to the *a priori* character of constitutional analyses. Since Professor Mohanty (among others) has shown that Husserl, with his concept of horizon, has gone far beyond this naive empiricistic position, I shall here simply reject this interpretation without further argument, though I shall say something more about it later on.

(2) It might serve the purpose of uncovering (by asking for the invariant structures of the *life-world*) the necessary conditions for the possibility of any coherent system of experience which makes up a world. This interpretation is the one I take Professor Mohanty to be advancing. I shall call this the 'transcendental interpretation'.

(3) Lastly, it might serve the purpose of eliciting and validating the presuppositions of any given theory (or any given world) (a) by matching it against the practical *life-world* out of which it arose; or (b) by matching it against other theories (or worlds) of the same level of naivete. In one reading of (3a) we simply have the foundations interpretation in a different guise, for the *life-world* as the ordinary pre-theoretical everyday world is here absolutized as an in-itself, i.e., as an absolute foundation. But as Gerhard Funke has argued,[2] if we are to be faithful to the phenomenological method, we cannot absolutize any domain of experience as experience of an in-itself. It is always relative to a consciousness. Thus any tenable reading of (3a) would necessitate our abandoning the idea of a '*letztmöglicher Grund*'. Let us call this the open regress interpretation.

(3b), which we might call the coherence interpretation, simply forestalls the question of justification (in the sense of grounding) by 'justifying' the adoption of any given framework (horizon) in terms of its coherence to others of the same level of theoreticity. But surely this interpretation is not faithful to Husserl's line of inquiry. Thus, unless we are to accept the open regress interpretation, we seem to be committed to defend the transcendental interpretation.

Let us then turn to a possible criticism of such a transcendental interpretation. An objection to this interpretation may be roughly stated as follows: The findings of a transcendental investigation are expressible in form of a statement as the conclusion of an argument. However, in order

for the premisses to be true, we must presuppose the truth of the conclusion. Thus a transcendental argument commits a *petitio principii* and
cannot possibly succeed. (See, e.g., Ayer's argument against Descartes'
existence 'proof' and Gram's argument against Kant's refutation of
idealism.[3])

There are at least two replies to that kind of an objection. A transcendental argument is not a species of deductive argument, but rather uses
a third form of inference based on the notion of presupposition. (This
is the argument Hintikka uses in his defense of Descartes' 'proof'.[4])
Aside from the difficulties (re-stated by Gram) of making a clear distinction between presupposition and implication, it is clear that Husserl
could not accept this reply; an appeal to a third form of inference would
still be a move within the framework of a naive logic.

The second reply would be to argue that rather than be a refutation
of transcendental arguments, the objection demonstrates the force of
their being *transcendental*. That is to say, it would be an objection if the
picture of the relationship between fact and theory or premiss and conclusion is one of independence, such that from neutral facts (premisses
to be established by appeal to those facts) a theory could be abstracted.
But that is the wrong picture. The investigation is transcendental precisely because the facts are facts only within a certain theory or framework, and the investigation is designed to elicit from what we *take* to
be facts the necessary conditions for their being facts (i.e., to elicit the
framework or horizon within which they are facts). But in that sense the
investigation transcends the naive acceptance of the alleged facts (or even
theories) as facts and rather sees them (through the reduction) in their
conditional character, in their constitutional dependence on the framework. In this way, I take it, Professor Mohanty argues that LW_2 is the
framework necessary for the possibility of experiencing the world that
we do experience.

As good a reply as this seems to me to be, we must still concede the
main point of the objection. The conclusion of a transcendental argument (or the result of a transcendental investigation) is not in itself
necessarily true, it is rather a conditional necessity. To see this more
clearly, let us look at Professor Mohanty's suggestion regarding the
status of LW_2. He says that "LW_2 is wholly *a priori*", in that it "pre-
determines the general pattern of any experience". But what grounds do

we have   t this is the *only* possible pattern of (that these
are the o   ries for) any experience? It won't do to appeal
to pre-prec   ce as an independent neutral testing ground,
because 'ex   finition of the lower order *life-world*, which
presupposes th   *life-world* as its framework, i.e., it is *dependent* on that fra   to say that Husserl wanted "to return to
the world of pre   -logical and pre-predicative experience"
must not be taken a   pure foundations interpretation. Hence
Professor Mohanty   swer in *Edmund Husserl's Theory of
Meaning*,[5] that "Huss   t raise the problem of *justifying* this
uniformity of nature, f   ity... is not a matter of theoretical
belief but *a phenomenolo   n of our primitive experience of the
environment*" cannot answe   ery, since, as we have seen, "our
primitive experience of the en   nt" is already tainted by its dependence on a framework within w   is given as experience.

The only other way to answer ou   uestion would be to argue that we
could not have anything that we might call experience unless it followed
this pattern. This would have to be a purely *a priori* argument, but I have
no idea what this argument could possibly look like. On what grounds
could we establish the premisses of this argument? It seems to me, then,
that we have to accept the conditional character of transcendental investigations.

What then is the purpose of a transcendental investigation? Let us ask
whether the transcendental interpretation is a genuine *alternative* to the
open regress interpretation. It seems to me that it is not, for only if the
transcendental investigations were to disclose an absolute domain of an
unconditionally necessary framework, could we avoid the regress. But
because of the mutual dependence of experience and framework, we
shall not arrive at a pure experience devoid of a framework (as the foundations interpretation would wish), nor at a pure framework devoid of
experience through which it is constituted (as a Kantian transcendentalism would wish). Thus we can see that transcendental as well as foundational arguments gain their strength only within a certain level of an
open regress, and that the appeal to a *life-world* serves the purpose of
elucidating and clarifying the implicit presuppositions that remain unthematic in a naive attitude within that *life-world*. It is for that reason
that Husserl speaks of clarification and understanding rather than justi-

fication in the strict sense. This situation is echoed by Neurath's metaphor of repairing a ship while being afloat, and by John Austin when he remarks about his method that appeal to ordinary language is the first word (i.e., by choosing this as our starting point) but certainly not the last word (i.e., the absolute ground of justification).

Though it might well be, then, that we shall not be able to uphold a sharp distinction between empirical and *a priori* investigations, we still want to preserve at each level a distinction between transcendental or conceptual inquiries about a domain and naive factual inquiries within a domain, however relative this distinction might be. It is with respect to this latter distinction, then, that we must understand Husserl's appeal to a *life-world* and the *a priori* character of its horizons.

Gendlin: It seems to me that Mohanty's careful analysis leads to the central question which, I think, you are very much aware of: to the question of this 'conundrum' of something, which is neither clearly conceptualized or patterned, nor is it sheer mush or sheer indeterminacy. I like the way you put it, as the conundrum of 'vague patterns'. On the one hand pattern is not 'vague' and 'vague' means 'not patterned', so we must distinguish between the function that this nuance serves in the argument which is quite sound. It seems that there must be something, such that it is not the same as the variety of patterns that we give to it, and yet, it is not anything completely arbitrary. That is, there seems to be a function to be fulfilled and something must serve this function. On the other hand, it seems to me, we cannot be satisfied simply to call this *tertium quid* 'vague typicalities'. In other words, the whole notion of a transcendental analysis implies this question because 'transcendental analysis' means undercutting that level after one has already reached further, and various thinkers who set up *a priori* categories do not agree with each other. So, I believe, that a transcendental analysis should not skip too quickly over the preconstitutive level of experience. From my point of view, you went hastily through this crucial level of preconceptualized experience, something, that is not arbitrary and yet is not structural. You called it 'vague forms', which alone would not suffice to account for the various forms that we can arrive at at the successive levels of constitution. If 'vague forms' or 'vague patterns' do not suffice to account for them, I think it is a mistake to assume that whatever would be the basis for 'clear patterns' should itself in some way be some type of patterns. The basis for the later structuration, idealization, conceptualization, assumes that it is of the same nature except that whereas they are 'clear', it is 'vague'. In fact, what is 'clear' is the final set of categories. But there seems to be several sets of categories possible. It seems to me that all philosophers following Husserl have this difficulty. On the one hand, they assert within the constitutive system the function of a level that is neither already conceptualized nor is it anything altogether undetermined and arbitrary, the pre-conceptual, pre-constituted, pre-thematic level, as it is usually called. Then they move directly toward perfectly rational set of categories. What is missing would be a study of this process of moving gradually from what is 'vague' to what is 'clearly conceptualized'; the variety of possible ways of conceptualization would be the first law of this process. In this respect, the first thing we might say is that the process of moving from the vague *life-world* to a set of categories has a variety of open possibilities. Secondly, one might venture, which I think Husserl has, an assertion that it is possible to have *vague experience*. Perhaps, thirdly, I am proposing here *a priori* categories of process – as you very well stated – that the process of structuring or conceptualizing is in some way objective, in the sense that it is a way of continuing what is possible with respect to the *world-horizon*. One takes a distance from one's own experience and says: these structures which I now establish for the first time have in fact always been there, which is correct in the sense that these structures were always possible. Other structures would never have worked. This would be an illustration of the sort of transcendental analysis that I think is necessary, which is an analysis of the process of moving from 'vague' to 'patterned', rather than asserting that some arbitrary set of patterns is the *a priori* structure. What is *a priori* here are the rules of the process of moving from the 'vague' to the 'structured'. Would you accept that as a Husserlian step?

Mohanty: I would prefer to respond at the end of the discussion. Meanwhile, I am going to gather my thoughts.

Morin: I was wondering why are you, Mr Gendlin, referring to the *life-world*? Why are you talking of 'vague'? If you are referring to the world-horizons and the system of constitution, why use the concept of 'vague' here? Does it even occur in this context?

Gendlin: I think you are perfectly right. 'Vague' is a word that means nothing else than something other than 'structured patterns'. It means 'non-structured', 'non-patterned', so that *life-world* or preconceptual experience is called 'vague' or 'ambiguous' by those who think chiefly in terms of patterns, or forms, or structures, and you are right that I should not call it 'vague'.

Morin: What I am concerned with is that the first paper and also the second paper refer to *life-world* conceived as a kind of a 'habitual-world', that we are used to live in, to something that in Ricoeur's thought is 'a reflection on habit'. In one passage in *Le Volontaire et l'Involontaire*, he mentions the horizon as the *monde de l'habitude*, the world of habit; it is more than the horizon of the world as possibilities of constitutive patterns, it is an established way or a system of acquired habits, from which you go on to something new, more interesting, like learning to skate, for example. You acquire the skill of skating and then you may move on to something more interesting, like playing hockey, entering competitions, etc. The horizon would be understood as the established system of life already acquired upon which one is working, further establishing more complex modes. This is why I don't want to use in this connection the expression 'vague'. It is not at all 'vague'. The life-horizon is the already established background.

Gendlin: I think you are right. It would be silly to say that living is vague, giving an absolute priority to concepts. It is only called 'vague' with respect to the way we would like to grasp it, or with respect to the steps of thought, or structures we would have to apply in order to do so and which its nature eludes; but what we want to grasp is, in fact, inherent in it. By our attempts to grasp it we unfold this virtual structure and structurize it further on higher levels.

Kuspit: I just wanted to ask whether what we are really talking about here – and I was struck particularly with the last presentation by Dr Meyn, and with the manner in which it circled back to what Professor Tymieniecka was talking about – is, that we seem to be reaching a kind of barrier, a boundary line of philosophy itself. It strikes me that in Husserl, and in the whole transcendental analysis, philosophy becomes much more tentative all the time; it becomes a *tentative enterprise* which cannot offer justification, which cannot come to any final ground, which cannot really work in that old sure footed way that it used to operate giving us a system in which everything was nicely laid out. This seems impossible on transcendental presuppositions, and what we are really asking for now is *poetry*, what we constantly come up against is that we seek something different than philosophical assumptions, principles, convictions; we want something that is life, something that is poetry, something that inherently is prior to the constructivism of philosophy.

Mohanty: Well, I think I see it in the following way. The problem of Husserl was not primarily the problem of *justification*, but his appeal to *life-world* intuition or experience was always a problem of *clarifying* meaning. In other words, he did make provision that the problem of justification can be within a certain theory or system, for example, mathematics may have its own internal problems and internal method of justifying the truth of a mathematical statement within its theory. It seems to me, I am not quite sure, that this is all he was doing, and that in his view any kind of concern or problem occurring within a theory would remain unclear, its meaning remains unclarified unless we in some way refer it to a pre-theoretical base outside this theory. That means that one can interpret, for example, his familiar method of *eidetic* variation not as arriving at a new truth by establishing the truth of a proposition through *eidetic* variation, but by intuitively *clarifying* the meaning of a proposition which we will be able to arrive at. In a peculiar way Husserl tried to bypass the problem of justification by emphasizing the problem of meaning. If that is so, one can possibly have an interpretation of the transcendental argument – take the problem of infinite regress interpretation which Dr Meyn gave – which would be an interpretation relative to justification. It means that one justifies each level by another level, and goes on in infinite regress. Or do you think that the process of clarifying meaning also would be involved in infinite regress?

Meyn: Yes, I think so. I think it would involve a regress here, too. I am reminded of Carnap's distinction between 'internal' and 'external' questions, and Husserl certainly doesn't want to ask 'internal' questions. But Carnap's answer to the 'external' questions are simply pragmatic, whereas Husserl claims that it is not enough; we have to have clarification, we have to understand what we are asking. You are quite right, it is not justification that he is seeking but clarification. But this clarification is still supposed to come from outside the framework and itself is not on any fundamental ground. It requires itself a further clarification or can be reiterated.

Mohanty: Yes, you have brought up the question of reiteration, which adds an important point. There is a kind of endlessness in the process of clarifying meaning, which, as you say, is the process of reiteration, but it does not seem to me that from this necessarily follows that you are always going behind to another level of experience. If I take Husserl's example of clarifying the meaning of a perfect geometrical concept, I might go through arranged sensible circular shapes in an infinite series. That means I can intuitively always go on through the series; it belongs actually to the meaning of the perfect shape that there is a possible infinite series through which you can traverse, with the perfection as its limiting form, but it belongs also to the concept of the perfect form that you do not go through it and imagine as if it has been gone through. That means, in clarifying the meaning of a construct or an idealized form, we find that it implies a sudden assumption, a sudden 'as if' we have gone through it. Husserl always realized, I am sure, that there is an infinity involved in any clarification of meaning, but he also realized that there is a peculiarly 'human' way of overcoming this infinity, precisely by this assumption 'as if'. So, if I am familiar with this 'transcendental argument', as he calls it, I think his version of the transcendental argument would be very different from the Kantian version. The Kantian version was a question of how this type of true judgments is possible. It was, as Kant has formulated it, a question of the conditions of the possibility of the truth of such *synthetic a priori* judgments and their meaning. Husserl's concern was not with the truth of certain scientific beliefs but with the possibility of having such scientific beliefs with evidence of meaning.

Meyn: If we ask about the beliefs – the possibility of scientific beliefs – this involves the concept of knowing, and then we are right back in the same problems as with the concept of truth. How do we get about to knowing these propositions? It seems to me that your answer really won't get us out of it.

Mohanty: Yes, yes. When Husserl spoke of how we know these propositions he seemed extremely dogmatic. He, for example, appealed to apodictic evidence, and there, it would seem that he was actually operating within an internal framework, e.g., formulating the question: "How does a mathematician arrive at a mathematical truth?" It would seem very dogmatic actually, but then as soon as he raises the question: "How do I make this clear, this meaning, this mathematical concept?" he reaches outside the framework to what he calls the *lived-world*. I do not think that he really keeps these two kinds of inquiries separate, and I am not even sure whether it is possible, as you say, to keep them separate. I confess, I cannot give an exact answer to the questions you have raised, but I think what is necessary at all to answer your question would be to be able to distinguish clearly between the Kantian transcendental argument and Husserl's transcendental argument.

Loverde: I would like to make a suggestion that may help to clarify Meyn's problem, because, I think, it has to do with two ways in which whatever it is that is constituted is going to be conserved. If you talk about something like an 'eternal truth' you presuppose it is conserved forever. For instance, when we talk about scientific laws we take for granted that even though the contrary is empirically possible, there are and will be empirical instances that confirm the law. There is then the sense of eternal conservation, and we have to distinguish two different things: one is a scientific law which we think of as an atemporal truth, and the other is what we were talking about before: instituted habits, ways of existing. Maybe we should then distinguish between the *a priori* of constitution as such and the *a priori* of its various levels. When you talk about the constituted *a prioris* of a given culture they might not be preserved throughout the future phases of cultural constitution of the *life-world*. Some of the *a priori* principles of the constitutions of a given *life-world* in the past may be preserved and others may fall by the wayside. There was, for instance, a time when slavery belonged to the *a priori* of the pattern of civilization. It has been rejected and its validity denied, which is a different way of progressing than what science at least claims to adopt. But I don't think when you are talking about what *a prioris* of constitution can be conserved, either of them has to have absolute foundations. I think science attempts to conserve well validated investigations, and I don't think the personal life and the cultural life needs to have that presumption. If this distinction is made we don't have to batter back and forth so much between what is an absolute eternal principle and what is relative. In may opinion they are very much related, and once you abandon the criterion of absolutism you can talk about different ways in which they are conserved.

Eng: I begin to see some strands of style forming themselves here. I think Donald Kuspit touched a very live nerve in our debate. To be philosophically an eternal beginner, as Husserl was, seems to lead to this critical situation; we see it in Heidegger's turning to Hölderlin, and Merleau-Ponty's last, very poetic phase. Incidentally, Socrates somewhere equates philosophy with music. I also think of course of Dr Tymieniecka's musical introduction. She used the term "orchestration" frequently, and this touches, of course, upon the guiding sensory metaphor in phenomenology. In suggesting that the philosopher returns to poetry brings to my mind a number of things, especially a topic I am still to write on called the *Suffering of Poets*. It includes Lorcus' recounting of the poet opening his veins in public, and Kierkegaard's telling imagery of the poet as being sacrificed in a kind of a Moloch, and his far-off cries – as he is perishing – reaching us as sounds of music. All this is related to the element of *pathos*; to the very deep relation of *pathos* to *Eros*, the suffering of the monad in the presence of *Eros*. Here we are in the highest Platonic realms, and this touches upon the element of 'self sacrifice' in the *epoché*. In may paper at the first Waterloo Conference on *Body Consciousness and Violence*, I talked about the element of *pathos* – which Aristotle discusses in his *Poetics* – as being integral and inseparable from the phenomenological reduction. I am not engaging in diagnostics when I mention the problem of personal depression of Husserl throughout his life, and which I am convinced is tied up with the difficulties in *Entsagung* or renunciation, in embodying *phantasia*, in embodying imagination. It is the renunciation of real advantages that the poet very gradually discovers as he goes farther and farther in giving body to the *logos*: he discovers that the *logos* has a life of its own, and he, the poet, surrenders to it. Hence, the 'self destruction' of poets, the 'madness' of poets is in some way, we sense, intrinsically bound up with what they have accomplished. Maybe it is in the struggle, in the attempt to come to terms with this conflict by really giving up one's own individual life and abandoning one's personal self to the transcendental subjectivity, that the personal meaning of the creative act lies. Out of this all imagination develops.

Loverde: What you said reminded me of the sort of correlation I made between the two notions of 'scientific' and 'lived', and of 'scientific' and 'poetic'. The idea of 'renunciation' on the one side seems to correspond to the idea of what you might call 'being able to stand back in a detached form' of scientific objectivity. The poet who leaves behind a habitual embodiment by taking a distance from it or a form of 'renunciation' is like a detached observer. On the other side, he gets embodied in – what was just referred to now as – a sort of a primordial *logos*, which has its own way of existing to which he gives himself up. This movement, I think, is something that is perhaps often ignored in the creative process, this ability to get disembodied for another purpose, and I am not sure that it is all that different than the kind of thing that a scientist has to do when he is trying to become objective. When Husserl talked about taking greater and greater distances of formalization from the concrete sensory experience of circle drawn on the ground, and successively rising to a much higher sense of an absolutely perfect circle, that is a kind of 'disembodiment' of forms or transcendental suspension. It seems to me that they are comparable with each other; in one case we say it is scientific law, and in the other we can poetically call it a kind of *logos*. But, I think, there are corollaries that can be made on all of these various alternative ways of talking about what perhaps might be pretty much similar.

Bove: We all agree with what is being said here: that the critical point in philosophy is to thematize the concrete texture of experience. A poet or an artist, a painter or a musician work with a concrete fabric. Part of the fabric is in his mind and part of the fabric is in the concrete elements of experience he handles. I think Professor Tymieniecka hit on a very important point for philosophers. We should neither succumb to the impact of *pathos*; we should not feel that the artist's inebriation with what he does has to make us become artists, nor should the philosopher become inebriated with philosophy the same way that an artist is with his feelings. Now, I think, that one of the reasons why an artist does not lose the intimate contact with concrete experience is that he deals with sensory things; it could be his emotions, his passions, his love. Whereas a philosopher, in this day and age, finds it very difficult to deal with the concrete facts of life, precisely, because he sees the disparity between the human experience of life and man's conceptual thematization of it. Husserl was keenly aware of this disparity between the two. I think then, that the practical problem of a philosopher today is, as Professor Tymieniecka said, to make these non-verbal elements, these 'vague patterns' that are shooting through the world play their role in the philosophical reconstruction of the human universe which necessarily operates with abstract rational notions. Actually, there is a great fracture among artists concerning the role and dimension of passion or *pathos* in contemporary art. We don't have many great artists of the stature of a Rembrandt who are whole persons. I have seen many artists who are trying to use art to *become* whole persons. The situation is the same in philosophy. But an artist suffers and eventually learns something. What he learns are not the kind of things that a philosopher usually deals with directly, but I think – and again Professor Tymieniecka was pointing in that direction – a philosopher should take and bring all his experiences into a system of rational expression. That is necessarily the craft of the philsoper.

Glyn: What has been said by the last three or four speakers about passion and living is either the total bankruptcy of philosophy or else it is a real step forward, which I think it is. It is a real step forward to assert the role of concrete experience for philosophy, and then to institute a really careful study about the central question which is: what is that progress from 'living' – I will call it now more correctly 'living' rather than 'vague patterns' – to 'conceptualizing'? What is that process that would accomplish the passing from living to something conceptualized: thinkable, speakable?

Pax: I am not satisfied yet with the discussion between Professor Meyn and Professor Mohanty on the question of clarification and justification. If I understand you correctly, Professor Meyn, you were saying that the transcendental argument leads back to a conditional type of position, arising from the fact, or because of the fact, that it is the condition necessary for our experience and is not therefore a justification of the transcendental position. What I would like to suggest is that in the Husserlian position the meaning of 'justification' is precisely 'clarification', and there is no way to ask for justification of the transcendental position: the clarification is itself the justification. I think this is also the explanation as to how clarification (or justification) of the constituted world proceeds by retracing the steps of the transcendental genesis from the most complex structuration of meanings back reductively to the *lived-world*, back to the life of passions, and so forth. It seems to me, if you are asking for a different meaning of justification it would have to be a justification which would in some way be 'outside' of the transcendental viewpoint which is all inclusive of experience.

Ver Eecke: When Dr Tymieniecka gave her inaugural lecture she spoke of two heroes of contemporary philosophy: Kant and Husserl. She claimed that, contrary to their assumption of the absolute sovereignty of reason, she proposes to establish the limits of reason. Hegel and Freud have already indicated that reason has its limits, although they also indicated that it is a task of man to widen the territory that reason can cover, but in expanding that territory, they had to pay a price, namely, that reason could not be the sovereign master. They have, furthermore, indicated that allowing reason to be dethroned does not mean that all 'possible worlds' become possible. It means that even the irrational forces have limits and that the limits of irrational forces can be clearly understood. Therefore, I agree fully with the basic thesis of Miss Tymieniecka, but I would not accept the last conclusion that the problem of all 'possible worlds' is fully open to debate. I would have, naturally, very great difficulties in attempting to show where exactly these limits are, I simply wish to refer to these two alternative heroes of the philosophical tradition, if you can call Freud in that tradition, to say, that they have already tried to indicate the limits of the irrational within the philosophical reflection.

Tymieniecka: I agree entirely with Professor Ver Eecke that Freud and Hegel have tried to show the limits of reason, I would even add to it that as far back as Plato we have already seen a most serious attempt to disentangle the roles which various functions of reason play in men, but I have chosen Kant and Husserl because what I propose to improve lies in their line of thought. The reasons for it are very clear. As we all know, no matter what previous philosophers have contributed to the task of analyzing man's faculties, the nature of the world and its origin, it is especially Kant and Husserl who have developed the most thorough analysis of the human consciousness as the engineer of the world. They seem to have synthesized the wisdom of history at that point, and they have done an enormous amount of detailed analyses on their own. Since I have been attacking this problem in its very roots, it is quite natural that they became the target of my criticism. There is also another matter which probably justifies my choice of Kant and Husserl. A scholarly pursuit in distinction to, let us say, a strictly artistic pursuit, always takes into account in a more or less precise way the *direct* inheritance of the issue in question. The artist when he is composing a symphony or writing a poem is only *indirectly* considering the fact that someone else has already done something similar, or that he is trying to do something new.

This relation between the artist and his predecessors in artistic creations is a loose one. In philosophy, taken as a *scholarly pursuit*, the situation is different. What we are ourselves accomplishing is to add a tiny element to the already established common good of philosophical reflection. It takes its meaning from being incorporated into what has been done before in the field, and by its distinctness it adds to the great work which has been done before. Otherwise, we are incapable to establish in a scholarly way neither the validity of what we are doing nor its contribution to the common treasury of philosophical speculation. We could profit then if you would bring out what Freud, Hegel, and other phisophers had to say on the issue concerning the origin of the human world.

Ver Eecke: I agree entirely with the second justification where you locate yourself against your predecessors. I accept immediately your invitation to say that when the results of someone else's study is on the table one might then refer to similar things said by other philosophers from another tradition; and it is because of that invitation that I want to insist that the title of one of the important works of Hegel is *Phenomenology of the Mind*, which means that you cannot pretend that you have the right to choose Kant and Husserl over Hegel, because both claim to study consciousness. I would like to say – and maybe then that is my prejudice – that the *Phenomenology of the Mind* does more, gives more light concerning the constitution of consciousness than the *Crisis* of Husserl. But that might be just my prejudice, and at this point we see that we have different origins and a particular kind of debate is going to originate when we confront our mutual results. What I respect deeply in Hegel, that Husserl has not, is the tragic dimension in consciousness and in the mind; that kind of dimension is more absent than present in the Husserlian and the Kantian analysis. That is the contribution, I believe, that Hegel can make whenever a scholar begins his investigations with the results of Husserl as his own, and then discovers a form of tragic dimension in consciousness.

Tymieniecka: I am gratified by what you are pointing out, and I must admit with a certain embarrassment that, of recent, I have not read Hegel indeed. Your remarks have importance because the 'tragic dimension' you are emphasizing has been in fact the gist of my research concerning the 'creative function'. If you happen to read my book, *Eros and Logos*, you will note that I am starting precisely with the premise that phenomenology of the transcendental constitution in the Husserlian vein is unfolding the *automatized* function of consciousness without taking into account that what man is *really dealing with* in his life. In other words, man's life does not proceed by universally predelineated processes, rather it unfolds and advances by *conflicts*. In fact my conception of the creative experience centers on the analysis of *this* crucial conflict. My 'phenomenology of the creative experience' is oriented toward the discovery of this tragic dimension and the specification of its role in the 'creative function'. Consequently, I am equally gratified by the remarks of Dr Eng concerning *pathos* and the tragic in human experience. As a matter of fact this is the focal point in the context of the creative experience or function in distinction to the constitutive framework which makes it so radically different from the simple unfolding of the cognitive function. It is related to the actual experience of living. This is the perspective which is so far missing in phenomenology, and it is being overlooked in the one-dimensional pursuit of the constitutive function. My perspective of the phenomenological study of man is based on the recognition of the crucial role of conflicts and the tragic, as real factors of life. This is the issue I endeavoured to bring to the fore in *Eros and Logos*. A final word about Dr Ver Eecke's remark on the idea of 'possible worlds. In my view the problem refers to the role of *creative imagination*, upon which I have dwelt considerably in my lecture. In fact, the freedom of the worlds of the *creative imagination* – as I conceive of it – proves the *a priori* of the constitutive function as the vehicle of *originality* in the creative process and consequently in the formation of 'possible worlds'. The all important difference is, however, that in my view creative imagination is not a *faculty*, as it is customary to treat it in philosophy from Plato to Sartre, rather it is a *function* emerging from the *creative orchestration* of man's faculties.

Straus: The situation of a chairman is like being of a guillotine; if he indulges in the others he may lose his own head, so I would like Professor Matczak to say a final word.

Matczak: From the discussion which we had, especially this morning, I gather that we have come to the fundamental problem of phenomenology. It cannot be answered in a few words, but I think it is of paramount importance that we discuss it. Today the discussions focused primarily on the problem of justification. This is an essential problem, namely, how to justify the 'phenomenological reductions'. The answer given was that justification lies in clarification. I think, however, that justification and clarification are two different things: clarification answers the question "*how* the thing develops". Justification answers the question "*why* it is so". With this is connected the further problem: if phenomenology could offer only clarification then it would be merely a kind of positive science, a science that tries to describe and clarify in a certain way things and beings. But philosophy goes to the final answer: *why* things and beings are as they are. If, as Professor Mohanty says, the world is infinite, then there is an infinite progress of both, the world and its clarification, and the problem still remains: where is the difference between science and philosophy? Science sets limits to its progress, but should philosophy set itself limits or rather search as far as human reason, feelings, and strivings permit it to go? If not the latter, philosophy will be a science like mathematics, chemistry, and so on. Another problem: the problem of *a priori* and facts. What is the justification for *a priori*? Isn't it so that when we discuss various problems we find the decisive certitude in facts and not in *a priori* principles? And this is the objection against Kant, against Hegel, against Husserl: that it is not the *a priori* which decides. Transcendental reduction will have its proof only in factual statements. It seems to me that in the line of justification there is a complete reversal of phenomenology, which demands to be taken into account. Fact is the decisive element and not *a priori* abstractions and speculations. The next issue I wish to bring up is the question of the relation between passion and reason. Professor Tymieniecka introduced this morning an extremely useful and clarifying distinction of a great value for phenomenology, and yet, the problem remains: what is the difference between passion and reason? Creativity, of which Dr Tymieniecka spoke, reminds me of the *élan vital* of Bergson. What is creativity? What is it and how do you know it? For Bergson: by a kind of feeling. But what about reason? Dr Tymieniecka has proposed a wholly functional system to account for creative activity comprised of both. Creative activity would be a process that develops and has a line of unfolding. But since she opposes the creative activity to the system of the functioning of reason, how do we know about this process, by feeling? And yet, when we analyze the creative progress as a functional system it is done by reason. How does reason stand, then, to this intuitive grasp of feeling? Especially since it seems that reason in its own function aims at something else. It was first the question of the distinction between phenomenology and a particular science. Now the question is of the distinction between phenomenology and literature, phenomenology and spirituality. Does a philosophical analysis of creativity oscillating between the intuition of feeling and the ordering of reason provide a basis for such distinction? Or is feeling and reason the same?

NOTES

* Research for this paper was supported by a summer grant from the SUNYA Committee on Institutional Funds.
[1] Barry Stroud, 'Transcendental Arguments', *The Journal of Philosophy* **65** (1968) 241–256; Moltke S. Gram, 'Transcendental Arguments', *Noûs* **5** (1971) 15–26.
[2] *Phänomenologie – Metaphysik oder Methode?*, H. Bouvier & Co., Bonn, 1966, esp. Chapter B6.
[3] A. J. Ayer, *The Problem of Knowledge*, Penguin Books, Inc., Baltimore, 1962, Chapter 2 (iii), esp. pp. 50–52. Gram, *loc. cit.*
[4] Jaako Hintikka, '*Cogito, Ergo Sum:* Inference or Performance?', *The Philosophical Review* **71** (1962) 3–32.
[5] 2nd ed.; Martinus Nijhoff, The Hague, 1969, p. 142, italics mine.

DISCUSSION

Special Contribution to the Debate

DONALD KUSPIT

The 'A Priori' of Taste

In a rather informal manner I would like to reflect on the phenomenological aspect of Kant's *a priori* of taste. It seems to me that, when
Kant is talking of the *a priori* of taste in the third *Critique*, he is talking
really about his sytematic presentation of cognitive materials. He is not
talking about beauty at all, as he affirms he does. His real concern is with
the nature of representation which seems to be at the core of this thinking,
what he calls in the first *Critique*: 'the matrix of representation'. I will
limit myself to a few quotations in his presentation of the *a priori* of taste
in the third *Critique* and will attempt to show how these ideas have to do
with the phenomenological nature, so to say, of representation. Before
I do that, let me briefly remind you what Kant claims to be doing in his
discussion of the *a priori of taste*. Firstly, he defines taste in general as
"an examination of what is required to call an object beautiful". Throughout his discussion, he emphasizes constantly the contemplative nature of
what he calls 'the judgment of taste'. He repeats this again in the introduction to the *Metaphysics of Morals*, where he emphasizes, for example, that
the judgment of taste is 'passive', it is 'contemplative', and he insists that
it involves a kind of elimination of the outflow of vital force. Secondly,
he indicates that eventually it is insignificant whether or not the object of
the representation, that we obtain in the judgment of taste, exists or not.
In other words, he is not concerned with ontological questions here. You
notice, that the very operation of the 'passive, contemplative' judgment
of taste, as Kant sets it up, is eliminating any attention to the object itself,
and instead, seems to be a focussing upon the idea of representation.
Furthermore, the judgment of taste, as he says, reaches its climax in the
"third moment of the judgment". The climax involves: "the consciousness of the mere formal purposiveness in the play of the subject's cognitive
powers". There is no mention strictly speaking of any *perception* of
beauty. Kant emphasizes that it "is a discovery of – what he calls – inner
causality", which is purposive with respect to cognition in general;
furthermore, this inner causality involves, what he calls, "a maintaining

Tymieniecka (ed.), Analecta Husserliana, Vol. III, 116–123. *All Rights Reserved.*
Copyright © 1974 by D. Reidel Publishing Company, Dordrecht-Holland.

of the state of the representation itself and the occupation of the cognitive powers"; and finally "we linger over the contemplation of the beautiful because this contemplation strengthens and reproduces itself". Now, let us put all this together and see what we have. The quotes I want to concentrate on, the three – we might say – ideas I want to concentrate on, are as follows: (1) what we have in the state of judgment, let us call it that, the state of the judgment of taste is basically a passive state of consciousness, (2) in this passive state of consciousness we have consciousness in some way calling attention to itself, maintaining itself, maintaining its own representations, and (3) this maintaining its own representations involves a kind of lingering sensation where we have a sense of lingering of conciousness, and this lingering consciousness is the butt of the contemplation of the beautiful. Hegel says in the *Encyclopedia* of 1839 about Kant that Kant's nature of the *a priori*, which I think brings us to the core of the matter, is simply "the spontaneity of thinking, nothing else". The spontaneity of thinking, says Hegel further, is the taking of the whole of experience into the subject. It seems to me that this remark by Hegel in the *Encyclopedia* applies to Kant's discussion of the *a priori* of taste. In fact, Kant is seeing consciousness in itself as nothing but the activity of representing without any concern whatsoever about *what* is being represented. This is why Kant speaks of a purposive 'freeplay' as operational in representation. This 'free play' – as Kant specifically says – is not connected with the categories of understanding. Thus, we have, on the one hand, the removal of the categories of understanding to one side, and on the other hand, an emphasis on consciousness as a spontaneous purposive 'free play', which is concerned with the production of representation which does not represent any object in reality at all. We have here, it seems to me, a very peculiar description of the phenomenological reduction in a kind of primitive form, pre-Husserlian form. Namely, first, in this initial passive state of 'contemplation', which Kant calls the 'dispassionate' or the 'disinterested state', consciousness discovers itself as nothing but consciousness in general, this means: spontaneous 'free' consciousness; in the next stage of lingering, what we have is consciousness displaying what its content is dealing with. The third stage is revealed by Kant's very interesting remarks in the conclusion of his discussion of the *a priori* of taste where he tries to illustrate his position by examples. Talking about delineation of forms, he says: "in paintings, sculpture, and in all the formative arts, in architecture, and horticulture so far as they are beautiful arts, the *delineation* – and he himself emphasizes this term – is the essential theme, and here it is not what gratifies in sensation but what pleases by means of its form that is fundamental for taste". And, incidentally, he talks about the superiority of delineated form over charm conveyed by emotion. (Schopenhauer employs this distinction in his discussion of the metaphysics of fine arts.) Delineated form is itself the reification, I want to contend, the reification in sensibility of the representation discovered by lingering. To summarise: first, consciousness in a passive state discovers that it is simply a free spontaneous activity; then, it finds this lingering consciousness of representation; finally, having no object, it reifies its own representation, that is, it discovers very simply that it has sensibility. In conclusion I would say that, from the point of view of Kant's Copernican revolution, viz., shifting the emphasis to the subject rather than to the object, under the auspices of discussion of the *a priori* of taste, Kant is concerned with the subject's desire for an esthetic form of its own. In other words, what Kant really says is that the delineated form, which presumably is apparent in the fine arts, is actually the form of consciousness itself as it is able to talk about itself. I, therefore, claim that Kant is only interested in consciousness's ability to be self sufficient and adequate to itself in its power of reasoning.

Straus: May I for a moment forget my position as a chairman, and may I ask is *representation* the word for the German word: *Vorstellung?* Furthermore, is it your position to maintain that in Kant there is no account of the object of a representation?

Kuspit: The question was, whether in Kant there is an account of seeing *something* or *someone*, is that correct? Yes. I would say there is, but I do not think that is what he is really talking about in the discussion of the *a priori* of taste. Because, I think, he is stripping down 'seeing' into this peculiar form which he himself calls 'esthetic', which is no longer 'seeing' like you or I see each other. The emphasis, it seems to me, is not on the objects seen but on the seeing itself; on the special phenomenon of seeing itself which goes right back to representation as, so to say, the phenomenon in itself which he is examining. He gives us the example of a palace. Rousseau, for instance, sees a beautiful palace and would like all beautiful places torn down, somebody else would prefer to live in a wigwam rather than in a beautiful palace. But, Kant says, these attitudes are in this case all irrelevant, because in this question we are not concerned with whether these things actually exist, whether a beautiful palace actually exists, we are interested in the idea of beauty. The idea of beauty is therefore not connected with the fact of seeing something in particular.

Straus: I look at a play in a theatre, I see a scene, and I see the people as I see you now in the surroundings here.

Kuspit: I say that there is more to esthetic perception than to ordinary perception. When I 'see' on the stage actors, am I seeing simply people in make-believe clothes? I feel that the question about esthetic 'seeing' is radically different from the question of whether I see you now. It is a different issue and a more complicated issue than we may realize.

Ver Eecke: I wanted to take up the issue of objectivity in Kant. If my understanding of Kant is correct, he has two steps for the constitution of the object. The first step was underlined by the previous speaker, namely, the part that is contributed by the senses and the result that is phenomenal. When I look at the same chair from three angles, I have three 'phenomena'. My senses give me three aspects of that same chair, but I confirm these three aspects as belonging to one object. The second step is the act confirming the three phenomena as beings aspects of *one* object, and this is an activity of the mind, whereby – as Kant says – I elevate the phenomenon to the status of an object. I believe that in this interpretation of Kant you can explain your difficulty with the esthetic perception. Namely, the role of the mind, or the emotional life, or the time dimension involved in constituting the object from the sensuous data is different from the manner in which they operate in the judgment of taste. You can have the differentiation between the real object and the esthetic object or theatrical object because between the senses and the object there is the act of constitution and contribution of the mind.

Tymieniecka: I am sorry but I have to disagree with the three previous speakers concerning the way in which Kant conceives of 'constitution', to use a Husserlian term, both in cognition and in art. If Kant would speak the way Dr Ver Eecke maintains, he does – but he does not – namely, that we see the chair from three sides and then from the level of seeing it thusly we pass into, what Husserl calls, the 'intentional object' of the chair, then Dr Ver Eecke suggests that this passage would be due to the 'activity' of consciousness whereas in the sensuous stage consciousness was 'passive'. Let us recall that what Kant distinguishes in the first place is the manifold of experience which does not yield cognition until it has fallen into the forms of space and time; still, this esthetic stage does not deliver any cognition in the strict sense either. However, on this point you are right: the spatio-temporal system of the receptive stage must be immediately submitted to the active stage of the constitution, viz., to the categories. On the other hand, the point on which – I think – you are mistaken, and from which Dr Kuspit fails to draw the conclusion is the fact that there is a *hiatus* between this concrete, completely unshaped received data – which the forms of space and time try somewhat to grasp but alone cannot bring to objectivity – and the categories. Kant introduces then – as I have shown in my lecture – something which is absolutely crucial for the Kantian system, and also for all the debates here, namely, the *Einbildungskraft*. It enters into play as a blind force of what Kant calls – the 'empirical soul'.

By 'empirical soul' Kant – as well as Husserl after him – means the elemental dimension of human consciousness or of the mind-body situation which for both, Kant and Husserl, remains completely in the shadow for the sake of the intellect. From this blind background of the empirical soul emerges the *Einbildungskraft* which is supposed to mediate first between the manifold of experience – by getting into the esthetic forms of space and time – and the categorial schemata.

Concerning the controversy between Dr Kuspit and Dr Straus, the crucial point of difference which Kant sees between the object of arts and esthetic judgment in general, on the one hand, and the object of cognition, on the other, lies – it seems to me – precisely in the role of the *Einbildungskraft*, namely, in what Husserl calls the 'constitution of an object of cognition', or the 'passive genesis', where the *Einbildungskraft* would be simply an agent of his own mediating between the passive side of subjectivity and the active side of the category of schemata. Kant realized – whereas Husserl did not – the role of the esthetic or artistic object in the constitution of the world as something absolutely crucial for the explanation of the world, but he realized this – as you well know – only after the composition of the *First Critique* where in the whole constitutive cognitive apparatus is set up. Then to account for the originality of the esthetic object, this object of creation, he takes recourse in the
· *Einbildungskraft* and its elemental force of nature in a very special way: the notion of the genius is introduced with respect to whom the *Ein-*

bildungskraft is taking the leading role; namely, the schemata are thrown aside, the works of reason are suspended completely, the *Einbildungskraft* is liberated from their jurisdiction and works in the genius on its own. Originality in the work of art is due to the genius having the unimpeded use of the *Einbildungskraft*.

At this juncture the question arises for Kant: under what auspices, according to what criteria is the *Einbildungskraft* operating? In order to answer his own question Kant introduces the notion of taste. Taste is the principle of the *Einbildungskraft*. Dr Kuspit has shown in an interesting way how a person in the actual living of an esthetic experience is simply 'contemplating', he is set in front of an object to see it in a special way when his consciousness is altogether freed from the schemata. But the reason why I did not agree with Dr Kuspit is that the meaning of this phenomenon – to use your term – is that consciousness freed from the schemata is left entirely overpowered by the *Einbildungskraft*, which then applies to the emotive-affective self-made, so to speak, taste for criterion.

Here lies the crux of the matter. We see that both Kant and Husserl failed on this point concerning the debate about the *a priori*. Both of them tried to reconstruct the world first in terms of the structuration of the workings of reason and intellect, and both came to the conclusion that it alone did not work. As we very well know the partition of conscious activities into the 'empirical consciousness' and the 'transcendental consciousness' is a terrible burden for the later Husserl which he tried to overcome the best he could.

Beginning with *Ideas* II and III, the notion of the 'empirical soul' appears in his reflections in such a light that it *could* mediate – as Dr Kuypers pointed out – between the two but he never managed to break through.[1] Furthermore, as I have emphasized in my lecture, Husserl could not account for originality along the line of the automatized rationality of constitution. Kant succeeded in making this breakthrough by the *Einbildungskraft*, but it seems to me, that he has made a major error: granted, he can in this manner retrieve the source of originality but then he has lost the ground for the determination of the criteria of taste. It is true that he tries to salvage the situation by suggesting that the *Einbildungskraft* receives the constructive principles of creativity from nature, it is nature that guides the genius in the organization of taste. But this interpretation is evasive and inadequate. For if this were the case we can legitimately ask: "Wherefrom did nature acquire these principles? Why would nature establish first the transcendental consciousness which would work on a routine basis, and them make a break in it and install other principles"?

It seems to me that the failure of both, Kant and Husserl, to account for creativity, originality, and thereby to find a way out of the *transcendental* and unique world to the *realism* of possible worlds lies in their approach to the origin of the lived world in terms of faculties. I have proposed to you my approach, which I believe is the truly phenomenological one, and which seeks to resolve the issue in terms of the *functional context of creativity*.

It appears then that imagination is indeed the key to originality and the way out of the dead-end street of transcendental idealism with respect to the origin of the world. But 'creative imagination', which I propose, is not a faculty; it is a functional system emerging from the whole *orchestration of the creative system*. As such it is free from the *constitutive a priori* and operates on its own.

Matczak: In order to fully comprehend Kant's mind, I believe, we have to add a *fourth* stage to the previously discussed three stages of his constitutive scheme, namely, that of *practical reason*. In the first three stages, that is in sensibility, understanding, and pure reason we establish the objective reality or the level of objectivity of structures and ideas; but then the question arises concerning the existential state of this structural scheme, not in the interiority of conscious faculties but in the actuality of the 'external world'. Dr Tymieniecka has formulated the problem of realism in terms of the issue of 'possible worlds', and presented it as a breakthrough in the idealistic position which encompasses man in the monolithic unicity of one transcendental world. But there still remains the problem of the existence of the world. It is precisely at this juncture that the fourth stage comes into play, namely, practical reason postulates that objects within space have to exist 'outside' of our mind.

Straus: In the *Critique of Pure Reason*, Kant speaks of the "esthetics of space and time", and I think he never went beyond it; he never treated 'esthetics' in the sense in which it was begun by Schiller and then developed in the nineteenth century.

Kuspit: Your remark confirms my interpretation of Kant. You are merely saying that an esthetic experience is reproducible because it is a representation; and I want to add that representation in fact is reproducible in a way reality is not. The point I am making is that Kant becomes fascinated with the suspended state of representation, with lingering, with the maintaining of representation. Professor Tymieniecka brought out very well in her remarks on my presentation that perhaps Kant does not fully know what to do with representation. I believe this is so and partly for two reasons: first, Kant was much more Humean and conventional than one might care to admit, witness Hume's essay on taste. Kant was too concerned with 'good' taste which remains with the rational approach and neglects the role of the *Einbildungskraft*. Secondly, it seems to me that in the issue of representation the Kantian emphasis is on the exposition of 'what it means to be a subject', thus on the analysis of consciousness, and not on the status of the object contemplated in esthetic representation.

Eng: I want to take up Dr Matczak's request for a discussion on the meaning of 'justification' in phenomenology, and combine it with the issue of Kant at hand. Incidentally, it is regrettable that Kant's *Opus Posthumus* was not mentioned by anyone here with its emphasis upon the lived body, a topic that is rarely touched on in discussing Kant. When he writes on *Leib* and *Leiblichkeit* he comes very close in his ideas to the later Husserl.

Concerning justification in phenomenology, it seems to me, that phenomenology finds its unique justification by beginning with the fundamental situation of man as holding within himself both: birth and death. This recognition is expressedly the radical character of temporality in Husserl. The central role of temporality, particularly the importance that temporality takes in the form of 'self-temporalization' in the inner time consciousness, is paramount in Husserl's thought. Man is a creature who is aware of both: his birth and his death; we carry within ourselves by our capacity for procreation birth, likewise we carry death, a beginning and an end.

Straus: This session is dedicated to Roman Ingarden. I had the pleasure of having met him in the year of 1969. Our first speaker is Professor Półtawski.

NOTE

[1] Cf. K. Kuypers, 'The Sciences of Man and the Theory of Husserl's Two Attitudes', *Analecta Husserliana* **II** (1972), pp. 186–195.

ANDRZEJ PÓŁTAWSKI

CONSCIOUSNESS AND ACTION
IN INGARDEN'S THOUGHT

In this paper I shall speak about the place of consciousness and action in the structure of man according to the views of Roman Ingarden. The paper consists of four parts. In Part I I shall try to show the development of Ingarden's views on pure consciousness as it has been described by Edmund Husserl, and to ask some questions about the actual possibility of using the concept of pure consciousness in the framework of Ingarden's ontology.

In the second part of the paper I shall discuss Ingarden's way to a consideration of the problem of action, of *praxis*, according to his essay 'Man and Time' which is particularly important in this respect and only a small part of which is actually available in a language other than Polish.

In Part III I shall give a short account of the situation of consciousness and action in the structure of man as it is presented in some passages of his main work, *The Controversy about the Existence of the World*,[1] and in his last treatise on responsibility.

In the concluding, fourth part of the paper some questions concerning the actual structures of consciousness, as they seem to be necessary from the point of view of the presented conception of human nature shall be asked and some consideration shall be given to a certain ambiguity and certain lacunae which seem to be present in Ingarden's analyses of consciousness. In this part I shall also mention some works which supplement the ensuing picture of man by elucidating problems which have been left open or not explained in a satisfactory way by Ingarden himself.

I. INGARDEN'S VIEWS ON PURE CONSCIOUSNESS

Roman Ingarden's many-sided and fruitful philosophical life was dominated by one idea: to clarify and to solve the realism – idealism issue. This idea originated from the opposition of the young Goettingen student of Husserl to the transcendental idealism of his master.

Nevertheless, in his main work, *The Controversy about the Existence*

Tymieniecka (ed.), Analecta Husserliana, Vol. III, 124–137. *All Rights Reserved.*
Copyright © 1974 by D. Reidel Publishing Company, Dordrecht-Holland.

of the World, Ingarden assumes as a provisional starting-point Husserl's formulation of the question at issue. Thus, initially, he accepts the conceptual framework of transcendental phenomenology. But, as his investigations proceed, he realizes more and more clearly the fundamental difference existing between his own approach and that of his master.

The transcendental way of putting the problem of the external world presupposes an opposition between the real world and pure consciousness. We may express this shortly (but rather crudely) by saying that, assuming the latter as existing beyond doubt, we ask whether the world given in our conscious acts exists also. And, if it does, how – what is its mode of existence. Thus, the concept of pure consciousness is crucial in transcendental investigation.

Husserl's position was mainly epistemological. He wanted to establish, in the first place, a philosophical foundation of knowledge, a *Wissenschaftslehre*. As Ernst Tugendhat writes, his ontology was, in fact, an *aletheiology*, a descriptive analysis of truth in which the main ontological concepts are regarded as derived from the respective forms of experience.[2]

Consequently, the way in which an object of some region is given to us has for Husserl an ontological priority to the structure or form of the object itself and determines its mode of existence.

This seems to show itself in a particularly striking way in his notion of pure consciousness. According to *Ideas* I, pure consciousness is:

(1) Adequately, indubitably *given* in immanent perception.

(2) A self-sufficient, *closed region* of being, "which receives nothing and from which nothing can escape".

(3) An *absolute being*, which *"nulla 'res' indiget ad existendum"* – does not need anything in order to exist.

(4) A sphere of *'irreality'* in which all reality is constituted.

In opposition to this, Ingarden's extensive analyses of the structure of single objects and, as distinguished from them, of regions of being seem to show that pure consciousness might, at most, be regarded as a single object and not as a whole region of being.

As to (3), his description of temporal existence, in particular his analysis of the consequences of the fact that only the present exists actually, being like a crevice between the past and the future – and his contention that our *Erlebnisse* do exist in time, speak against ascribing to pure consciousness an absolute, wholly self-sufficient being.

Concerning (4): According to Ingarden, pure consciousness and the pure *Ego* form with the real psychological subject an indivisible unity: the human monad. And both this subject and the respective stream of consciousness are very closely connected with a living human body. This seems to put into doubt the possibility of regarding pure consciousness as exterior to the real world and as having another – and, so to speak, stronger – mode of existence than this world (in particular: than the real human person whose consciousness it is). If there is any difference at all, then the stronger mode of existence would rather be that of the real person.

Consequently, from among the four main features of pure consciousness as stated in Husserl's *Ideas*, it is only the first – that of adequate (or, better, if we take into account the later development of Husserl's thought, *apodictic*) and indubitable givenness – that has been left.

But, as Ingarden stresses, ontology and theory of knowledge are mutually quite independent, and it is a mistake to deduce ontological statements from the epistemological ones. This seems to be particularly obvious when applied to any attempt at making the form or mode of existence of a real object (i.e., one which is, by its very notion, independent of its experience) dependent on the way in which it is given.[3] Ingarden also states explicitly that the problem of the ontological structure of a human person is not the same as the question of what is indubitably given, and that the line separating the actual strata of being is not the same as that which divides what is indubitable from what is not.

As a result of this development, we are left with the problem of what – according to Ingarden – *pure* consciousness, as such, could possibly be. Because he does not abandon the term altogether and admits that, in theory of knowledge, in order to avoid begging the question at issue, we must proceed in a transcendental way. In the additions to the third, German edition of the *Controversy*,[4] he still explicitly approves the procedure of basing an investigation of the problem of the external world on the ontological insight into the essence of pure consciousness – the insight from which we learn that it is given in an indubitable way in immanent perception. On the other hand, when writing *Controversy*, he does not attribute to our conscious acts any other mode of existence than reality (as he did as late as in the first edition of *Das literarische Kunstwerk* in the early thirties). And he states explicitly that pure consciousness is concretely contained in the innermost core of the real self and can be discerned as such "only abstractively, rather in thought only, and merely to some degree" [5].

But we may ask: is this delimitation of pure consciousness really a result of an *ontological insight* into the essence of consciousness as it is given in experience, or is it merely an epistemologically motivated construction of a sphere of *apodictic* givenness, i.e. an abstract epistemological idealization? The question, in other words, is whether what is called *pure* consciousness may be regarded as something individual and, in its way, concrete – or whether, on the contrary, it is only an abstract, ideal pattern which is often mistaken, if not for a whole region of being, then for an actual stratum in the structure of man (and probably of other living beings).

The already sketched line of development is, so to speak, the negative side of Ingarden's investigations concerning human consciousness. But there exists another, positive and more interesting line of his thought which leads to the problem of action, of *praxis*. I shall now proceed to an account of it.

II. REALISM, TIME AND HUMAN NATURE

There has appeared recently in Poland a little book, entitled *A Small Book on Man*,[6] containing Ingarden's papers on human nature.

Taking very seriously his master's ideas of a *Philosophie als strenge Wissenschaft*, of philosophy as a really well-founded, responsible investigation, and trying to establish it on the basis of a strict and detailed ontology, Ingarden was very cautious in expressing general views and in drawing general conclusions from his analyses. Consequently, the only technically written item in this publication (and the only comparatively long one) is his last and important treatise on responsibility. The other five are short essays, amounting together to less than half of the volume.[7]

One of these essays, 'Man and Time', in six parts (just thirtyone small pages long) seems to be of particular interest in tracing the development of its author. It took Ingarden almost a decade to write it, and it can indeed be regarded as a most striking indicator of the direction in which his philosophy was moving; and even, together with the already mentioned treatise on responsibility, as his philosophical testament. The first two parts of this essay were orginally Ingarden's contribution to the IXth

International Congress of Philosophy in Paris (*Congrès Decartes*) in 1937. The third part appeared in Polish in 1938. The concluding parts (V and VI) were written during the war and published in 1946, together with the rest of the essay, in a Polish literary magazine.

In his contribution to the *Congrès Descartes*, Ingarden describes two different ways of experiencing time. Their diversity, he says, is the deepest source of the fundamental opposition in metaphysics between the Heraclitean conception of being and that of the Eleats. And, in modern times, of the controversy between realism and transcendental idealism. But its most important consequences show themselves in the problem of human nature, making it the main query of philosophy.

In the first type of time-experience I start from the feeling of my own identity with myself. Identity means here two things: (1) That I am numerically the same, and (2). That I have one constant, individual nature which constitutes me as a particular real person. As given in this mode of experience, my *Erlebnisse* are only transient manifestations of my self which is transcendent in relation to them. That I am conscious of myself and of anything existing or going on 'in' myself – and that, in consequence of this, I actively shape my personality in a certain way – says Ingarden, is not simply because some of my *Erlebnisse* have taken place, but because of my actual human activity. This activity does not consist in experiencing anything, but leads to an impact of my real forces on some reality. Similarly, I owe my awareness of people, things and events in the world, and the possibility of my living together with them – my living in which I really change and transform myself – to the fact that my experiences are caused by real forces. Conscious acts constitute merely the mode of my life, and they do not even contain everything that takes place 'in' me and in my life. I can even exist when none of my experiences are actual. I transcend them, i.e., I am not any part or abstract feature of my consciousness.

Thus, in this mode of experiencing, time is not given as existing in itself but, at best, as a secondary appearance which owes its occurrence to some particular features of real being. Consequently, I perceive myself as *the same* as I was before. I live, says Ingarden, under the burden of my past which determines to a great extent my actual life, and I am almost constantly directed towards my future in which my plans are to be realized. And so the past and the future appear as actual factors of my life, and my self is given as transcending the boundaries of the present.

But there exists also another mode of experiencing, in which time and myself appear as the opposite of what they seemed to be in the first mode. It is, says Ingarden, in situations of two different kinds that an experience of this second mode may arise. This occurs, in the first place, when we perceive the fragility of a human being and the destructive power which time exerts upon it. Or, when we realize that, as human persons, we do not just exist, but that we constitute ourselves in various temporal perspectives.

In this attitude we see – or seem to see – that a temporal being exists essentially at the actual moment only and that it cannot really last, because every moment must necessarily pass and give place to the next one. And the present which, in the first form of experience, seemed to be a univocally determined phase of my history, becomes now but a point in which nothing can possibly be contained. And so it seems that nothing can really transcend the actual moment – and nothing can find its place within it. Were this true, I would have to regard myself as being just the actual phase of my ('pure') conscious processes or as the pure, transcendental subject whose existence and qualities are exhausted in saying that it is the subject of these acts. And we would have to conceive it as vanishing and coming again into being at every single moment of the period of time during which it is supposed to exist. But this would really mean that, as a man, I do not exist at all. And I am not, says Ingarden, prepared to admit this readily. Therefore, I appeal to my individual, immutable nature whose existence seemed so obvious in the first mode of experience. But when I try to ascertain what this nature is really like, I realize that it is itself necessarily immersed in time. I can only perceive myself in temporal perspectives which necessarily change with the passage of time, and I can only learn who I am at the moment when the actual present is already past.

Concluding his Paris contribution, Ingarden quotes Descartes' demand of inquiring who I am – the I who, according to the second *Meditation on First Philosophy* does necessarily exist. This inquiry, says Ingarden, must be started once again on a new basis: that of the different modes of our time experience. And, on this basis, we must seek a new solution for the question.

And so the Paris lecture was but a formulation of the problem that is to be solved. In the third part of 'Man and Time', Ingarden discusses some consequences of assuming the second mode of time experience to be true. Living in this mode, we consider the present as a point dividing two regions of inexistence: the past and the future. And within this point not only the actual subject and his life, but also time itself must be constituted. But if my real self is nothing more than, in a sense, a fictitious creation of my conscious processes – why should I not, asks Ingarden, deem the past and the future and the passage of time itself as merely an intentional creation of my acts? This has been indeed attempted in several different ways, for example by Kant, Bergson and Husserl. Nevertheless, it seems difficult to understand how such a constitution in the present, conceived as just a point in time, could be possible at all. And this seems to put into doubt this whole conception of time.

To avoid abandoning it altogether, we can assume that our acts transcend the actual present. But this is tantamount to admitting that, conceiving the present as a temporal point, we may assume an identity of an *Erlebnis* which transcends this point and is extended in time. And if we concede this, why not admit as well that all the transcendent beings, and in particular the self which was given in the first type of experience, do really exist in the way they seemed to exist, i.e. as independent of our conscious acts, and not merely as their intentional constructs? Moreover, if the past and the future were pure nothing, speaking of the present would become meaningless. And if the three '*extases*' of time were but intentional constructs, why should we regard the passage of time as something absolute, deeming transcendent realities as purely intentional beings?

And so the first mode of experience seems more trustworthy then the second. But Ingarden does not want to assert this yet. To solve the involved problems, he says, we need a more detailed inquiry into both: the content of these experiences and the way in which they unfold themselves.

In Part IV of 'Man and Time', some practical consequences of living in the second mode of experiencing time are considered and it is shown that, if we take it seriously, this way of looking at time makes us lose our real identity as men endowed with a definite personality. On the other hand, if we are able to stay true to our own nature – by self-control and by building ourselves up in the constant struggle against fate and against ourselves – this identity is saved. When we adopt the attitude of mastering time, it changes its look and becomes for us an occasion to establish our spiritual power and identity.

I shall now proceed to the conclusion of 'Man and Time' (Parts V and VI), written during World War II when Ingarden was working on the *Controversy about the Existence of the World*.

There exist, he says at the beginning of this conclusion, my free and responsible deeds. I perform them in the difficult moments of my life, sometimes in danger of death. They are rooted in the innermost depth of my self, such as it has become in the history of my life, and they are only possible inasmuch as I am still really staying the same. The existence of a transcendent real self is, then, a necessary condition of their possibility. On the other hand, in performing them I actually experience this self and am actually building myself up, becoming more and more independent of time. But if I am not mature enough to perform them – if, by wasting my powers, I am unfaithful to myself – I begin to disintegrate, to decay in time. And so it seems that there are different possibilities of existing in time – Ingarden says tentatively: different varieties of time – and that they are derivative of the modes of our behaviour, as mere expressions of the mode of existence of temporal beings.

However, says Ingarden, this conception of time and of myself may only be accepted if the self is distinct from its stream of consciousness, if this stream is but a manifestation of the real self. And he concludes by stating that, as a man, I am a power which is constantly building itself up and transcending itself. Placed in a body, I am marked by it and, although often dominated by it, I can also master my body and use all its possibilities to further my growth. As this power, thrown into an alien world, I make this world my own and create new worlds which are necessary to my existence. I am a power which wants to be free, and I am ready to risk my life for my freedom. But I can only live and be free when I devote myself, of my own accord, to the production of goodness, beauty and truth. Only then can I exist.

III. MAN AND RESPONSIBILITY

And so Ingarden accepts as genuine the first variety of time experience, deeming the second one – which is, according to him, the actual root of the transcendental conception of reality – as not substantiated properly and, in the first place, contrary to the experience we have of ourselves while acting in a free, conscious and responsible way. We are actual powers, he says. This is indeed a very dynamic view of man and his reality. But it is only a sketch, a primary, almost poetical rendering of a fundamental insight into the nature of man and into his situation in the world. And it was not before the very end of his life that Ingarden undertook, in the already mentioned treatise on responsibility from 1970, a more technical and systematic investigation into the details of this view. In this work he surveys the most important ontological presuppositions which must be taken for granted if we want to assert that we can act in a responsible way.

I shall now give a short account of the situation of consciousness and action in the structure of man, referring to some passages of Vol. II of *Controversy* and to *Ueber die Verantwortung*.

As we have already seen, free human action is the starting-point of the whole investigation of responsibility. And it has been already stated in 'Man and Time' that, as it seems, such an action cannot be undertaken by a pure *Ego* of transcendental phenomenology. It is only a real human person, endowed with a definite ontological structure, that can undertake it. As Ingarden writes in *Controversy about the Existence of the World*, the pure *Ego* is but an abstraction which cannot be effectively separated from the actual texture of the real person. It is, he says, merely a characteristic feature of the *Gestalt* assumed by the soul when it gains self-consciousness and begins to live in its conscious acts and actions. It is a centre, an axis of the soul, of its forces and properties, the source from which the stream of *Erlebnisse* flows and in which all the powers of the soul are focussed. Without the real soul it would be just an incomprehensible skeleton. Owing to its crystallization, to the fact that it becomes the ordering, organizing and dominating factor in the soul, the soul is transformed into a person.[8]

Thus, as Ingarden writes in a passage which we find a little farther in *Controversy*, the soul, the pure *Ego*, the stream of consciousness and the human person which crystallizes owing to their mutual functional dependence and cooperation – are only aspects or features of one coherent conscious being: the human *monad*.[9] The stream of consciousness contains two sorts of *Erlebnisse*: passive experiences of the

Ego undergoing something that it must passively endure and, on the other hand, manifestations of active operations, performances in which something is produced or realized. Both are modes of behaviour of a subject which, in the former case, is affected by a real action of a transcendent being – and which, in the latter case, is the actual actor of his deeds. In performing them he endows his *Erlebnisse* with their actual shape and meaning; and his acts are, so to speak, his doings even when it is only cognition that he is aiming at.[10]

The main ontological foundations of responsibility, as indicated in *Ueber die Verantwortung*, are:

(1) The actual existence of values, the possibility of their concretization in real objects and situations. We shall not dwell on this problem here.

(2) Identity in time of the persons who are regarded as responsible for their actions.

(3) A particular substantial structure of the persons in question and

(4) Freedom of any action for which one may be responsible.

We have already discussed some aspects of items (2), (3) and (4). The last of them, freedom of action, presupposes a definite structure of reality and of man. Namely, to accept it, we must conceive ourselves as partly isolated systems which are not wholly dependent on the rest of reality, but are endowed with barriers separating them from some influences coming from outside; while, on the other hand, they must be partly opened and causally connected with the rest of the world in order to be able to act upon it. And so, according to Ingarden, man is precisely such a complicated system composed from a hierarchy of partly isolated systems of lower order. In fact, the human organism seems to be composed of two main systems of this sort:

(A) The system which keeps us alive, and

(B) The system of procreation.

In the former (A) he discerns four principal subsystems:

(i) The skeleton and the system of movement,

(ii) The system of metabolism,

(iii) The regulation system, and

(iv) The system of information.

It is a difficult question, says Ingarden, whether the soul may be regarded as a system which is partly isolated from the human body or whether it is directly connected with it. But he is rather disposed to think that it is partly isolated. There are, then, three main elements in man: the body with its systems (A) and (B), the stream of consciousness (C) and the soul (D). This conception of man presupposes, of course, a particular causal structure of the world and its temporality. This is discussed at some length in *Ueber die Verantwortung*, but I shall not speak about it in this paper.

And so man, according to Ingarden, is a partly bodily, partly spiritual creature centered upon an *Ego* (*ichhaft*) and endowed with consciousness. He is – or, at least, seems to be – a high-level system of functional systems comprising on the side of his body as well as on the side of his soul a hierarchy of partly isolated systems. And both the soul and the body express and reflect themselves (Ingarden writes: *finden ihren Ausdruck und ihre Auswirkung*) in his pure consciousness and in the way in which his pure *Ego* behaves. [11]

But what is the precise status of consciousness in this system? It owes its existence to the system of information in which we must, as it seems, assume a particular subsystem of the 'gate to consciousness', allowing different facts and events to become conscious. The stream of consciousness itself, as a pure happening, as a process, is not a system and needs some ontological foundation. And it is provided with such a foundation, on the one hand, by the body and, on the other hand, by the soul of man. The stream of consciousness, says Ingarden, is like a surface of contact between the living body and the soul. It consists partly of the information obtained by the *Ego* by means of the bodily information system and concerning, in the first place, the body and what is going on in it – and, further, the external things and what happens with them. On the other hand, there appear in it from time to time some manifestations of the soul and of the changes taking place in it. For, although it is essential to the soul to be conscious, everything that concerns it must also pass the 'gate to consciousness' in order to become known to the *Ego*. Ingarden stresses the fact that we are not conscious of everything that resides or is going on in the soul. The acts of thinking and, perhaps, those of the will, in particular its decisions seem indeed to be performed consciously. Both are ways of the *Ego's* behaviour or, better, its deeds. They are made possible by the powers of the soul, and to speak about an *Ego* without a soul would indeed be self-contradictory. But the soul may be, and in fact always is, in different respects closed to the conscious subject, and it often takes a considerable effort to be able to dominate over one's own soul.[12]

IV. THE PROBLEM OF THE STRUCTURE OF CONSCIOUSNESS

As we have seen, Ingarden speaks about consciousness as comprising pieces of information, in the first place about the living human body, and in connection with it about the external world and about the soul. Thus, it is merely information – merely, so to speak, the picture of the person and of his world that the stream of consciousness is made from. And the pure *Ego*, as the axis of the person and as the pole from which the person's actions shoot out does indeed transcend the conscious acts. The *Ego* is the actual pole both of these pieces of information which we call conscious acts and of the actions of the real person.

These actions, as we could describe them (but it is not Ingarden's term) are mirrored in consciousness. It does not seem that there are any reasons to object to this. But is it enough to assume this mirroring in order to explain conscious, deliberate action? And is it not a natural thing to assume that, at least in the primitive cognition of the material world around us, these pieces of information are inseparable from the corresponding actions or tendencies to act – that viewing the intentionality of acts as purely intellectual and, in a sense, unreal reactions to the object of thought is an abstraction which can only be achieved at a comparatively high level of consciousness. In other words, as the *Gestalt* psychologists used to stress, the system of movement and that of information are in fact functionally one system.

It should be noted here that Ingarden speaks about a very close connection between the functions of the system of bodily movement and those of the *kinaesthetic* sensations in our muscles.[13] But what about sensations in general? It seems that it can be shown – as I have tried to do it elsewhere [14] – that there exists an interesting ambiguity in Ingarden's conception of the actual structure of consciousness. Because, in the *Controversy about the Existence of the World*, he accepts, so to speak, officially the scheme "*Auffassungsinhalt – Auffassung als...*" as it is introduced in Husserl's *Logical Investigations*. He says, namely, that it is to the cooperation of our having a sensation (understood as a sense-datum of some sort) with an intentional aiming at a transcendent object that the so-called bodily presence of an object in sensory perception owes its occurrence.[15] But, if we look at his description of the primitive levels of consciousness in Section 46 of the German edition of Vol. II/2 of *Controversy* we may realize – and this is not surprising if we take into account what has already been said in this paper – that it is the actual, dynamic and *symbiotic* contact of the experiencing creature with his or its environment that Ingarden is actually describing here: that he cannot help laying stress on the actual dynamism of actions and reactions in the primary, sensory experience as it is given to our consciousness.

This aspect of our sensing has been shown in the analyses of some contemporary psychologists, psychiatrists and philosophers; in particular, in the most clear and comprehensive way in Erwin Straus' book *Vom Sinn der Sinne*.[16]

Let us now come back to the first question asked at the beginning of this part of my paper, namely to the problem of the particular structure of consciousness itself which is indispensable to make conscious, deliberate action possible. If we put it in this way, it seems indeed clear that such a structure must actually exist. In fact, this structure has already been described. As far as I know, this has been done most distinctly, taking into account the information furnished by psychiatric research into the disintegration of human consciousness in mental disorders, by Henri Ey in his book *La conscience*.[17]

Concluding: I have tried to sketch the development of Ingarden's views on consciousness and action and their place in man. This development seems to be of particular interest from the point of view of his attitude towards Husserl and his transcendental phenomenology as well as from the point of view of his own, highly original and comprehensive analyses which have been reported here very shortly and fragmentarily. But it seems that analyses of consciousness itself were not the domain of his main achievements and that his insights concerning its nature have not been followed by an actual elucidation of its structures corresponding to the ontological states of affairs described by him. In this respect his results seem to be in need of some supplements.

NOTES

[1] Roman Ingarden, *Spór o istnienie świata*, Vol. I–II, P.A.U., Kraków, 1947–48; 2nd edition P. W. N., Warszawa, 1960–61; 3rd German edition: *Der Streit um die Existenz der Welt*, Niemeyer, Tübingen, 1964–65.

[2] Cf. Ernst Tugendhat, *Der Wahrheitsbegriff bei Husserl und Heidegger*, De Gruyter, Berlin, 1968, 2nd edition 1970, pp. 178ff.

[3] Cf. 'Constitutive Phenomenology and Intentional Objects', by the present author in *Analecta Husserliana* II (1972), pp. 90ff.

[4] It is interesting to note that the most important passages stating the difference between Ingarden's position and Husserl's conception of philosophy, philosophical method and consciousness have been added in the German edition of *Controversy* (1964–65) and do not appear in the previous Polish editions of this work.

[5] Cf. Ingarden, *Der Streit*, Vol. II/2, pp. 370ff.

[6] Roman Ingarden, *Ksiażeczka o człowieku*, Wydawnictwo Literackie, Kraków, 1972.

[7] The book contains: (1) Ingarden's speech at a discussion at the XII International Congress in Philosophy in Venice in 1958 *'L'homme et la nature'* (Man and Nature) – in *Atti del XII Congresso Internazionale di Filosofia, Venezia 12–18 settembre 1958*, Vol. 2, Firenze 1960, pp. 209–13. (2) *'La nature humaine'* (On Human Nature) – a contribution to the *XI Congrès des Sociétés de Philosophie de langue française*, in *Nature humaine, Actes du…, Montpellier 2–7 Septembre 1961*, Paris 1961, pp. 220–25. (3) A short essay (primarily a radio speech) *'Człowiek i jego rzeczywistość'* (Man and his Reality). (4) *'Człowiek i czas'* (Man and Time), published as a whole in the monthly magazine *Twórczość*, Warszawa, Vol. II (1946) No. 2. (5) A Polish translation of the book *Ueber die Verantwortung. Ihre ontischen Fundamente* (On Responsibility: Its Ontological Foundations), Reclam, Stuttgart, 1970. (6) 'O dyskusji owocnej słów kilka' (Remarks on Fruitful Discussion), a short article in the newspaper *Przeglad Kulturalny*, Warszawa, Vol. X (1961), No. 48 (483).

[8] Cf. *Der Streit*, Vol. II/2, p. 321.

[9] Cf. *ibid.*, p. 325.

[10] Cf. *ibid.*, p. 300f.

[11] Cf. *Ueber die Verantwortung*, p. 97f.

[12] Cf. *ibid.*, pp. 91ff.

[13] Cf. *ibid.*, p. 188.

[14] Cf. an article by the present author, *'Dane wrażeniowe a doświadczenie pierwotne'* (Sensory Data and Primary Experience), in the special volume of *Studia Filozoficzne*, Warszawa, entitled *Fenomenologia Romana Ingardena* (Roman Ingarden's Phenomenology) and his book *Świat, spostrzeżenie, świadomość. Fenomenologiczna koncepcja świadomości a realizm* (World, Perception, Consciousness: The Phenomenological Conception of Consciousness and Realism), P.W.N., Warszawa 1973.

[15] Cf. *Der Streit*, Vol. II/1, p. 197.

[16] Erwin Straus, *Vom Sinn der Sinne*, Springer, Berlin, 1933, 2nd (enlarged) edition 1956. English transl.: *The Primary World of Senses: A Vindication*, The Free Press, Glencoe, Ill., 1963.

[17] Henri Ey, *La conscience*, P.U.F., Paris, 1963, 2nd (enlarged) ed. 1968. German transl.: *Das Bewusstsein*, De Gruyter, Berlin, 1967 (there exists also a Spanish and a Japanese translation).

AUGUSTIN RISKA

THE 'A PRIORI' IN INGARDEN'S THEORY
OF MEANING

The theory of meaning which is so closely connected with the problems of *a priori*, is one of the most discussed themes in the philosophy of our century. Roman Ingarden contributed to this topic in a very significant way, going deeper than a typical positivistically minded philosopher of his time. In Ingarden's major work, *Das literarische Kunstwerk*,[1] the problem of meaning seems to dominate the author's concern with the nature and structure of a literary work of art. It is well known that Ingarden distinguished four basic levels (*Schichten*) which in their inter-relatedness constitute a literary work of art as a polyphonic organic unity. These levels are: (i) the level of a word sound, or generally, the level of the material dress of a language in which the work is expressed; (ii) the level of the meaning of linguistic expressions; (iii) the level of objects represented and/or described by the work (*die Schicht der darge-stellten Gegenständlichkeiten*); and (iv) the level of schematic views (*die Schicht der schematisierten Ansichten*). No doubt the second level, the level of meaning, has been emphasized by Ingarden as being the key-stone as to the remaining basic components of a literary work of art. This observation may also be supported by his lengthy Chapters 5 and 6 that have been devoted to the problems of meaning composing almost one half of the entire treatise.

Let us now attempt to review the basic features of Ingarden's philosophy of meaning and language as expressed in *Das literarische Kunstwerk* and in his later works.[2] Ingarden's point of departure is the category of word, although he admits, in accord with Frege or Wittgenstein,[3] that "a really independent linguistic entity is not the single word but the sentence,"[4] and moreover, even a sentence is only relatively independent, if regarded as a member of the entire complex of sentences. Despite these assertions, the investigation of the meaning of a word precedes the concern with the meaning of a sentence and complex of sentences. In this respect, Ingarden follows the traditional path, marked by, for instance, Berkeley or J. S. Mill. However, his theory of meaning is radically op-

Tymieniecka (ed.), Analecta Husserliana, Vol. III, 138–146. *All Rights Reserved.*
Copyright © 1974 *by D. Reidel Publishing Company, Dordrecht-Holland.*

posed to any psychologistic interpretation which identifies meaning with a kind of a psychological, subjective experience or impression; an interpretation so prevalent in British empiricism and the logic of the nineteenth century. Words (and phrases) are classified by Ingarden into two basic groups: (i) names which possess full meaning, so-called nominal meaning[5]; and (ii) functioning words (in the traditional terminology: syncategorematic), which we could call, following the terminology of Kotarbiński and Leśniewski, functors. This distinction resembles Pfänder's division of concepts[6]; although Ingarden avoids the use of the term 'concept' in this context, apparently being afraid of a possible psychologistic interpretation, which could lose hold of intersubjectivity. It is otherwise interesting how frequently Pfänder's conception of meaning is approvingly referred to in Ingarden's work, in spite of slight disagreements between them. It appears that Pfänder's contribution to the theory of syntactic categories, as expressed in his *Logik*, has not yet been appreciated by the students of this field whose attention so far has been paid to Husserl, Leśniewski or Ajdukiewicz.

The cornerstone of Ingarden's philosophy of meaning is his theory of nominal meaning or meaning of a name. Five basic aspects have been distinguished by him: (i) intentional direction factor (*der intentionale Richtungsfaktor*); (ii) material content; (iii) formal content; (iv) existential characterization; and possibly, (v) existential position or location. These aspects are characteristic of any nominal meaning; if the word which possesses this meaning is taken in isolation from the sentence-context. Otherwise, a new factor may arrive, (vi) apophantic-syntactic aspects, which modify the meaning of the word in question, and contribute to the meaning of the sentence or complex of sentences in which our word occurs. In order to explain what Ingarden meant under the above aspects of meaning, let us quote some of his relevant formulations: "The aspects... which determine an object as to its qualitative nature are called the *material* content of the meaning of the word, whereas that aspect by which the word 'refers' to this, and no other, object, or – in other cases – to one of such objects, is called the *intentional direction factor*".[7] He further states that the material content 'assigns' to the intentional object in question certain material characteristics, and, together with the formal content, literally 'creates' ('*schafft*') the object. Thus, as to its qualities, the intentional object is completely determined by the material aspect.

However, Ingarden also appreciates the role of the third, formal aspect without which no object could be completely determined. According to his view, philosophers had, in the past, overlooked the radical differences between the formal and material content, and had failed to incorporate formal content into meaning altogether. In general, formal contents deal with the formal structures or aspects of objects, stating that something is, for instance, a thing, or a process, or a state of affairs, or something else. On the other hand, Ingarden does not say too much about the existential factors (iv) and (v). He merely claims that the existential characterization is always contained in the nominal meaning, whether implicitly, or, less frequently, explicitly. Using as an example the word 'Hamlet', he shows that while the corresponding fictitious object is existentially characterized, its existential location, in reality, is not determined. This would mean that the fifth aspect of the nominal meaning is only optional.

At any rate, the aspects (i) and (ii), i.e. intentional direction factor and material content, have been exposed in detail, in a very interesting manner. If the material content may remind us of Mill's notion of connotation, deprived of any trace of a psychologistic coloring, or of the Aristotelian essentialism, the intentional direction factor is conspicuously akin to the semantic relation of reference, denotation or designation (the choice of the term may be a matter of a terminological preference). It is to be stressed that here we have in mind the very semantic relation, not its members, which are, according to modern semantics, the name or phrase as a linguistic expression, on the one hand, and the referent, denotatum or designatum (again, a matter of a terminological preference), on the other. As known, the reference-relation is not the only semantic relation, and it is even not fully accepted in all semantic theories, especially not in those which are governed by the maxim 'use-instead-meaning'. Another important semantic relation, which is again not accepted in all semantic theories, is the relation between the sense of a linguistic expression and the object referred to by the expression.[8]

Now, in order to set Ingarden's highly original notions into a proper perspective, let us review his detailed characterization of the intentional direction factor and the material content. At first, there is a two-fold classification assigned to the notion of intentional direction. On the one hand, the intentional direction may be either unidirectional (*einstrahlig*), or polydirectional (*mehrstrahlig*). The polydirectional one may be either certain (as in a phrase 'my two sons'), or uncertain (as in a word 'people'). On the other hand, the intentional direction may be either constant and actualized (an example, the phrase 'the centre of the Earth'), or variable and potential (as in the word 'a table'). Both classifications are very illuminating and lend themselves to be compared with other elaborated semantic positions. There are three philosophically important pairs of notions which are backing the above classifications: one–many, constant–variable, actual–potential. The pair: one–many, has been significantly used for clarifying the issues of individual and general, and also, plural names. Ingarden fights the traditional belief, spread from Berkeley to Kotarbiński, among others, that general names refer to more than one object (take, e.g., a word 'table'). With Pfänder, he assigns this right only to plural names, like 'my two sons'. General names or phrases, on the contrary, refer to any individual object from the class of objects which is unambiguously determined by the material content of the name or phrase. This is a conception which resembles very closely the semantic relation of multiple denotation, re-established by R. M. Martin in pursuing nominalistic objectives.[9] Altogether, there are three possible approaches to the question of general names: they refer either to: (1) the entire class of objects, but not to the individual members of the class (collectively, but not distributively); or, (2) any individual member of the class, as well as to their sum (i.e., both distributively and collectively); or, (3) any one individual member at one time (distributively only). Ingarden took the third position without committing himself to nominalism; since to him it is only a problem of the intentional direction, not of the existence of the class as an abstract entity. All these remarks would be incomplete without considering the other two pairs of philosophical notions. The pair: constant–variable, is of extraordinary importance in the languages of mathematics and logic. But while in these symbolic languages the property of 'being a constant' or 'being a variable' is attributed to linguistic expressions themselves (at least in the time of Frege and A. Church), Ingarden assigns them to both the intentional direction factor and the material content of a nominal meaning. As a matter of fact, there is a correlation between the appearance of variability in the material content and the variability of the intentional direction. In other words, if all components of the material content are constant, then the corresponding intentional object will be unambiguously specified, individualized, and thus, the intentional direction will be unilateral and constant as well. Any variability is accompanied by a kind of uncertain description, openness, and thus, ambiguity. Suppose that we specify a thing by saying that it is colored, but we leave unspecified what kind of color it should be. In the material content of the respective word, there is a constant aspect (we know that the thing is colored), and also a variable aspect (uncertainty which color, connected with a possibility of a further specification).[10] In distinguishing constant and variable aspects, Ingarden has helped to turn the attention to a very important problem – to recognizing variable factors in the expressions of ordinary languages, where they are not as conspicuous, to an unexperienced eye, as in the symbolic languages. Enjoying the fact that literary works of art are usually not written in the language of Boolean algebra, we can also see the import of this procedure for the efficient analysis of the works of art.

Ingarden was also concerned with the limits of variability which, as he warns, should not be identified with the traditional notion of the 'extension' of a 'concept'. He thinks that the 'extension' of a 'concept' has been erroneously identified with the scope of the objects falling under the 'concept'. Likewise severe objections have been raised against the traditional notion of the 'content' of a 'concept' which suffers, according to his view, by disregarding variables and by considering alleged 'common characteristics' of the objects which fall under the 'concept'. Here the reasons are not as well founded as one would expect.

The remaining pair: actual–potential, completes the story. Here, potential goes hand by hand with variable, and actualized with constant. If we take a word 'table', both its material content and the intentional direction factor are variable and potential, but they could be specified, i.e., made constant and actual, if an individual table is sufficiently characterized and singled out of the class of tables. A rich stock of potential components of a nominal meaning is thus the source for the actual meaning specifications which seem to be mined from the golden vein of an implicitly complete meaning. Ingarden is aware of different degrees of potentiality as to its chance of being actualized. All this is very interesting, and supplemented further by an illuminating discussion of the difference between the nominal meaning and the meaning of a functor, as well as by the discussion of a verbal meaning, i.e., the meaning of phrases like 'he writes', 'we go', etc.

However, a philosopher of language will be bothered by several problems which are not easily answered by employing Ingarden's explicit remarks. One of such problems is how do we know that the meaning specification has been completed and we have the right to delineate the corresponding intentional (or other) object from, say, its environment. The assignment of the meaning to a word or, generally, an expression, is not a mere conventional act, and even if it were, we would be committed to keep intact the, already assigned, meanings in a broad linguistic context, to avoid complete anarchy. Although Ingarden does not explore the questions of definitions in technical sense,[11] he seems to be aware of the problem of a mutual consistency of meaning components, particularly when explaining in detail the nature and properties of pure intentional objects as opposed to ideal or real objects – if there are such objects at all (to use Ingarden's popular phrase). He is also concerned with the modifications and variations of meanings of words and phrases, when they are members of sentences and complexes of sentences. Of course, such modifications are quite natural and, only due to them, new and higher meaning units, corresponding to complex phrases, sentences and complexes of sentences, arise. Ingarden elaborates these points very nicely, in almost a phenomenological anticipation of Chomskean 'generative grammar'.

Another problem which would worry a philosopher of language is closely connected with the problem of *a priori*. It is the danger of a vicious circle which is contained in the relationship between a nominal (or other) meaning and its corresponding intentional (or other) object, which could take place of the referent in the semantic relation of reference. What is determining what? The meaning, its corresponding intentional (or other) object, or vice versa? Apparently the first holds; especially if we imagine the intentional direction factor as a kind of a vector or arrow pointing to the object in a style of an ostensive definition. Then, the constant and actualized intentional direction factor unambiguously focuses on the object which is fully determined by the material and formal contents of the meaning. However, this 'created' intentional object presupposes its forerunners – the intentional objects, that are hidden beyond those words, or, generally, expressions which have been used for the specification, or, if you like, definition, of our 'created' intentional object. This is an old trouble with any definition, as Pascal cleverly noticed. If we ask what is in this determination *a priori* given, we may conclude that it is positively the formal structure of the 'created' intentional object which directly results from the grammatical category of the word, or in general, expression referring to this object.[12] On the other hand, the material content and the feature of existential characterization – though necessary components of a nominal meaning – may be internally divided into *a priori* and *non-a priori* elements, the fate of which could be decided by some Kantian, Husserlian, or Quinean criteria. Personally, I would rather leave this question to the specialists in Ingarden's ontology.

There are too many other issues which could be discussed in the context of Ingarden's theory of meaning. In particular, I have set aside his paragraphs concerned with the meaning of different kind of sentences and complexes of sentences. However, our paradigm cases could be extended to those domains as well. In this connection, I have at least to mention that entities, which are ascribed to sentences as their, say, referents, are called *Sachverhalt* (this troublesome term appears also in Wittgenstein's language and is usually translated as 'state of affairs'). The intentional (and other) objects, being components of such states of affairs, are actually determined by a set of states of affairs in which they participate. Although the Leibnizian combinatorial techniques are not explicitly present in Ingarden's approach, his consideration of the possibility of infinitely many sentences and states of affairs, touches upon points which are so vividly discussed in modern semantics connected with the notion of state-description, analyticity, synonymy, modality, probability, etc. However, since the intentional (or other) objects are determined by their appearance in the finite set of respective states of affairs – as being the intersections or overlaps of them – the problem of the determining factors which are hidden in the meaning of the corresponding sentences, re-appears in a critical way. Now, the atomized meanings of the single words or phrases, definitely become only methodological abstractions, and the 'higher meaning units' – using Ingarden's phrase – which correspond to the complexes of sentences, and are the products of the merging meaning of the lower linguistic units, will contextually determine the intentional (or other) objects in question. If someone would like to push things so far as to attempt to replace the former 'referential' theory of meaning by a kind of 'use-in-context' conception (say, late-Wittgensteinian or Austinian), he might do so; but I think that it is the first that holds here. Moreover, Ingarden deals with a semantic relation having three members: (i) linguistic expression (presented by its sound or shape); (ii) its meaning or sense; (iii) the corresponding intentional (or other) entity, such as object, state of affairs, etc. The intentional direction factor seems to appear between the meaning and the corresponding entity; being, at the same time, a part of the meaning in the broader sense; whereas the corresponding entity, i.e. intentional object, etc., transcends both the linguistic expression and the meaning. We could even speak about various degrees of this transcendence, according to whether it is a pure intentional object, 'created' by a literary work of art by virtue of so-called 'quasi-judgments', or a real object, possessing autonomous being, or something else.

Before I finish this essay, I have to return to the problem of *a priori* in Ingarden's theory of meaning. Evidently, I have only posed the problem without giving a satisfactory answer. As I have suggested, I could transfer this burden to the students of Ingarden's ontology, and so lose the responsibility for it in the realm of the philosophy of language. Such a deliberation could be well supported by the lack of Ingarden's explicit remarks concerning the *a priori* components in the meaning of a linguistic expression. Yet, the problem is too challenging to dismiss it so easily. I presume that the clues to further advancements lie in answering the following questions:

What are the criteria for stating the necessary connections and dependencies between elements of the material content of a linguistic expression?

What is the invariant core of the changing meaning, if the meaning, according to Ingarden, is not an ideal entity, but is formed by our intentions to fulfill extra-linguistic goals and functions?

How do the functors like 'and', 'is', 'every', 'if..., then...', etc., determine or modify the emerging complex meanings of compound phrases and sentences?

What sufficient and necessary conditions must be satisfied to compel an intentional direction factor to refer unequivocally to its corresponding object?

And so forth.

I realize that to impose upon Ingarden's conceptions some of the answers that have been provided by contemporary semantic or semiotic theories, could lead to an inaccurate, distorted picture of his attempts. Thus, the only alternative is to analyze Ingarden's work in a broader context and perspective.

NOTES

[1] Roman Ingarden, *Das literarische Kunstwerk* (The Literary Work of Art), 3rd ed., M. Niemeyer, Tübingen, 1965; first published in 1931. Also appeared in Polish under the title *O dziele literackim* (PWN, Warszawa, 1960).

[2] See his selected lectures, 'O języku ij rego roli w nauce' (On Language and its Role in Science), in Roman Ingarden, *Z teorii języka i filozoficznych podstaw logiki* (From the Theory of Language and Philosophical Foundations of Logic) PWN, Warszawa, 1972, pp. 29–119.

[3] Gottlob Frege, 'Sinn und Bedeutung', in *Funktion, Begriff, Bedeutung* (ed. by G. Patzig), Vandenhoeck and Ruprecht, Göttingen, 1962, pp. 38–63; first published in 1892. Ludwig Wittgenstein, *Tractatus Logico-Philosophicus* (transl. by D. F. Pears and B. F. Mc-Guinness), Routledge and Kegan Paul, London, 1961, par. 4.23: "It is only in the nexus of an elementary proposition that a name occurs in a proposition", etc.

[4] Ingarden, *Das literarische Kunstwerk*, p. 43.

[5] There actually are two basic categories of the terms possessing full meaning: nominal and verbal (action-oriented).

[6] Alexander Pfänder, *Logik*, 3rd ed., M. Niemeyer, Tübingen, 1963, Part ii: 'The Theory of Concepts', particularly pp. 156–179 (first published in *Jahrbuch für Philosophie* IV (1921)).

[7] Ingarden, *Das literarische Kunstwerk*, p. 63.

[8] There are interesting contributions to this topic by G. Frege, C. I. Lewis, K. Ajdukiewicz, A. Church, R. M. Martin, W. V. Quine, and many others.

[9] R. M. Martin, *Truth and Denotation*, University of Chicago Press, Chicago, 1958, Chapters iv and v.

[10] It is noteworthy that R. Carnap is discussing similar problems, also using color-examples, in his *Meaning and Necessity*, University of Chicago Press, Chicago, 1956, p. 28. See also C. J. Ducasse *Truth, Knowledge and Causation*, Routledge and Kegan Paul, London, 1968, p. 32f.

[11] Roman Ingarden, 'Essentiale Fragen', in *Jahrbuch für Philosophie und phänomenologische Forschung* VII (1924/25); especially Part V. The Polish translation appeared in Roman Ingarden, *Z teorii języka i filozoficznych podstaw logiki* (PWN, Warszawa, 1972), pp. 327–482.

[12] This feature may remind us of Carnap's 'formal mode of speech'.

DISCUSSION

Straus: Thank you very much for this excellent presentation. I also want to express my gratitude to the program chairman for the thoughtful idea of having dedicated one session of the Congress to Roman Ingarden. In listening to your presentation I found myself reminded of the relationship between Goethe and Schopenhauer, at least in the field of color-doctrine. As you know, Goethe has developed the manual of color-doctrine, not on the basis of scientific experimentation in physics rather simply from seeing, from observing. Scientists at large were dubious as to its value but some individuals, among them Schopenhauer, accepted Goethe's theory. He remarked to Goethe that he had started the majestic work of building a pyramid of which only the tip was missing, and he, Schopenhauer, added the tip to it. Without a tip the pyramid is incomplete, but if it has a tip you can pull away the base and it still remains a pyramid. Incidentally, it is not known whether Goethe took Schopenhauer's remark as a compliment or as an insult. At any rate, the story can be applied in some way to the relationship between Husserl and Ingarden.

Tymieniecka: Being responsible for the program and the scholarly procedure of our Conference I take the liberty to emphasize the themes and issues our debates have so far concentrated upon. But first of all, allow me to express my joy at having Professor Półtawski with us here. I remember him well from our seminar with Ingarden twenty-five years ago.

As I have indicated in my inaugural lecture the themes of our Conference – although they are very specific in themselves – fall into the general framework of the great issues with which the phenomenology of Husserl and the other phenomenological thinkers, like every great philosopher, have been always profoundly occupied, namely; *the origin of the world.* This issue has received a very special formulation in phenomenology, like in Kant, regarding its relation to human consciousness. In the lecture of Professor Mohanty we have seen how the problem of the world in

Tymieniecka (ed.), Analecta Husserliana, Vol. III, 147–158. *All Rights Reserved.*
Copyright © 1974 by D. Reidel Publishing Company, Dordrecht-Holland.

Husserl is differentiating into various *life-world* problems at various levels, and as a result we are left with the unanswered riddle: whether the differentiated *life-world* changes anything in Husserl's views concerning the constitution of the *fundamental* world. With Dr Sweeney we have recaptured the realm of passion against 'sovereign reason'. Dealing with certain aspects of the life of passion he clarified some pivotal points regarding the *a priori* constitution of the affective realm. Dr Murphy gave a comparative exposé of the problem of the *a priori* in constitution as it is seen by Kant and Husserl. In Dr Półtawski's lecture on Ingarden we were shown the problem of constitution at various levels in his philosophy. The presentation clearly brought to the fore the precarious, if not ambivalent, status of *a priori* regulations for organizing experience in the philosophy of Ingarden. Lastly, Dr Riska has presented the role of the *a priori* in Ingarden's theory of meaning. Here again a further controversy opens up to which neither Husserl nor Ingarden gave a definite answer. Namely, "What is the structurizing, regulating system in the theory of meaning"? Ingarden spoke of 'ideas and their interconnections' to which he has devoted a great deal of attention, whereas Husserl is wanting on this point. This brings us back to Husserl again, and the issue we intend to pursue is to see whether he could have ever handled adequately the problem of constitution without spelling out clearly the organizing principles.

Gumpel: I merely wish to add some of my thoughts about Ingarden to the ongoing discussion. As far as the field of language is concerned I consider him one of the greatest critics of the twentieth century; in fact, I would align him to the Humboldian tradition of the nineteenth century. Ingarden was the first one who has shown systematically how language as a creative medium has to be distinguished from all other media of communication. I intend to mention here only one point that I regard to be Ingarden's lasting contribution to the analysis of the structure of language, namely, his distinction between the 'pure intentional object' and the, so-called, 'ideal' and 'real autonomous object'. Let me illustrate. The 'real autonomous object' for instance, is the tree in nature, the natural object; the 'pure intentional object' differs from this kind of object and also from the 'ideal object', like a triangle which has an 'ideal objective' existence. The 'pure intentional object' differs from both in that it has arisen through man's own intentional, subjective processes.

A natural object can be analyzed scientifically, and an ideal object can be analyzed mathematically, but it is through this latter type of meaning, i.e., insofar as the pure intentional objects being 'pure' of the 'ideal' and of the 'real' exist in themselves, that man has really shown himself to be a creative being, a basically spontaneous creative being.[1] It is exactly at this point where the Chomsky School of the 'generative grammar' has run aground. That every meaning has in itself a constitutive essence and existence is obvious from the fact that we can communicate intersubjectively, that when you utter a word and I utter a word we are able to communicate without having the actual referent in front of us. If Ingarden had not distinguished these features of the pure intentional object we could never have arrived at a real understanding of how literary language differs from the language we use in everyday communication, where we are constantly referring by means of – using Ingarden's terminology – 'adequation', '*Anpassung*', and '*Identifikation*'. In other words, in ordinary language the meaning of words always refers to the reality around us, but in literature meanings exist only as a totality, as an entity invented for a specific purpose, yet they have existence. If Ingarden had not clearly differentiated the pure intentional object as a purely intentional entity, we would never have been able to prove the identity of literary language. As far as I know, Ingarden is the first and only one who ever pursued this study systematically. As Dr Riska has mentioned, from the point of view of ontology certain serious problems remain. I mean, if this object is neither ideal nor real, how are we to understand it to be 'objective'? And yet, the very fact that we intersubjectively communicate through the convention of word-sounds gives us some basic stability even though this 'object' is not as 'objectively' real as a mathematical triangle or an actual thing. On this point the Chomsky School falls short because they discuss meaning only from the aspect of reality. In their interpretation words like 'piggy' or 'bunny' is merely child's talk and has no meaning, the proper word and proper meaning is pig and rabbit respectively. Ingarden would say that *qua* meaning, *qua* 'pure intentional object' the words 'piggy' and 'bunny' have meaning even if the actual referential object is a pig or a rabbit. So when T. S. Eliot uses in his poem a childish word we immediately give it a certain relevance; we know, for instance, that he is being sarcastic or flippant for certain reasons. And that is the decisive factor, not whether 'bunny' corresponds actually to an autonomous real object: the rabbit. 'Bunny' is a pure intentional object, and it has just as much significance and purpose as the meaning 'rabbit', insofar as in literature we are able to gain a certain significance from it concerning the intention and state of mind of the author, that is, ironical, sarcastic, and so on.

Pax: My question is directed to Dr Półtawski. In talking about the bases of responsibility, I believe you meant it from the ontological viewpoint, the first one you mentioned was the *actual existence of values*. You did not sufficiently develop this topic according to my expectations. It appears to me, that it is on this point that the question of the *a priori* and of constitution, especially with respect to the activity of consciousness, has to be pursued. I wonder whether you could elaborate further on the question of how Ingarden perceived these values, and what kind of status they had in reference to human activity.

Tymieniecka: I would like to add another question to the previous one. You have shown Dr Półtawski, how Ingarden's concern with action in his later years seems to have enlarged, or even questioned his world view of earlier years, that of the strictly ontological world view. It seems to me, that the question of Mr Pax touches exactly on the issue: whether the regulative principles which apply to the will and to action, would not have changed a great deal his view upon the constitution of the whole *life-world*, as Husserl would say, with respect to the role of reason in this constitution, as it does for Kant. The function of the will in Ingarden's philosophy is one of the major issues that must be clarified.

Półtawski: My first answer refers to the question of values. As it is almost proverbial by now, Ingarden was very careful not to pretend to know more than he actually thought he did. He would rather probe, search, and query than make dogmatic pronouncements, but on one point he was as unwavering as was Kant: in order to show that we act responsibly, that there is such a thing as responsibility, we cannot begin with a relativistic conception of values. However, in my opinion, the determination of the actual status of values remains a vexing problem for him. He says as much that values actually, really exist, even though we cannot say that they are real in the sense in which chairs and tables are real.

The question about the *a priori* is a difficult and complex one. Let me begin by saying that, for Ingarden, the fact of our freedom of action is not something that would undermine the actual *a priori* structure of the world. He adhered to the theory of the ontological structure of the world and our ability to apprehend it by our ideas, but the precarious problem is: how are these ideas related to actual reality. This is the problem of participation, which is left unanswered in Ingarden's philosophy.

With this is connected Ingarden's understanding of the distinction between ontology and metaphysics. Ontology, for him, deals with pure possibilities, whereas metaphysics studies the structure of what actually exists in some way.

Tymieniecka: That means that the matter of will and of action would belong to metaphysics.

Półtawski: Well, that is the point. I am convinced that, when Ingarden thought he was doing ontology he did much more metaphysics than he would be willing to admit. As I have pointed out in my lecture, when he undertakes the analysis of the relation between man and time, he is not only an ontologist, he is, in fact, a metaphysician. There is the crux of the matter: he ignored his own line of demarcation between ontology and metaphysics.

Levine: But does one have to be conditioned by the other? Would it not be possible to gain some freedom in the realm of metaphysics which the ontological structures otherwise would not allow?

Półtawski: I don't think so, because the domain of ontology is more encompassing for Ingarden than metaphysics. In ontology you may deal with things which do not actually exist, only as universal ideas; metaphysics on the other hand, is that part of ontology which proves the things to be present in reality. Because his starting point was Husserl in the transcendental approach to the problem of the external world, it made Ingarden extremely cautious in asserting that something applies to reality, and the issue is indeed metaphysical.

Silverman: I have a further question for Professor Półtawski. In an earlier part of your paper you discussed the problem of constitution, or, to be more precise, the problem of personal identity. It seems that, as you have noted in the section on man and time, I remain the same and I am conscious of myself throughout the entire process of my world constitution, but not of any of the things in myself, so that the self continues as before. Later you talked about the experience of time, which is directly related to the continuity of the self, and maintained that the experience of time makes us lose our identity. I would like you to explain this point, because it seems that experience itself includes, through constitution, all of the self, so that we do not fall into the Humean problem of a succession of impressions, where one cannot be sure of a totality of the self.

Półtawski: I don't think I have said what you are imputing to me. In the first place, I did not say that "through constitution we become conscious of our identity", for two reasons. Firstly, there is one mode of experiencing time in which I am aware of my own identity in time. It is a problem which requires further detailed investigation, but we can say that it is an experience in which I am conscious of my identity continuously from the beginning to the end. Secondly, there is another mode of experiencing time: I am building myself up, I am constituting myself as a real person, as a power perhaps – as Ingarden would say it, or accept us to say it. As this gradual, step by step constitution takes place I am concomitantly conscious of time in which it takes place.

Your other question was whether the second mode of experiencing time does not make us lose our identity. That is not quite correct. Living in a way in which we would seriously accept this mode of experience as allowing us to structure our consciousness, we cannot lose our identity just because we might resort to commanding and mastering time to this effect. There is some connection between Hume's problem and ours, of course, but the point is that experience itself cannot make you lose or gain anything, only experience itself may be modified.

Matczak: When you speak about 'polarization' of man, and the strict connection of man with time, I am very much interested to know whether there is any close similarity between Heidegger and Ingarden in this regard? Moreover, as you presented Ingarden's thoughts on action, and man's reality through this action, I wondered how does it relate to the position of Nicolai Hartmann?

Maydole: The clarity of Dr Riska's exposition, and the remarks of Dr Gumpel made me very curious about Ingarden's philosophy of language, particularly his investigations about 'intentional objects'. I am wondering, exactly what problem in the meaning of expressions is solved by reference to intentional objects. One point that comes to mind, independent of the proliferation of entities beyond need, is the question of synonymy. Traditionally reference to intentional entities is brought about in order to help us with the problem when two expressions – be they terms or longer grammatical units – are synonymous. This answer, however, immediately raises other questions, namely, under what conditions are these intentional entities identical; according to what principles are they individuated? I wonder, whether the answer to these questions is any easier to come about than the answer to the question of synonymy. The answer to the latter might be found on a much more mundane level than that of intentional entities, such as reference to usage, convention, or perhaps something even more naturalistic.

Riska: Unlike Frege, who starts his semantic exposition with the problem of identity, Ingarden does not employ the same starting point despite the fact that he was well aware of the problem and discussed it several times. I would say his theory is a straightforward referential theory of linguistic expressions. This could be a methodological approach. Then, we should not forget that his theory of language was elaborated primarily for the purpose of analyzing the language of literary works, which contains special problems. I refer in this regard to the problem, for instance, of being confronted with mythical figures and imaginary objects in literature which are produced by the language of the literary work, but they neither depend for their existence on reality nor do they refer to reality. As a consequence, Ingarden's foremost concern is not with types of sentences, rather with attitudes toward a sentence. He distinguished three different attitudes. First, the mere assertion made by the sentence, like in Frege, with the sign being horizontal, but without the vertical stroke. The second could be exemplified by the attitude contained in judgment (using the term in its traditional sense), in which we are not only expressing a sentence, but also approving or disapproving its content (in a way that we assign, or we are committed to the truth or falsity of that sentence); Ingarden accepts here explicitly Russell's 'sign of assertion'. The third is found in the language of the literary work of art and it is an attitude in-between: it is neither a mere propositional sentence with its correlated intentional entity, nor an assertion which would be committed to truth or falsity, but something in-between, which Ingarden calls the '*quasi judgment*'. The '*quasi judgments*', although not mere assertions nor commitment to truth or falsity, produce an organized 'appearance' which makes simple things to be true or false, albeit it is not truth or falsity in the logical sense. All the while we have to keep in mind that this was an analysis of language for a very specific aim: for the literary work of art. On the other hand, I have just read recently his lectures from 1948, published under the title *On Language and its Role in Science*, and even though the context is different, his views do not differ radically from those expressed in *The Literary Work of Art*. The single noticeable difference is that in the former work, instead of referring to intentional objects he resorts frequently to the use of the semantic term 'designation'.

Nonetheless, I am of the opinion that pure intentional objects in literary works, in the analysis of languages of art are important indeed, for what kind of entity should be assigned to Pegasus or to the 'quadratic circle', an issue so intensely debated at the turn of the century. We must give credit to Ingarden for introducing the notion of pure intentional object for the analysis of the language of art, even if we cannot credit him with solving the problem of intentional context, or the problem of synonymy.

Having said all this, there still remains another question to ask: "What else is there in the intentional object above and beyond what has been given to it by the meaning of the corresponding linguistic entity"? It is at this juncture where ontology and free linguistic experience enter; to elaborate upon it, however, is beyond my competence.[2]

Gumpel: The problem of synonymy touches upon the very heart of the matter, because to understand the pure intentional object one has to preclude the synonym. Wilhelm von Humboldt made this quite clear a hundred years ago and, parenthetically, Chomsky fully agrees with him, when von Humboldt expounded on his theory of the 'inner form' to show that each linguistic entity exists in itself, insofar as it is a 'creation of the spirit', or in Ingarden's terminology, a 'pure intentional object'. It has to be 'in itself', otherwise it could not be truly 'created', and it can never be absolutely correlated with another entity. To illustrate our point let us use the well worn example of the words 'king' and 'monarch'. When we talk about the autonomous entity, the actual person – suppose we were to be introduced to the King of Sikkim – it does not make too much difference referentially whether we say 'king' or 'monarch', because we mean the same autonomous, existing person. However, in case of a poem or a work of fiction we choose one word over against the other, we decide on 'king' instead of 'monarch' and we never use them interchangeably. The reason why we cannot paraphrase a literary work is because the words in themselves constitute the being of the literary work of art. Apart from the matter of rhyme, meter, etc., in the case of a poem, the words specifically chosen by the author constitute the very being of every type of literature. Only the words of the author *qua* author have relevance, have existence, not autonomous real existence, but 'literary existence', 'artistic existence'. Insofar as it is so chosen by the author nobody has the right to exchange, for instance, the word 'king' for 'monarch' as if they were identical.

The situation is entirely different in the case of ordinary language. If a historian is talking about King *x* and interchanges 'king' with 'monarch', he is adequating his language with the autonomous reality and it is the actual person who becomes important. But we are no longer in literature and it might be permissible to paraphrase and to say 'king' instead of 'monarch'. Inasmuch as Ingarden has affirmed the unique creative significance of the word and its function in language, *qua* language – not *qua* reality – he is the first critic in modern times who has really been able to prove *why* it is that in literature we cannot paraphrase. Modern literary critics constantly orate about the 'heresy of paraphrasing', but they never explain 'why'. Ingarden's pure intentional object in its relevance, not to philosophy or to logic, but to the literary domain shows why it is a heresy to paraphrase, for each meaning *qua* meaning has significance and this significance is created not for objective relevance, as in history or the sciences, but solely for the purpose of literary communication; consequently, words exist in themselves and are not synonyms or relational, they are absolute. By this standard alone can we truly judge the value of a literary work of art.

Martin: My intention is to continue this discussion about the literary work of art. To this end I would like to have some clarification concerning the ontological status of the work of art in Ingarden's philosophy. His treatment of the problem had considerable influence on many literary men. I am thinking of René Wellek in particular, whose work on the theory of literature, even his language, bears a striking similarity to Ingarden. Yet, in René Wellek's work there is a quite definite notion as to what the literary work of art is: it is a kind of ordered structure of norms; and he expends a great deal of effort to clarify just what that is. Now, I did not get any indication from the preceding debate or from

Dr Półtawski's lecture this morning as to *how* would you top this pyramid, as it were, of Ingarden's language in order to give some very definite statement concerning the ontic status of the literary work of art in particular.

May I just comment further on one significant point in Dr Gumpel's remarks, namely, on the issue of paraphrasing a work of art. I am inclined to agree with her fully that to paraphrase is to distort, but the problem of what the intentional object is still remains and it is a hard one to crack. However, I would like to venture a suggestion. In order not to proliferate objects without necessity, why don't we employ the notion of 'null entity' which has a great tradition in both philosophical and scientific literature. Frege and Leśniewski use it, and so do many others. If we can describe the 'null entity' in different ways, we have different linguistic expressions all of which can be taken to designate or stand for any meaning you like. With the 'null entity' we could perhaps get at the problem of intentional objects; in clear cases, where it is not necessary to state specifically what they are, we can use the 'null entity' taken under the given linguistic description in the particular way we describe it linguistically, and we can have all the linguistic diversity we want. I propose this as a possible approach to the problem of intentional objects.

Półtawski: I have several questions directed to me and I want to take them in the following order: first, Dr Matczak's questions about Ingarden's relation to Heidegger and Hartmann; and then, answer Professor Martin who asked me about Ingarden's influence on Wellek.

Let me preface the first answer by saying that I am not an expert on Heidegger, and neither was Ingarden. He read some of Heidegger's work, but he did not consider them to be genuine phenomenological analyses. Consequently, there is hardly any ground to claim a direct influence of Heidegger on Ingarden. They may have concurred in their reaction against Husserl's approach to consciousness, to temporality, to being-in-the-world, as Heidegger would say, but, there again, we are immediately faced with the difference between the two. Heidegger's being-in-the-world is just a structure of the pure self, and that is quite contrary to what Ingarden was aiming at. It seems that Heidegger's conception of being-in-the-world isn't really 'real'; it is, perhaps, 'esthetic', but definitely not 'realistic' in the sense that Ingarden was aiming at reality. In his *actual realism* being-in-the-world is not just a structure of the self, but a structure in which the self is and so is the real world, as we regard it in life, and not in philosophy. These interpretative ideas of mine echo the thoughts of a French philosopher who contends that the problem of reality lies in the present just as much as it does in the past and future. This seems to be a good criterion to apply in order to differentiate on their respective positions and inter-relations. Husserl, in fact, was interested in constitution, which shows how the *past* begins the present, whereas Heidegger was interested in the *future*. The present as a crucial factor for a really 'realistic' point of view was neglected by both of them, and to an extent, I think, by Ingarden as well. Concerning the second question of Dr Matczak, I would prefer that Professor Tymieniecka answer that, as she wrote a book comparing Ingarden's relation to Nicolai Hartmann.

Lastly, Professor Martin asked for some clarification on the issue of the ontological status of a work of art in Ingarden and René Wellek. Ingarden's approach to the mode of existence of the work of art is that it is a purely intentional form of existence, and it consists in being spoken of, or in being thought of. On the other hand, René Wellek, following the inspiration of the Prague School, spoke of 'norms' as the structuring principles of a work of art. Ingarden was dissatisfied with the vagueness and ambiguity of Wellek's ideas and they have engaged in extensive polemics on the issue.[3]

Questioner: I have only one question to Dr Półtawski. In your lecture you have pointed out an interesting distinction between ontology and metaphysics in Ingarden's philosophy. On the basis of that distinction Dr Levin wanted to know the ontological state of a work of art in Ingarden. My question is now: could you elaborate on what you see to be Ingarden's position of the metaphysical status of the work of art?

Półtawski: As far as I can tell Ingarden applies the distinction between metaphysics and ontology when he speaks about reality, but he does not apply it to the work of art. He speaks only of the ontology of the work of art.

Maydole: My remarks have to do with a distinction in the meaning of two synonymous words we have neglected heretofore. This distinction ought to be taken into consideration when a preferential choice is being made between two words in a literary work. Let us take again the example of the words 'king' and 'monarch'. Cognitively they may well be synonymous, however, in ordinary language words have more than just cognitive meaning. They have what we may call 'emotive meaning', and on the emotive-level the meaning of 'king' and 'monarch' may well differ. The difference lies in the different ways in which they are used. To mention only one distinguishing feature, the word 'monarch' has a much more honorific aura than does the word 'king'. This may well explain the reason for the difference in choice. Independent of the fact that rhyme and meter are all very important in a poem, one has to distinguish between cognitive and emotive meaning. This consideration may well take us a long way and help us to dispense with the appeal to intentional entities.

Riska: In my understanding of Ingarden he is fully aware of the 'emotive components' in the meaning of words; but again, he was not concerned with the problem of synonymy. In modern semantics we are introduced to concepts like 'emotive-synonymy', 'emotive-meaning', and so on, but to repeat, these are linguistic issues that lay outside Ingarden's interests in the structure of literary language.

NOTES

[1] Cf. *Analecta Husserliana* II, pp. 68–75, 90–95, and 114–117, for an exposition of the notion of intentional object. [The Editor].
[2] Cf. *Analecta Husserliana* II, pp. 68–75, 90–95, and 114–117. [The Editor].
[3] In the first edition of the *Theory of Literature,* R. Wellek has adopted Ingarden's *purely intentional mode of existence,* and has openly acknowledged his debt to Ingarden. Cf. A.-T. Tymieniecka: *Phenomenology and Science in Contemporary European Thought,* Noonday Press, 1962. [The Editor].

PART II

ACTIVITY AND PASSIVITY OF CONSCIOUSNESS

MARIO SANCIPRIANO

THE ACTIVITY OF CONSCIOUSNESS:
HUSSERL AND BERGSON

A speculative philosophy, which opens up horizons of pure thought and offers luminous vistas to the originary consciousness, cannot exist by itself or even separated from a philosophy of life which returns to the obscure origins of our *élan vital*. In this way the *eidos* becomes the light in the depth of the *pathos* and illuminates the remote extremities of primordial life. The 'visual distance', whereby consciousness defines itself in order to recognize the objective truth of things, is overwhelmed by consciousness's own native, immediate rapport with the life around it. Objective vision is not to be confused with indifference or apathy. Knowledge translates itself into possession. Intelligible perception impinges on vital perception and is interpenetrated by it. Husserl's *identisches Ich* resolves into Bergson's *moi fondamental* which can be brought into the internal flux of conscious life by means of a process which avoids the 'alteration' and 'solidification' of external structures. The *eidos-ego* of reductive analysis lived in the synthetic intuition of the 'fundamental I', or rather of the *moi fondamental, tel qu'une conscience inaltérée l'apercevrait* (as an unaltered consciousness would perceive it).[1]

But it is the task of phenomenology to safeguard the identity of the conscious in the face of whatever process which tends – as this of Bergson does – to break down consciousness itself into its components out of which its development arises. The category of unity is necessary to ensure continuity of the life of the conscious, as Roman Ingarden illustrates in the critique of Bergson, written in the *Jahrbuch* of Husserl (1928). Such unity is in no way imposed on consciousness, as an *a priori* form, from the gnoseological plane, but emerges on the ontological plane in the process of organization of the life of consciousness.

Bergson's *moi fondamental* needs Husserl's *identisches Ich* in order to continue to be fundamental in its interior movements. In the same way, the vital impulse (*élan vital*) has intimately a need for unity if it is to be organic and functional. In order to be 'unaltered', consciousness liberates itself from external symbolic superstructures and phenomenological

Tymieniecka (ed.), Analecta Husserliana, Vol. III, 161–167. *All Rights Reserved.*
Copyright © 1974 by D. Reidel Publishing Company, Dordrecht-Holland.

methodology can be used in order to endeavor to discover the original springs of life.

The inwardness inherent in the philosophical intuition comprehends an activity of consciousness which does not resolve itself in terms of experience of life nor is it affected by duration (*durée*); since consciousness, in its own essential existence, is identical with itself and in its internal perception transcends any temporal development. If it is true that there can never be two identical, successive states in the flow of consciousness, inasmuch as the second state must in addition possess the memory of the first, then memory is unique in its omni-temporal encompassing of the present. 'Consciousness' is a concrete term which denotes a subsistent reality, and we cannot attain its concept as an abstraction from the actions of the *psyche* as one could with the abstract term, *coscienzialità*, or 'to be conscious'. In the evidence of the *cogito*, consciousness is a living reality, diverse and multiple in its acts, yet one in its intentional structure. These actions can be contrasted in the dialectic of life and death and can resist unification; but basically their selfrevealing capacity is one.

Thus the *time of consciousness*, which is heterogeneous in duration (*durée réelle*) refers back to a *consciousness of time* (*inneres Zeitbewusstsein*) which transcends psychological time, in the homogeneous act of its constitution. From its point of observation consciousness comprehends also extra-temporal reality, as is illustrated in the manuscripts of Husserl, published by Gerd Brand: in fact, "in concrete *durée*, (*sachlich*) the permanent I remains an I devoid of extension, identical, motionless and permanent in its evolutional actualization".[2]

The basic constitutional unity, which realises itself in the *Ich-Mensch* of Husserl resolves in a unitary way also the dualism of 'matter and memory' which the French philosopher has above all never mastered in the development of his thought. In reality one single development reconfirms again the concrete unity which unifies the *io-uomo* wit the *io-transcendentale*: "*Dem konkreten transzendentalen ego entspricht dann das Menschen-Ich, konkret als rein in sich für sich gefasste Seele, mit der seelischen Polarisierung*".[3] It is however also true, in ensuring the empirical unity of the *Ich-Mensch*, that Husserl, parting from the *aistheta*, aims at a purity of spiritual life in which the ultimate essence of the *eidetische reine Seelenlehre* consists. It is in no way fortuitous that in the development of the doctrine many significant operations in corporeal and social life are deducible from the transcendent I which lives in the first person in every person, operations successively evaluated in the *corps propre* of Merleau-Ponty and in the intersubjective social rapport of E. Paci.

A synthesis of *objective intentionality and operational as well as vital intentionality* is possible at the moment when an *eidetic* reduction operates in the area which is vital, not only to draw to itself the origins of a unified impulse of life in consciousness, in its instinctual purity, but also to make from this a deduction in the transcendental field. That is made possible by a return of consciousness into itself, living in its own fulness and capable of comprehending within itself its own original impulses and its most genuine operations of life which emanate from consciousness itself without being identified with it. As Tymieniecka observes, "experience, as it arises in the world. stands out in profile against the background of the fulness of conscious life, as indivisible fulness which is not given but submerges us in its pre-sence." [5]

Now the indivisible life of consciousness which has its centre in the *Ego-Pol* exhibits itself in the multiple experiences of life (*Erlebnisse*); but, because they are always incomplete and refer back to the other, fulness itself is conceivable only as an intelligible unity which demonstrates itself in the process of unification and which comprehends the diverse movements of the life of consciousness, insofar as it is possible, in their accordance and harmonious fusion. And this done in virtue of one sole unifying act, which, being such, a sovereign act of consciousness, has in itself the fulness of total unity (*uni-totalità*).

In fact the *io identico* performs in itself the unity which rules the whole life of consciousness and, by reflex, the unconscious, in the following ways:

(1) as a *categorical unity* which operates in the unity of apperception and in the essential connection of the events in life;

(2) as an absolute unity through which consciousness does not identify itself, with the *Erlebnisse*, nor with their sum total, but in the ego, consciousness transcends them.

The return of consciousness into itself, into its own essential unity leads us back to the immediate research of a fundamental sense in the origins of the *life-world* which precede any particular scientific superstructure or any rational construction. This reduction becomes necessary not only in order to seek the genuine origins of our formations of sense but to ensure also the very intelligibility of the vital impulse, immanent in consciousness in its original directions, at the level of pure instinct. This is arrived at after having neutralized any natural attitude, any existential affirmation on the 'inhuman' violence of instinct, so that the bio-theoretic significance of life emerges. Similarly, in inter-subjective rapport, a liberation of instinct, reduced to its pure state and made intelligible as such, may facilitate immediate communication through empathy (*Einfühlung*), thus reviving the understanding between hearts in an active way and preserving it from any deviation.

Consequently, the activity of consciousness, in the unity of each perception, makes the vitality of sensation adhere to the rational intention, and this vitality is destined to flow back to the intellectual structures, even to the most reduced ones. In fact, if it is true that thought makes intelligible the origins of the *life-world* as they are and not only insofar as they are predisposed to life in a generic sense (biological, theoretical, religious life, etc.), it is also reciprocally true that a vital impulse is immanent in pure thought as well as in the passive *genesis* of perceptive receptivity and it also accompanies the *active genesis* of logical formations in their intentional dynamics.

The formations pertaining to the attentions, to the interests and to the genesis of the propositions have been amply illustrated by the late Husserl's document.[6] Thus the revelation of an 'intelligible' moment of a vital event and reciprocally the discovery of a 'life' of thought convince us of the profound unity of the *bios* and the *logos* which reveal in their origins a unique activity of the transcendental subject which *thinks in life* in its creative expressions and *lives in thought* in its proliferations and in its fertile and originary formations of sense. Husserl's work on the 'genealogy of logic' (*Erfahrung und Urteil*), collected by Landgrebe and revised by the author in the last years of his life, is undoubtedly a most original document provided that it is not understood simply as an *eidetic* result, therefore pre-theoretical, in opposition to a historical, concrete and vital philosophy, but also understood for what it is and for what ultimately it implies, as a work in which the 'vital' processes are studied, the kinaesthetic processes of the formation of thought (insofar as they are originary processes). As I have previously indicated in my book, *Il Logos di Husserl*[7] the originality of Husserl rests on his way of utilizing the logical instruments he possesses by going beyond the rigidity of Kantian forms and by conceiving the flow of conscious life as a Heraclitean view where the permanence of the *Logos* (which is truly present in this divine river) does not absolve us from researching into the kinaesthetic motives which solicit the very life of thought in forms which are always new. Very willingly Husserl utilizes all the terms which contain within themselves an implicit conception of a dynanism in constant progress, from the constant reference to the function of the ego to the development of his *Habitualitäten* and ideal forms: nor is there lacking an analysis of vital movement in which each perception carries within itself a development of potentiality already contained in the simple response to stimulus, as the disciples of Husserl have demonstrated, in the phenomenology of perception.

The central chapter of *Erfahrung und Urteil* (Section II, Chapter I) is dedicated to the general structure of the predication and to the genesis of the principal categorical forms. The detailed examination of the 'relation' (which recalls the *Urkategorie* of E. Von Hartmann) has predisposed the reader to the most intimate process whereby consciousness rises from interests in perception to analysis of predicative activity in general in its diverse modalities and in its objective reference, to the comprehension of the same modalities (departing from their constituent origins) as modes of defining to themselves the ego. This analysis brings us nearer to a more accurate approximation to things, insofar as it may be conceded by the static methodology of the analysis which emerged from *Ideas* I, where the object was placed at a certain 'visual distance'. The full resolution of the cognitive tensions is thereby brought about (*Erfüllung des erkennenden Strebens*).

This activism in general emerges also from the pre-categorical origins of the *life-world*[8] and from the return to history as the 'genesis of mankind' and appears likewise in the manuscript published by me for the first time on the pre-conscious activity of early infancy apart from the 'proto-child' (*Urkind*). This child was studied in its unactualized situation and in its vital experience with the body of the mother (*das mutterleiblich Kind*) which constitutes with it a primordial unity (*primordiale körperliche Einigkeit*) and therefore a point of departure for all perspectives. Thus the ego of the baby awakens to life from a corpuscular stage which is premundane. Husserl recalls in clearly intelligible terms the simple and vital fact of birth or rather wakening to life (in those who are already living) and studies their historical-social connections and associations.

Das Ich von diesem Erwachen, das Vor-Ich, das noch nicht lebendige, hat doch in seiner Weise schon Welt, in der Vor-Weise, seine inaktuelle Welt "in" der es unlebendig ist, für die es nicht wach ist. Es wird affiziert, es bekommt Hyle als erste Fülle, ersten Anteil an der Welt der wachen, der lebendigen Ichsubjekte, die miteinander schon in lebendigem Konnex sind und mit denen es damit in einen ersten geburtlichen Konnex tritt: es hat Eltern, und diese sind in einer Allgemeinschaft von lebendigen Ich in der historischen Allzeitlichkeit, der sie angehören. Die Lebendigen wecken den Unlebendigen.[9]

Alternatively, that which is not rendered intelligible to the origins of life and to the same stream of *Erlebnisse*, may be considered obscure and conducive to involutions in the individual consciousness thereby resolving itself in the negation of its own life – or rather in death; that which instead appears in itself following its own teleological direction towards the formation of psycho-physical structures may justly be termed intelligible and comprehensible even in the terms of the statistical and mathematical language used to express, as long as we speak of the origin of phenomena, the proportional increase of their parts, with recourse also to irrational numbers. Reciprocally, for an analogy which inheres in the basic unity of intelligible and vital reality, a speculative thought which does not appear rich in motivation, nor fertile in development (delineating new planes of intellectual, moral and social life according to valid directions), one may justly term dead. For example, B. Croce in his critique of Hegel asked what was still alive and what dead in Hegelian thought.

In conclusion, consciousness acts according to intelligible and operative patterns in such a way as to enable the intelligible patterns to illuminate from the inside the operative ones, right to origins of the vital impulse, immanent in the life of consciousness considered in its primordial directions. *Life is thought* out in the philosophic intuitions of its origins and development; and, reciprocally, *thought is lived* in its active genesis, in its creations and ideal formations. Bergson did not think that scientific constructions, before being 'solidified' in the space-time-conceptuality are lived in the inventive experience of the researcher: while Husserl appears better disposed to search in the *life-world* the originary *Sinnesfundament* of the science of nature.[10] Consequently, the speculative philosophy of an evolution of the life of consciousness, towards the actualization of its fulness, renders philosophically intelligible the process of psycho-physical life right to its originary instructive purity. Such ideal comprehension does not limit itself to establishing the fact of the origins and the progress of the vital impulse, but endeavors to deduct these impulses in the transcendental field. This ideal comprehension observes all vital acts in their primary objections on the plane of rational teleology. In their turn, vital energies, with the genuine freshness of originary emotions, revive the motivations and immanent interests of speculative thought.

NOTES

[1] E. Bergson, *Essai sur les données immédiates de la conscience*, 1889, p. 96.

[2] (Ms. C10, p. 28.)

[3] E. Husserl, *Cartesianische Meditationen*, 35, *Husserliana*, p. 107.

[4] *Ibid.*

[5] Anna-Teresa Tymieniecka: *Quelques notions de phénoménologie créationnelle de l'être*, atti del XII Congresso internaz, di Filosofia (Venezia), Firenze, 1960, vol. V. p. 526.

[6] *Erfahrung und Urteil* (ed. by L. Landgrebe), Hamburg, 2nd ed., 1954.

[7] M. Sancipriano, *Il Logos di Husserl. Genealogia della logica e dinamica intenzionale*, Bottega d'Erasmo, Torino, 1962.

[8] *Krisis der europäischen Wissenschaften* (1936), *Husserliana*, 1954.

[9] Husserl Archives, Louvain, Ms K III 11, pp. 14 seg. 1935, published in part by M. Sancipriano *Aut Aut* 1965, No. 86, pp. 21ff.

[10] (*Krisis* 8, h).

MARY ROSE BARRAL

PROBLEMS OF CONTINUITY IN THE
PERCEPTUAL PROCESS

This paper will probably have the unique and doubtful distinction of asking more questions than it tries to answer. In fact, it is meant precisely to be an exercise in questioning phenomenology so as to discover how much it can be expected to give us.

The occasion for this is the thought-provoking article by A.-T. Tymieniecka, 'Ideas as the *A Priori* of Phenomenological Constitution'.[1] In it the author takes a new approach to the problem of the continuity of the perceptual process, positing the ideas as the active regulative factor in the dynamism of consciousness. It might be profitable to pose the fundamental questions of the possibility of the phenomenological investigation itself in every circumstance and in any field, and to pose such questions from a natural and even naive point of view, so as to leave no doubt in the mind of the questioners.

At times it will be necessary to reiterate a question and to belabor a point: all this, for the purpose of discovering *if* and *how* the phenomenological investigation does indeed provide us with a trustworthy tool for the discovery of the world of beings as well as the world of knowledge, and, more to the point, to learn whether a presuppositionless approach is ever possible on our part.

The problem of continuity in the perceptual process can be viewed and analysed under several headings and from various points of view. It can be viewed genetically or historically; from the viewpoint of experimental psychology or from that of phenomenology; and if phenomenologically, then either in relation to the act of perception from the natural standpoint, or from the more sophisticated level of the transcendental constitution.

Whatever be the choice, it will be necessary first to investigate perception genetically, in order to uncover the roots of the phenomenon under discussion; then to distinguish between original perception (of the child, for instance), and that perception which occurs within the conscious subject whose experience encompasses may levels of awareness. Thus it

Tymieniecka (ed.), Analecta Husserliana, Vol. III, 168–182. All Rights Reserved.
Copyright © 1974 by D. Reidel Publishing Company, Dordrecht-Holland.

may not be possible to treat these two phenomena on a par. The pheno-menological aspect of the one may not at all be comparable with that of the other, although the essence of perception itself must be the same in both.

There then appears to be two sets of phenomena to investigate in order to discover, possibly, (1) the continuity of the perceptual process in the adult whose experience of the world is rich and extends to many kinds and levels of being; (2) the continuity of the perceptual process in the child, whose experience of the world is merely incipient.

In the first instance the phenomenological investigation would have to consider a wealth of experiences; this would have a bearing on the process of continuity in that previous experiences would illuminate and thus facilitate subsequent ones, although the process itself does not depend on any one particular experience.

In the second, there would be as yet no experiences, or at least very few, for the perceiving subject to investigate. Moreover, there seems to be another question here: one should determine whether one wants to speak of the child's perception from the child's own experience, i.e., describe it as he himself might, or better, gather the child's own account of the perception in question – assuming he were capable of revealing it; or else, whether one wants to speak of the child's perception as viewed by a transcendental consciousness. But before a 'transcendental *Ego*' could have a child's perception as object of phenomenological investigation, the perceptual experience itself of the child should be 'lived', that is, it should be interiorized by this 'transcendental *Ego*'. But how does one 'live' the experience of another?

Unless one takes from the data of psychology the very structure of the original perception, one is faced with a difficult, if not impossible task. It would be relatively easy, of course, to provide oneself with an object of consciousness for investigation by accepting the language expression of the child's perception. Still, without questioning the absolute validity of this procedure (for the language expression will already have modified the perceptive experience [2]) the difficulty remains: the language expression would present the complete perception, not reveal the process by which one had arrived at it. But the question here is not the perception as such, rather the continuity of the process, that is to say, the conscious or unconscious steps by which the subject arrives at the perception of the total object – granting, of course, that perception is by profiles. The problems referred to have been investigated in various ways, philosophically, and under different aspects, also psychologically. However, as Tymieniecka points out in her article, 'Ideas as the *A Priori* of the Phenomenological Constitution', it seems that the crucial matter of the perceptive process has not been fully investigated, phenomenologically; or, at least, the puzzling aspect of it has not been resolved – not even by Husserl, Merleau-Ponty or Gurwitsch. In other words, though through their elaborate theories of perception, they have developed the fundamental question of the constitution of the perceptual object, they have not given a full account of the way in which the partial perceptions coalesce to form the total object.

Although the question here proposed deals with perception, the underlying puzzle refers to the deeper problem of the relation of ideas to things. Therefore, before we consider this question, it is necessary to refer to another article by Tymieniecka, in which she gives a detailed analysis of the nature of ideas, as she sees them, in a role of constitutive principles of concrete beings.

More precisely, in this article, '*Eidos*, Ideas, and Participation – The Phenomenological Approach',[3] she outlines the function of ideas as (1) *eidos* and (2) as *arche*. The idea as *eidos* would reflect the "universal features of a concrete structure" and would show how particulars can have a common nature; it would also serve as a medium for the cognition of concrete beings. "The function of the idea as *arche* [would be] the principle of the origin and career, such as is prescribed by the idea as *eidos*."[4]

The problem of participation is here reformulated following Roman Ingarden's analysis of ideas. This analysis basically affirms the autonomous status of ideas, which can be described as 'strictly determined' possessing (a) their own nature *qua ideas* (characterized by universality) and (b) having within that nature a specific content.[5]

In virtue of this twofold structure of universality and specificity (of content) "an idea can play towards a concrete being the role of prescribing their entire structure, from the most universal component to the most particular".[6] Thus the ideas are seen as principles of the constitution of concrete beings precisely because of their content, while, *qua ideas* they are identical – "the idea of the tree concerning its structure *qua idea* is exactly similar to the idea of a triangle".[7] Tymieniecka describes this content as 'pure qualities' specifying universal content of concrete beings in their universality, or indicating the necessity (hence still universal) of singular qualities in the structure of the concrete. It follows that, as the content of an idea is strictly bound up with the ideal structure which it is, there is no transposition between the two realms – the ideal and the concrete – but, referring to the ontical structure of ideas in terms of 'matter, form, and mode of existence', the relation between the two is thus made clear: each being can be considered, abstractly, from these three points of view. Then "the ontical structure contained within an idea is there in the mode of the principle of constitution of something other than the idea itself".[8]

Summing up, the role an idea can play towards the constitution of a concrete being is to describe its entire structure, in its universal as well as in its particular aspects.[9]

Further, in the article we find the development of the theme of ideas as principles of motion, and the description, in some detail, of the elements previously discussed. Limiting our study to points more pertinent to our topic, we learn, from some of the conclusions reached, that the ideas in the role of *eidos* are "the entire structure of the being as present in the content of ideas"[10]; and that "the ideas in whose content a specific essence is present is the principle of the origin, constitution and motion of concrete beings".[11]

The summation of the whole paper at once introduces us to the article on the 'Ideas as the *A Priori*...' and gives us a point of departure for our projected discussion:

(1) While ideas function as a point of reference for rationality and the structure of concrete things and beings, they themselves appear as absolute, that is, autonomous and independent of them. (2) Their relation to concrete beings seem to consist of a special kind of structural hysomorphism. (3) This hysomorphic structure common to both the universal idea and the heterogeneous concrete being does not, however, consist of static features, but is grounded in a structural mechanism of the idea as such; this mechanism consists of two interrelated factors, *constants* and *variables*, and serves to explain the structural 'constitution' of corresponding concrete beings, thus operating the crucial 'jump' from the universal to the concrete.[12]

Reverting now to our specific topic, the continuity of the perceptual process, we find that, in the words of Tymieniecka, two questions remain unanswered:

(1) "What is the nature of the continuity of the different phases in the constitutive process of perception and by virtue of what is this continuity possible?"

(2) "How is it that after a series of partial, momentary, and disconnected perceptual glimpses or instances, a unified, consistently structured and strictly determined construct is achieved, rather than an incongruous, incoherent, formless shape?" [13]

Leaving aside the interesting point of the exact nature of these questions, that is, whether they demand a phenomenological, a metaphysical, or even a transcendental investigation, they are undoubtedly legitimate and well taken.

According to Tymieniecka, the Husserlian theory of perception, as developed by Merleau-Ponty and Gurwitsch, does not account for the *how* of the continuity of the perceptual process adequately. Having accepted Husserl's notion of perception as basically a process by which several partial stages converge to the production of a coherent 'object', both Merleau-Ponty and Gurwitsch attempt to resolve all problems of perception on a structural level.

The synthesis of the *noematic* fragments is attributed to the nature of representation and to the internal rules of structuration of the perceptual context, that is, of the field of consciousness.... The perceptive process itself is consequently assumed to be autodetermined, self-sufficient, self-organized and moving exclusively through its intrinsic potencies. [14]

And the coalescing of the partial perceptive phases is understood as a passive synthesis.

But as Tymieniecka sees it, the process of identification and verification of the partial perceptual data resulting in the perceptual object will turn out to be an active function instead. Further, what guarantees the correct unfolding of the *noematic* synthesis points to a *finality* in the perceptual process transcending its unfolding.

This finality would then consist of the ideas, as the *a priori* of the phenomenological constitution. In this sense, ideas would be irreducible to the field of consciousness; they would be regulative factors of the constitutive process as well as the structuration of the constituted object itself. Now, the regulative factor of the ideas would be fulfilled by a constitutive variation of the ideas themselves (the term does not refer to Husserlian variation) which, as active function of consciousness would be the condition of the passive syntheses itself which Gurwitsch describes.

It is precisely this 'originary variation' which constitutes the transcendental *a priori* of consciousness.

Tymieniecka analyses in detail the workings of this originary variation of the idea as regulative principle, and shows that all the steps necessary to the completion of the perceptual process are performed according to a pre-established order. Thus the idea becomes the point of reference according to which the partial perceptive moments are organized. However, the idea is not merely a static model of the object of perception to-be; its invariant content is also unfolding; its variant content is flexible enough to organize the specific features of the types of objects constituted according to it.

Ideas are thus seen to fulfill their traditional role of "rules for the constitution of objects and principles of cognition".[15] This *a priori* of cognition is neither a metaphysical nor a Kantian *a priori*.

Without digressing on the meaning of ideas in this sense, versus ideas in the Kantian sense, one may say that, ultimately, ideas as the transcendental *a priori* conditions with relation to constituted objects are limited to providing the laws according to which a structure aimed at must be constituted. They constitute a regulative principle of the very mechanism of constitutive operations.

Thus, briefly, the main lines of argument for the ideas as the *a priori* of the perceptual constitution of objects, according to Tymieniecka.

Now, many questions arise regarding child perception and the perception of adults, the type of questions asked in relation to the constitution of being, and the ideas as regulative of such perception.

Beginning with that which may be considered the end, we may look at the perceptual process as it occurs at the adult level. The question is to determine *how* the perceptual object is achieved. To act somewhat as the devil's advocate — to stir up arguments — it may be pointed out that phenomenology has come on the scene well after many philosophical efforts had already been expanded in an attempt at discovering the essences of things, and that therefore, at the present level of sophisticated knowledge, it is never the case that we are faced with a totally new perceptive object, or at least one to which we do not — rightly or wrongly — already assign a definitive essence. This will therefore lessen the chance of witnessing a truly original perceptive experience which it may be possible to investigate phenomenologically, without presuppositions.

Merleau-Ponty, in *The Primacy of Perception*, affirms, "We can only think the world because we have already experienced it." [16] and,

Phenomenology could never have come about before all the other philosophical efforts of the rationalist tradition, nor prior to the construction of science. It measures the distance between our experience and this science. [17]

This fact might very well indicate that we are already so steeped in the knowledge of the world – and of individual beings in the world – that when presented with a partial view of an object, not only do we find very little difficulty in supplying the unperceived portions, but, also, we are already committed to a meaning. Therefore, speaking from a natural point of view, the activity of consciousness is taken for granted. Perceptions occur, and we do have ideas of things – which, presumably, we acquired in the course of our previous experience. [18]

It may be objected that the fact of having had many experiences is immaterial, because the real point at issue is the working of consciousness in *any* perception, even the perception of an object well known. Granted. But if this be the case, then, it seems there would not be much of a problem. If the object is already known, there is no question of anticipating, comparing, imagining, etc., to complement the profiles actually given – or, rather, these would occur, but without any difficulty, and, probably, instantaneously.

In order to discover, phenomenologically the *how* of the perception of the total object, it seems to be necessary that one be faced with a *new*, as yet unknown presentation.

This then refers us back to a child's perceptual experience as to one more suitable for investigation, since the child does not yet possess a patrimony of perceptual knowledge or of ideas culled from previous perceptions.

Incidentally, it would also have the advantage of clarifying how the intersubjectivity opted for by Husserl for the perception of the world is at work when the perceiver has not yet reached the predicative level. Does not the child perceive very much *with* the adult who, at the very least, offers him different views of the perceptual object together with the verbal expression?

Granting, then, the difficulty of a truly phenomenological investigation – that is one without presuppositions – because of the multitudes of objects which already are constituted in our consciousness, is it our problem to find the *how* of perceptual continuity phenomenologically or metaphysically? It seems that, if the investigation is to be phenomenological, then a great difficulty ensues: all that we are permitted to do is 'describe' the phenomenon as it appears in consciousness. But the questions posed are going beyond description: they are seeking that 'in virtue of which' the synthesis of the various partial perceptions results in the total object. If we are looking for the mechanism by which the multiple aspects of an object coalesce to form a perceptive, intelligible unit, then are we not looking for more than the investigation of the phenomenon may give?

Granted, the investigation of the phenomenon is supposed to reveal not the mere content of the experience in question, but, specifically, the characteristic traits without which such an experience could never be. But does the phenomenological investigation also reveal 'that in virtue of which' a phenomenon *is*?

Husserl was concerned with Being for consciousness. According to his own rules of phenomenological investigation – reductions, ideation, *epoché*, etc. – he seeks nothing more than that which Being reveals of itself to be – in consciousness; Husserl wants to avoid more than anything else any construction whatsoever; certainly, he does not want to presume or interpret the possible hidden 'causes' of the phenomena he investigates. It is necessary to recall also that Husserl's real concern was with the *noema* and the *noesis*; that for him the phenomenology of the perception of bodies did not consist in a study of actual perceptions, or of those which one could anticipate, but of those invariable 'structures' outside of which no perception of bodies, simple or complex, could be had.[19]

This invariable structure he also called 'essential style', a term which could apply as well to the perceived object as to the constituting consciousness:

The multiplicity of possible perceptions, memories, and, indeed, intentional processes of whatever sort, that relate, or can relate, 'harmoniously' to one and the same physical thing has (in all its tremendous complications) a quite definite *essential style* which is identical in the case of any physical thing whatever and is particularized only according to the different individual things constituted in different cases. In the same fashion, the modes of consciousness that can make us aware of some ideal objectivity or other... have a definite style, essential to this sort of objectivity.[20]

This, it seems, can very well be the reason why his description of the perceptual phenomenon admits a spontaneous synthesis – a passive synthesis – that is one which can be revealed to consciousness as such, but whose inner workings are not disclosed.

It seems that to ask *by what means does this occur*, and *why this result rather than another* is to ask for more than phenomenology is meant to disclose.

The question then arises: to discover the inner workings of the continuity of perception, must we rely purely on the findings of psychology?

There are really two ways of looking at this problem: (1) to attempt to investigate and then constitute, phenomenologically, the phenomenon of perception as it appears in consciousness; to attempt to discover its essence – and this requires that no construction of any kind be interjected of *how* the perception occurs, but only describe it, so that, following Husserl's own method, it may be possible to constitute the true and pure essences of both the perceiving itself and the pure consciousness. But since only the *given* can be taken into account the question is to see *if* and *how* that 'in virtue of which' the continuity of the process occurs can be or is given; and if the idea is the *a priori* of this perceptive process, then to see if it be perceived in the same sense in which a sensuous object is experienced (for the whole discussion here is about the perceived as sensuous, appearing in consciousness). And if not, then a different phenomenon would have to be investigated; that is, a judgment about the phenomenon in question. But this brings up the next possibility, which is (2) to submit to a phenomenological investigation the data supplied by psychology regarding the perceptual process. These data, as an object of consciousness, can very well become the subject of a phenomenological constitution and can furnish the phenomenologist with the key to the original perceptive experience, provided the investigations of the psychologist have truly discovered the inner workings of the process – be it only at the scientific and not yet at the philosophical level.

This second possibility would also open the way to a study of the perceptive process as it occurs at the child's level; here the investigation would promise to be very fruitful because of the more primitive and, hopefully, as yet unspoiled vision the child has of the world.[21] But the real problem is the difficulty of determining *that*, or *if*, the psychological explanation is valid. Psychological studies are, after all, techniques devised for discovering the workings of the psyche (of others); they are not – at least in the case of the child – aimed at one's own psychic activities.

Therefore, the whole question of a phenomenological investigation aimed at discovering the fact of the unification of the perceptive experience remains problematic.

But let us look again at the question of the perceptual process as presented by Tymieniecka, without trying to determine the nature of the questions posed.

The suggestion that the *a priori* of perception is idea is very interesting, and so is the notion that the perceptual process involves an active synthesis. We are of the same opinion, perhaps not for the same reasons.

To investigate this latter possibility, it is necessary, first of all, to recall the distinction between perception and sensation. Sensation could be taken as the impact of material objects on sense organs. This process could indeed result in a unitary impression without any activity yet on the part of a subject – theoretically, at least, for it is very doubtful that a conscious subject may experience pure sensation; one would at the very least have a vague awareness of an object being presented.

Perception, on the other hand, could never be passive, because it already involves a consciousness which, upon receiving a given sensuous impression immediately cognizes an object, albeit by profiles. This activity of the subject would be instantaneous. Merleau-Ponty speaks of the 'spontaneous method of... perception' as

that kind of living system of meanings which make the concrete essence of the object *immediately* recognizable, and allows its 'sensible properties' to appear only through that essence.... The object speaks and is significant, the arrangement of colors straight away means something... the subject's intentions are immediately reflected in the perceptual field, polarizing it, or placing their seal upon it, or setting up in it, effortlessly, a wave of significance.[22]

Perception, thus described, could hardly be labeled 'passive'.

But a further distinction must be made between perception of sense objects (this is usually the meaning intended when using the term) and the perception of nonsensible objects, for instance, the interior acts of a subject judging, reflecting, or becoming aware of relations. These purely immanent object of consciousness are objects of perception in a different way, but are no less perceived; i.e., they are seen, they appear in and are constituted by consciousness. Seen phenomenologically, perception is the act of seeing of the transcendental *Ego*; this seeing and consequently constituting, is a constituting of the essences which have revealed themselves in consciousness through various and different acts of perception. Here the activity of consciousness is even more evident. Gurwitsch expresses it thus:

> Through an act of consciousness an object appears and stands before the experiencing subject's mind in a certain manner of presentation. That object has to be taken such, exactly and only such as it actually presents itself through the act under discussion... to every act of consciousness there belongs the object of which the act conveys consciousness, such as the object appears, presents itself, is meant and intended through that act.[23]

Thus the object would give itself to the receiving consciousness in a variety of ways. The stress in the passage above is on the 'receiving' of consciousness: it refers to *noema* and *noesis*. Yet, this cannot be taken as a purely passive process. Insofar as Gurwitsch speaks of process, the notion of dynamism is involved. In that part of the discussion in *The Field of Consciousness*, dealing with this problem he hints at least to an activity which progresses towards actualization of the identity of an object:

> ...to the extent that the thing is actually perceived under aspects previously only anticipated, the identity of the thing is not only posited and intended, but fully actualized and fulfilled. Hence, the perceptual process is a process of fulfillment.[24]

Insofar as both Merleau-Ponty and Gurwitsch refer to certain pre-predicative contacts of the perceiving body with the world, the reference to passive synthesis could be considered, perhaps, also under a different aspect. Could it not be, in addition to expressing the basic fact of the receptivity of consciousness as opposed to constructive activity, a reference to a type of perception approximating, but not yet fully equivalent to perception as just described; we mean, a type of perception similar to sensation in that it does not express the full conscious activity of the subject, but results in a type of knowledge at the pre-objective level. In Merleau-Ponty especially, this is often referred to as that knowledge by which the body is placed in contact with the world. This could be possibly – a passive type of perception. The following passage in Merleau-Ponty seems to confirm our supposition:

> Perception is not a science of the world, it is not even an act, a deliberate taking up of a position; it is the background from which all acts stand out, and is presupposed by them. The world is not an object such that I have in my possession the laws of its making; it is the natural setting of, and field for all my thoughts and all my explicit perceptions.[25]

It is also worthy of note that in this passage there are two points of particular interest to our discussion:

(a) "The world is not an object such as I have in my possession the law of its making." Does this not hint at the fact that for Merleau-Ponty phenomenologically the *how* of the perceptive process is not given to me?

(b) The phrase 'my explicit perceptions' seems to indicate that Merleau-Ponty distinguishes between the perception which is described as a 'passive synthesis' and that which is 'active' because explicitly intended. Perhaps this second sense of perception is equivalent to that of which Tymieniecka speaks. It seems that in Gurwitsch the same supposition is at work. At some point at least both Merleau-Ponty and Gurwitsch must posit an 'active' consciousness, if they have to assign to it the tremendous task of constituting its objects.

Gurwitsch also points out that "the correlation between an object and the multiplicity of its modes of presentation is dominated by *a priori* laws: it is throughout a matter of essential necessity".[26] This necessity might be interpreted as arising from the nature of the object itself or as constituted by consciousness. In any case, in pointing out the *a priori* regulative laws, Gurwitsch does indicate that his thought is not resting with a passive synthesis simply. Of course, this is *not* equivalent to positing ideas as the *a priori* of the perceptive process.

But if ideas are indeed the *a priori* of the perceptive process, then a few questions need be asked in relation to them.

Are the ideas *in* consciousness, though not reducible to consciousness?

Are ideas *outside* of, or transcendent in relation to consciousness? But *how*?

Exactly, what is the origin of these ideas, if they have to fill their role as regulative of the perceptive process?

If ideas fill their traditional role of principles of cognition, in what sense can they be said to be instrumental in the perceptive process – assuming that perception is the basis of all knowledge.

Are ideas the nature or essence of things? And if so, are these essences created by the conscious self or constituted by the transcendental *Ego*? But if ideas are constituted, is this possible prior to perception?

If ideas are instrumental in the constitution of things and their origin is not in consciousness but in the things themselves, there is danger of being involved in a vicious circle. But if their origin is neither in consciousness nor in the 'things themselves', whence are they?

A further question insistently presents itself: Could one, from a pure description of that which transpires in any perceptive process, arrive at the discovery of a causal *a priori*? Or, is there not at the basis of any such attempt a psychological and/or a metaphysical presupposition on which the *a priori* rests?

If not only the nature, but the genesis of ideas could be elucidated further, possibly all these questions would be answered. Tymieniecka seems to give us a fleeting hint to a genesis when she writes: "It is the composition of beings which is a condition and guarantee of rational knowledge. Rationality, in turn, is the level at which concrete beings confront us with ideas".[27]

It might have been a very interesting discussion had Tymieniecka developed this aspect of the question. That is, is the idea arising from reason itself being confronted with a *nautre*, an *essence*, a *being* presented to the knowing subject for the first time? Or for the *n*th time, for that matter. But in this latter case, the then really concrete object would become clear to the rational subject involved precisely *because* of the idea developed from it. In other words, the real problem at issue here is the formation of the first idea – not the analysis of ideas we already have. Or, should one think that rationality itself confers a *nature* on the perceived object? But if so, we are brought back to the beginning: how do we differentiate things? Why confer 'this' *nature* or 'this' *essence*, rather than another, on this bit of perceived being?

This is indeed the crucial question, and precisely so because we are dealing with perception – at every level, but specifically at the child's level. And since we are investigating this occurrence phenomenologically, the problem is even more difficult to solve: we have only the given, nothing more, on which to base our conclusions.

At this point, it seems clear to us that much of Tymieniecka's investigation has greatly exceeded the purely phenomenological level and has gone further into the metaphysical realm. No pure phenomenological study could have yielded all that which is embodied in the two articles discussed.

And yet, the prospect of ideas as the *a priori* of the perceptive process is enticing. It is very ingenious to posit ideas as 'that in virtue of which' the total perceptive process is achieved: it appears a very satisfactory answer to the problem proposed. The ease with which perceptions occur seems to emphasize the point. But the crucial question is how to account for these ideas.

Evidently, if by definition they are the *a priori* of phenomenological constitution, it should not be necessary to account for them. But, on the other hand, is it sufficient to say, simply, 'they are', if one does not admit either a Platonic or a Kantian definition of such ideas?

We are inevitably brought back to the fact that our phenomenological enterprise is confronted by a wealth of accumulated experiences wherein ideas loom large. From this background of already possessed knowledge, how easy, or how difficult is it to approach our quest in a truly presuppositionless way?

Still, perception, even if admittedly instantaneous in its grasp of the object, albeit by profiles, of necessity must synthesize subordinate moments; we want, and ought to investigate all experience. But should we expect that phenomenology open to us every door? Or, are there questions which we cannot answer phenomenologically?

So, to return to the original question of the perceptual process in the child, it seems that this problem could be elucidated only by the aid of psychology. A phenomenological description could not be elicited from a child, and an outside observer could never discover the inner workings of a child's consciousness. Short of construction, it appears, a child's perceptive process cannot be described. But if one constructs a theory, it remains just that. It will have the intentional being of an object of consciousness, but never the givenness of the primordial experience of lived perception.

Concluding, the only contribution this paper may offer is to further the advance of phenomenology by retarding its course; that is to say, by insisting that in the phenomenological investigation nothing be admitted which is not given, and by suggesting that one ought to accept the fact that not every question can be answered phenomenologically. Indeed, phenomenology can and does utilize all other types of philosophical and scientific investigation, reserving for itself the right to accept nothing which has not passed its rigorous test of evidence. This would seem a good reason for phenomenologists to do the same, without apologies.

NOTES

[1] *Kant Studien*, Vol. 55, 3, 1964, pp. 368–83.
[2] In fact, perception is never possible without language; at the child's level the learning process is very much dependent on the language of the adult as well.
[3] *Kant Studien*, Vol. 52, 1960/61.
[4] *Ibid.*, p. 61.
[5] *Ibid.*, p. 63.
[6] *Ibid.*, p. 63.
[7] *Ibid.*, p. 63.
[8] *Ibid.*, pp. 64–65.
[9] *Ibid.*, p. 66.
[10] *Ibid.*, p. 83.
[11] *Ibid.*, p. 87.
[12] *Ibid.*, Vol. 55, p. 368.
[13] *Ibid.*, p. 370.
[14] *Ibid.*, p. 371.
[15] *Ibid.*, p. 378.
[16] Northwestern University Press, Evanston, Ill., 1964, p. 17.
[17] *Ibid.*, p. 29.
[18] In Husserl's *Logical Investigations* II (transl. by J. N. Findlay), Humanities Press, New York, 1970, p. 853, we read: "A division of perceptions is... mediated by a division among the objects of perception, though a difference in origin is likewise set beside it. In one case, perception arises from the effects of physical things operating through the senses on our spirits, in the other case, out of a reflection on the activities carried on by the mind on the basis of 'ideas' won through sensation." Husserl here indicates the danger involved by assuming this type of thing. The question is precisely to know how, having surpassed the investigation at the purely natural level, one can retain the notion of ideas as "fulfilling their traditional role of guides for the constitution of objects and principles of cognition," without assuming that which one wants to explicate.
[19] Cf. *Encyclopedia Britannica*, 14th ed., p. 700^b.
[20] Husserl, *Formal and Transcendental Logic* (transl. by D. Cairns), Nijhoff, The Hague, 1969, p. 246.
[21] It goes without saying that a child never reaches the age of perception proper – that is, the time when he is capable not only of sensing, but also of knowing, to a degree, the object appearing to his consciousness – without having already in some say become part of his milieu.
[22] *Phenomenology of Perception* (transl. by C. Smith), Humanities Press, New York, 1962, p. 131. Italics mine.
[23] A. Gurwitsch, 'The Last Works of Husserl, II: The Lebenswelt', *Philosophy and Phenomenological Research* 17 (1956/57), 390.
[24] Duquesne University Press, Pittsburg, 1964, p. 213.
[25] *Phenomenology of Perception*.
[26] Gurwitsch, 'The Last Works of Husserl', p. 383.
[27] *Kant Studien*, Vol. 52, p. 62.

HANS KÖCHLER

THE 'A PRIORI' MOMENT OF THE
SUBJECT-OBJECT DIALECTIC
IN TRANSCENDENTAL PHENOMENOLOGY:
THE RELATIONSHIP BETWEEN 'A PRIORI'
AND 'IDEALITY'

I. THE SYSTEMATIC RELEVANCE OF THE
SUBJECT-OBJECT-DIALECTIC

"Consciousness appears... differently depending upon the difference of the given object and its continuation *qua* continuation of the given object" (*"Das Bewußtsein erscheint... verschieden nach der Verschiedenheit des gegebenen Gegenstandes, und seine Fortbildung als eine Fortbildung des Objects...."* Hegel).[1] This statement expresses the essence of what is implied in the subject-object-dialectic of human consciousness: subject as well as object are feasible and definable only in a complementary or reciprocal distinction from each other. This position is supported especially in the phenomenological definition of the *phenomenon proper*, which presents itself in the mutual interchange of subjectivity and objectivity which in turn are to be understood as correlational concepts which cannot be defined as independent 'substances'.

"To conceive the universe of true being as something outside the universe of possible consciousness, and both of them only externally joined to each other by a rigid law, is nonsense." (*"Das Universum wahren Seins als etwas außerhalb des Universums möglichen Bewußtseins... fassen zu wollen, beides bloß äußerlich durch ein starres Gesetz aufeinander bezogen, ist ein Nonsens."* Husserl).[2] This programmatic presentation in *Cartesian Meditations* postulates a mutual relationship between consciousness and being (*'Welt'*) and shows the problematic issue which Husserl was attempting to adequately present and rationally resolve through perpetual reflection (notably in the period after the completion of *Ideas*). It leads then to the question: What is the relationship between subjectivity and objectivity? This same question also arises from further considerations in *Logical Investigations*[3] which are based on the insight into the intentionality of consciousness. This, in turn, leads to the following problem: Consciousness emanates through the process of contradistinction by which the outer world is mediated, and thus consciousness constitutes itself. This poses the question: What is the *mode of being* of this 'object' (*der Welt*)? Further themes for investigation also arise: the problem of the world as it is *per se* (*das 'Ansich' der Welt*), its 'transcendence', its material and concrete existence (*'Realexistenz'*). Another problem is the invalidation of the subject-object contraposition in a mutual correlation, the postulation of a realm 'in between', a mediating *'medium'* that does not allow a final emphasis toward either side. This proposition of a 'polarity'[4] between subject and object leads to the question of the character of this *medium*: Does this *medium* of itself tend toward one or the other side of this polarity?[5] Can the nature of this postulated 'other', the object realm, be regarded as something *distinct from* consciousness and therefore be understood as transcending consciousness? Or can it be defined as something distinct *within* consciousness in terms of 'the meant of a meaning' (*als das 'Gemeinte eines Meinens'*) which means the content of an act of consciousness, and would it thus be totally dependent upon the experiencing subject in this mode of being? Here we have the central problem of any controversy concerning idealism.

It is only against this background of all the aforementioned problems that the dialectic of the concept 'phenomenon' elucidates itself[6]: The experienced object can only 'appear' if there is an experiencing subject. The subject in its own consciousness is intentionally directed toward the object-world (*das Erscheinende*) as contradistinct to the experiencing subject. Is the subject contradistinct relative to itself or to something which transcends consciousness? Can we assume the existence of an object which is independent of consciousness in its mode of being (*einen Grund der 'Erscheinung'*)[7] and which lies behind the respective φαινόμενον which is only present in the medium of consciousness? Or is it possible that the postulated contraposition is not doing justice to reality since transcendental phenomenology relates to an altogether different level, to a level beyond idealism and realism?[8] Finally, we must ask if an interrelationship need be understood as being identical with mutual dependency. All these questions must be discussed in the context of the *'a priori problem'* in transcendental phenomenology. The problems discussed here center around the search for the *ontological substance* of what is experienced. They concentrate on the problem of the type of relationship between subject and object. Subject and object cannot be understood as two independent contrasting poles for they are linked by the transcendental structures of the experience of the subject proper.

However, there arises the question of the nature of the so-called 'other' viewed as contradistinct to subjectivity. A. de Waelhens, following Hegel and Husserl, conceives the essence of consciousness as "being this other in such a way as to still remain distinct from it" (*"d'être cet autre sur le mode de ne l'être pas"*).[9, 10, 11] Yet the question remains concerning the mode of being of the so-called other.[12] This is an essential question. The same holds true for Hegel's contention that "subject can be spoken of as subject if it is able to transcend itself and thus, through negation, in turn becomes able to constitute itself (*"Subjekt ist dies, daß es in ihm dies Anderssein sich gibt und durch Negation seiner zu sich zurückkehrt, sich hervorbringt"*).[13, 14]

According to Husserl, the existence of an essential difference between the world of objects and consciousness proper must be seen as one emphasizing pure consciousness (*besteht "eine essentielle Differenz zwischen der realen Welt und dem reinen Bewußtsein, und zwar zugunsten des reinen Bewußtseins"*).[15] This 'idealism', as expressed in *Cartesian Meditations*, is a natural consequence of the propositions we find in *Logical Investigations* where the intentional object as the phenomenal antipole *within* consciousness already has been postulated. This is not to say, however, that the assumption of an object as object would have to be excluded. In this study, Husserl conceives the aim of a universal science from an idealistic viewpoint: "*Pure* phenomenology... is the ontology of '*pure consciousness*', of '*pure ego*'. That is to say that it is not concerned with the level of psychophysical nature presented by transcendent apperception. It does not supply experiences or judgements that would refer to objects transcending consciousness" (*"Die reine Phänomenologie ist... die Wesenslehre von den 'reinen Phänomenen', denen des 'reinen Bewußtseins' eines 'reinen Ich' – das ist, sie stellt sich nicht auf den durch transzendente Apperzeption gegebenen Boden der physischen und animalischen, also psychophasischen Natur, sie vollzieht keinerlei Erfahrungssetzung und Urteilssetzung, die sich auf bewußtseinstranszendente Gegenstände beziehen..."*).[16] It should be emphasized, however, that the mere abstention from any statement on transcendence does not imply the negation of the meaningfulness of such a postulate. It is in *Ideas* that this negation was first made explicit. In *Cartesian Meditations*, Husserl defines transcendence as an *immanent mode of being constituting itself within the ego ("ein immanenter, innerhalb des ego sich konstituierender Seinscharakter")*.[17] The intentionality as "direction toward the 'other', ...openness toward meeting the other" (*"direction vers l'Autre, ...ouverture pour une rencontre"*), as J. Hyppolite formulates it,[18, 19] is thus only the mode of a pure encounter with one's own *ego* (*Selbstbegegnung*) by way of reflecting an 'effected' (*geleistetes*) νόημα, correlate of a νόησις;[20] thus, indeed, mind is only the One which is infinite and identical with itself, the pure Identity which separates itself from itself as the 'other' than itself, the '*Für-sich-sein*' and the '*In-sich-sein*' (*"das Eine, sich selbst gleiche Unendliche, die Reine Identität, welche sich von sich trennt, als das Andere ihrer selbst als das Fürsich- und Insichsein...."* Hegel).[21] If A. de Waelhens, as also J. Hyppolite, defines intentionality as being open (*'Offenheit'*) toward the other – the other that is experienced as something contradistinct to the *ego*, as something that is external to consciousness (*Außersichsein*) while at the same time being inherent (*Beisichsein*) in it – in this context Heidegger would speak of 'Eksistenz' and 'Transzendenz', then de Waelhens in his assumptions goes beyond those of Husserl. Husserl states: "Real being is not a necessary prerequisite for consciousness *per se*" (*"kein reales Sein [ist] für das Sein des Bewußtseins selbst notwendig"*)[23, 24] and proclaims 'transcendence' to be applicable only as a transcendence within the 'immanence' of consciousness (=meant to be a specific understanding of being in regard to intentional modification). For Husserl, transcendence has only 'transcendental' significance; it is to be understood as the manifestation of an operating subjective 'transcendentality' that, in turn, shows itself in a kind of intentional correlate νόημα; although intentionality, in this context, would imply nothingness, voidness; for the object-pole ('Gegenstandspol') depends on the subject-pole ('Ichpol'): thus "transcendentality means a subjectivity objectifying itself" (*"die Transzendentalität [bestimmt] sich... als sich objektivierende Subjektivität"*)[25]; the *mediation* (*Vermittlung*) of the transcendental pure subjectivity is effected by the 'act of objectifying itself' ('*Objektivierung*'); this self-objectifying subjectivity (*'sich-objektivierende Subjektivität'*)[26] is the final absolute which can be arrived at through a radical method of systematic reflection (*"in 'radikaler Selbstbesinnung'"*).

II. SHIFTING OF EMPHASIS TO THE 'A PRIORI'
OF SUBJECTIVITY

Thus, the meaning of *experience* (*Erfahrung*), as it is originally conceived in philosophical phenomenology, receives an essentially different significance: Experience, basically considered, presents itself as a subsequent realization of the structures of the effecting (*'leistend'*) transcendental subject. The receptive nature of experience is valid only at first sight: The 'phenomenon' as produced in the experiencing subject (*empirisches Subjekt*) finally presents nothing other than the manifestation of an absolute subject. Thus, each *a posteriori*, as encountered when experiencing the world, is made possible and deduced from a final *a priori* – namely that very *a priori* which constitutes '*pure*' subjectivity.

Thereby a shifting of emphasis within the subject-object-dialectic is obtained which leans toward a total subjectivity ('*All-Subjektivität*'); this total subjectivity invalidates the object-pole as a product of subjectivity. Thus, according to this theory, *empirical* experience is defined not only in its *mode* of experience but also in its *content* by the *a priori* of an effecting consciousness. This extension of the *a priori* concept to the *entire content* of experience signifies a phenomenological idealism, an idealism which, by means of analyzing the structures of an experiencing consciousness, arrives at a theory of an ontologically creative consciousness.

In this way, the *a priori* of experience, as compared with the transcendental starting point of Kant, constitutes itself in a fundamentally different and novel sense. The *a priori* no longer is a merely formal condition of experience, that is to say, a structure or a mode of experience by means of which the human mind is enabled to perceive the signals from outside (*das ihm 'Entgegenkommende', Transzendenz*). The *a priori* is rather a *material a priori* which seems to include all requirements of experience and thus cannot receive anything strictly empirical. Thus, the Kantian meaning of experience is invalidated (Kant understood experience as something lying between what does not emanate from the subject, meaning a transcendence that warrants empirical experience, and the structures of an experiencing subject proper). This mediating function constitutes the subject-object-dialectic of cognition which, at the same time, corresponds to an interaction of empirical and *a priori* elements. As has been shown, this interrelation receives, according to Husserl, a shifting of emphasis toward the *a priori* that proved to be a transcendental subjectivity. The *transcendence* of experience is thus invalidated in its empirical sense, but is, however, taken up again, *formaliter*, as a transcendence of the absolute subject. Hence the *a priori* of cognition proves to be the 'immanent transcendence' in the structures of subjectivity.

Thus, this final *a priori* which defines the transcendental subject also upholds the '*facticity*' ('*Faktizität*') of empirical experience as such. At this point, then, when speaking of *a priori* or *a posteriori*, we no longer mean the distinction between the constituting structure of the perceiving subject and something that comes from the outside, something not inherent in the subjective structures. The correlation between *a priori* and *a posteriori* is only applicable to the two 'immanent' realms of the pure ego and the empirical ego. Within this defined realm only, the possibility of *a priori* (in phenomenology) can be upheld. In the same way, the concepts of 'absolute' and 'relative' are to be located within the polarity of a pure and an empirical *ego*, thus establishing the significance of an *a priori* in transcendental phenomenology; for the *a priori*, in the end, is based on the pure ego.

III. TURNING POINT TOWARD IDEALISM

The *a priori* of any experience is manifested in transcendental phenomenology as the 'transcendence in the course of immanence' ('Transzendenz in den Spuren der Immanenz'), that is to say, as a kind of '*Nichteinholbarkeit*' of the ego (meaning: the ego is prior to any act of consciousness). The last aim of transcendental phenomenology is to base the perception of the surrounding world on an absolute, transcendental experience of the self and to found its empirical content in the *a priori* of subjectivity; further, the aim is to comprehend[27] the phenomena as emanating from functioning intentionality ('*fungierende Intentionalität*') and thus, through intentional analysis, supporting the fact that being in the sense of a conscious experience of the self ('*Für-sich-selbst-sein*')[28] is the only true being.[29]

In this way, according to Husserl's intentions, the bi-polarity of *ego* and external world (*Ich und Welt*), a gap which arose in the course of transcendental reflection, was to be overcome, an expectation that, in fact (see Husserl) could be realized only in an idealistic sense. The instance which unites subject and object thus presents itself as something principally subjective, a 'spiritualism' (*ein 'Geistmonismus'*) which should lead beyond the objectifying subject-object contraposition; seen in this way, reality '*per se*' proves to be a directly accessible 'subjectivity'. For as Husserl puts it, "it is neither the external nor the world that arrives at an existence *per se*; nor is it the *ego* experiencing the world; it is the transcendental subjectivity that through presumptive evidences both constitutes and sustains the world in its relativity" (*"Nicht das Weltliche und die Welt kommt zur fertigen Selbstgegebenheit, auch nicht das Ich als Mensch der Welt, sondern die transzendentale Subjektivität als Welt in präsumptiven Evidenzen zur Geltung bringend und ständig in Relativität in Geltung habend"*).[30]

In the final quest for the "infinity of transcendentally total subjectivity" (*"Unendlichkeit der transzendentalen Allsubjektivität"*)[31] comparable with Hegel's concept of 'absolute Spirit' (*'absoluter Geist'*), there is an attempt to overcome the antagonism between being and thinking caused by reflection. The starting point for an absolute and universal science which is based on transcendental phenomenology is thus the "nullification of the distinction between conceptions and reality" (*"Verschwinden des Unterschiedes von Vorstellung und Wirklichkeit"*)[32], the total invalidation of the subject-object dialectic[33] by way of a consummation of transcendental reflection which does not remain on the level of mere counterposition but succeeds in achieving a synthesis; for, "as long as I do not interpret transcendental subjectivity in its full range as a subjectivity which is to be conceived in a state of natural worldliness in a transcendental way, thereby interpreting myself transcendentally, the discrepancy between the ideas of the world and the world proper remains" (*"solange ich die transzendentale Subjektivität in ihrem vollen Umfange als die im Stande der... natürlichen Weltlichkeit lebende... nicht transzendental ausgelegt und mich... so transzendental verstanden habe – solange besteht die Spannung zwischen Weltvorstellung... und Welt selbst, seiender, wirklicher Welt"*).[34] This mode of thinking relates Husserl to Hegel's 'absolute idealism' (*'absolutem Idealismus'*) where the distinction between 'object' (*'Gegenstand'*) and 'certitude about a conscious self' (*'Gewißheit'/Selbstbewußtsein*) is annihilated by an act of absolute reflection.[35]

To cite Hegel a universal science based upon this contains "the ideas to the same extent that ideas represent the objects in themselves", or contains "the objects in themselves to the same extent that the objects correspond to the pure ideas" (*"den Gedanken, insofern er ebensosehr die Sache an sich selbst ist, oder die Sache an sich selbst, insofern sie ebensosehr der reine Gedanke ist"*).[36] In this, Hegel finds himself in agreement with Husserl. The above quotation reveals Hegel's original starting point, a point which constitutes the basis of any idealism, namely the theory that human consciousness rests 'with things' (*'bei den Dingen'*) and even knows itself to be identical with these things in that they are manifestations of that very consciousness. Thus, "the dialectic nature of the consciousness of a subjective being conscious of itself and also another such being, an objective one, is overcome, and being is conceived as pure thought which, in turn, is recognized as the only true mode of being" (*"der Gegensatz des Bewußtseins von einem subjektiv-für-sich-Seienden und einem zweiten solchen Seienden, einem Objektiven, als überwunden, und das Sein als reiner Begriff an sich selbst, und der reine Begriff als das wahrhafte Sein gewußt wird"*).[37] In this context Husserl states that 'truth' (*'Wahrheit'*) and 'true being' (*'wahres Sein'*) are events which both appear and achieve their realization in the perceiving consciousness (*"im erkennenden Bewußtsein auftretende, in ihm selbst sich realisierende Vorkommnisse"*).[38] This statement shows that for Husserl the exploration of transcendental subjectivity, and this is not because of a postulated parallelism between the structure of being and the structure of cognition,[39] but rather because subjectivity as such *contains and represents* the entire range of phenomena: therefore, "an exploration of subjectivity, an in-depth exploration of the being and activity achieving its fulness within the subject, leads to all understanding, all truth, and the true world" (*"eine Erforschung der Subjektivität selbst – eine Innenforschung des in ihr sich vollziehenden Seins und Lebens –[führt] zu aller Erkenntnis, aller Wahrheit*[40] *und der wahren Welt"*).[41] From an ontological viewpoint, mind is for Husserl prior to the world of things (*"vor der Welt"*), for mind is conceived as constituting world[42] and therein finding its self: "The encounter of the self in the other constitutes the very essence of subjectivity. Subjectivity can be spoken of as subjectivity... insofar it is able to reflect itself in respective phenomena" (*"Die Subjektivität selbst ist, ...indem sie für sich selbst 'erscheinende' ist..."*).[43] Thus the coherence of a postulated *a priorism* of consciousness with the fundamental claim of idealism becomes apparent: as the facticity of the reality (*'Geltungscharakter'*) perceived by the subject becomes integrated into the 'effecting' (*'leistend-fungierende'*) consciousness.

The subject-object dialèctic as manifested in the respective intentional acts reveals that the subject by its very nature is related to something clearly experienced as exterior (irregardless of its mode of being; in fact, real existence may even be lacking altogether!); the subject can only be feasible in that postulated contraposition. This contraposition on the first level is 'effected' (*'geleistet'*) by those contents of intentional acts that are conceived as 'real' in a naïve consciousness that does not reflect itself (*'naives' Bewußtsein*). On a higher level this contraposition takes place already *within* the *medium* of pure consciousness (On this level according to Husserl and Hegel, the postulated contrast between imagination and reality is presupposed to have been overcome). Consciousness *per se* becomes 'objective' through the objectification of its contents; this process is repeated on every level of reflection (*'aufsteigender' Reflexion*); however, there will always remain an incommensurateness which has already transcended every objectification in that it is this very incommensurateness which first makes the objectification possible.

Though the range of 'objects' be extended ad infinitum, even from an 'idealistic' viewpoint, there will always remain the subject, that is, the subject *'per se'* (*das 'Ansich' eines Subjekts*) which is incomprehensible and unmediated, just as there also remains for the realists, and here the basic similarity of the structure of the problem presents itself, a *per se* in the form of an object contradistinct to subject, a *per se* which, in principle, cannot be fully integrated into the subjective.

We can thus say that the intended *invalidation* of the contradistinction between imagination and reality (This conclusion is of primary importance for the discussion between idealism and realism within the context of transcendental phenomenology.) is *not* achieved in idealism, not even in its most radical type as in the case of Fichte, Hegel, and Husserl.[44] Although the experienced objects are integrated into consciousness since it is the consciousness (of the experiencing subject) that denotes, thinks, and intends these objects, intended in their whatness (*Sosein*) and thatness (*Daßsein*), and are thus 'effected',[45] – it is now, *within* consciousness, that the incommensurate gap between the reality proper of the pure *ego* on the one hand, and the concrete empirical reality as dependent upon that transcendental *ego* on the other hand, becomes apparent. This primary reality is elusive; the *imagination* as realized in the consciousness of the respective *ego* cannot actually comprehend reality. The reconciliation (*Versöhnung*) of imagination and reality, as programmatically intended by Husserl in his idealism,[46] does not prove to be realistic[47] since reality, even if conceived as effected by consciousness (as has been claimed by idealism), only 'appears' (*'erscheint'*) in the mode of simple subsequent acts of consciousness but is not fully absorbed by them. (An exception would be a philosophy, as for instance in the case of Hegel, which claims for itself *the* absolute point of view).

IV. DIALECTIC CONCEPTION OF THE 'A PRIORI' – BASIS FOR
EPISTEMOLOGICAL REALISM

If the aforesaid be true, the *a priori* of experience could be comprehended within its realm of validity. The *a priori* would no longer only be determined by the structures of subjectivity, but would also include the ontological premises in their entire systematic significance, premises which are not grounded in the subject proper, not even in a hypostatized transcendental subject. Thus this novel 'realistic' conception of the *a priori* would do justice to the dialectic of consciousness which conceived consciousness as "ecstatic interiority, that is, as an interiority which exhibits a movement toward the other than itself" (*"intériorité ekstatique, c'est-à-dire, comme une intériorité qui est ... un mouvement vers l'autre que soi."*)[48] If, however, in opposition to this view (as was shown in Husserl's transcendental phenomenology[49]) consciousness takes on 'monadic form' (*'monadologische Gestalt'*)[50] thus claiming to be the *final a priori* of any experience, then the subject-object dialectic is to be viewed on the basis of a common 'subjectivity' (*ein gemeinsames 'Subjektives'*)[51]; any objectivity is grounded on subjectivity.

In Husserl's terminology, however, the final absolute (meant to be the ultimate condition of a possibility of any intentional act), does not concur with that which we usually define as 'consciousness'. The final absolute, then, is to be conceived as being transcendent in relation to each respective act of consciousness and, at the same time, is dependent, if it is to possess being at all,[52] on the contrapositions of the empirical ego as manifested in its acts.

The final absolute can only return to its origin (*reditio, reflexio*) after having constituted the 'other' (*das 'Andere'*) as intended in those acts (*Beisichsein werden*). It is in this process that subjectivity is shown to be something absolute; hence, on this level nothing seems to be possible but subjectivity.[53, 54, 55] It can now be said that "any instance of direct empirical evidence turns into the respective transcendental empirical evidence, thus receiving its true significance" ("*jedes Stück natürlicher Empirie übersetzt sich in eine entsprechende transzendentale Empirie und weist in dieser seinen wahren Sinn aus'*").[56] For direct empirical evidence possesses only mediated being, in as much as it is outcome of, and therefore something *contradistinct to*, 'effecting Subjectivity' (*Produkt und Gegensatz der 'urquellend-fungierenden Subjektivität'*).[57, 58, 59, 60] This 'effecting subjectivity' will also be realized as the true nature of a philosophizing mind, that is to say, as its 'pure *ego*' that by its very nature cannot be conceived as being separated from the *ego* of the philosophizing mind[61] although transcendent to its empirical act of gaining consciousness.

Taking into consideration Husserl's statement on the subject-object dialectic, this dialectic has to be defined as a purely *immanent* one and hence, if a 'realist' asks: "In what way can one go beyond that which is imagined to be 'world' while at the same time remaining within ((transcendental)) *epoché?*" ("*Wie soll man, wenn man in der ((transzendentalen)) ἐποχή verbleibt, je über Weltvorstellungen hinauskommen?*")[62], this is putting forward the question in the wrong manner and as such this question must be called 'naive',[63] since it has not surmounted the contradistinction "between that which is imagined to be world and the world *per se*" ("*zwischen Weltvorstellung... und der Welt selbst'*")[64], but rather remains in a kind of static antithesis. Realism seems to refuse a consequent realization of transcendental reduction by way of which the motive of an '*in te redi*' receives its full and phenomenologically adequate completion. This completion would signify a return of the mind from its most extreme point of objective manifestation (*extremste Entäußerung*) in modern scientific 'objectivism' to the absolute subjectivity of true being. For it is absolute subjectivity which allows the manifold of the finite world, finite world here being again understood as identical with absolute subjectivity.

By the act of annihilating the objective realm in pure subjectivity through a monistic propagation of a primarily dualistic concept of transcendental philosophy, the subject-object dialectic of the 'phenomenon' has arrived at a dead end.

In analyzing the phenomenon and the experience of the respectively given environment (*Analyse des In-der-Welt-seins*), the subject-object dialectic proves to be a structure of mutual correlation between subject-pole and object-pole; this subject-object dialectic, if comprehended in its full impact, demonstrates the *a priori* preconditions of cognition to be valid for both sides of this subject-object-polarity. On the part of the subject, the *a priori* conditions consist of the *formal* structures of experience which, however, may change in the course of empirical processes and thus do not represent something final or absolute. On the part of the object, we find the *a priori* moment in the transcendence of the outer world to be such that in relation to consciousness it constitutes a permanent other.

The empirical *a posteriori* moment of cognition which constitutes the means through which cognition gains its creative nature is founded in the mediation of a reality as expressed in the individual course of life which provides the possibility of a manifold of events which cannot be foreseen and in which reality is reflected correspondent to an ever changing perspective of the perceiving subject. Integration of this manifold of experiences into the 'effecting' (*das 'fungierende Leisten'*) intentionality, understood to be transcendental subjectivity, would invalidate the thatness (*Faktizität*) of experience. In the last instance, that would be nothing but a realization of the *a priori* on the part of the discerning subject. This is the central argument of transcendental idealism as presented in Husserl's papers. An essential side effect of this kind of idealism is the overrating of the *a priori* nature of subjectivity, which thus becomes *the* ontological category. Only a full recognition of the subject-object dialectic in the phenomenon will turn phenomenology into a methodological preamble to a critical, as well as realistic, ontology, such as is provided by Heidegger's existential ontology where the final *a priori* of cognition and human existence is not fully absorbed by subjectivity. Thus, only a phenomenology in which '*a priori*' and '*subjectivity*' do not reach full correspondence can do justice to the dialectic between subject and object, a dialectic which does not allow a shifting of emphasis toward either the subject or object pole.

It is only in this way that phenomenology can be enabled to mediate between the extreme positions in epistemology. For, on the one hand, idealism is not acceptable from a critical viewpoint, and on the other hand, realism too should not be uncritically accepted. Along these lines, Merleau-Ponty, when referring to the nature of phenomenological dialectic, says that: "the essential merit of phenomenology is, in its conception of the world or of rationality, to have reconciliated the extreme positions of subjectivism and objectivism" (*"La plus importante acquisition de la phénoménologie est sans doute d'avoir joint l'extrême subjectivisme et l'extrême objectivisme dans sa notion du monde ou de la rationalité."*)[65]

NOTES

[1] *Enzyklopädie, Coll. Works* VI (ed. by Glockner), p. 246.

[2] *Ha.* I, p. 32ff.

[3] Cf. L. Landgrebe, *Phänomenologie und Metaphysik*, Hamburg, 1949, p. 59ff.

[4] Cf. *Ha.* IV, p. 105ff.

[5] Cf. T. W. Adorno, *Zur Metakritik der Erkenntnistheorie*, Stuttgart, 1956, p. 192.

[6] Cf. E. Fink, *Sein, Wahrheit, Welt (Vor-Fragen zum Problem des Phänomen-Begriffes)*, The Hague, 1958 (*Phaenomenologica* I).

[7] Cf. *ibid.*, espec. p. 88ff.

[8] This view is held by several French phenomenologists.

[9] 'L'idée phénoménologique d'intentionnalité', p. 119.

[10] Cf. 'Réflexions sur une problématique husserlienne de l'inconscient', 221ff.; *Existence et Signification*, p. 7ff.

[11] Similarly J. Hyppolite also in dialectical formulation: "l'essence de l'homme c'est... d'être soi dans l'autre, être soi par cette altérité même." ('Phénoménologie de Hegel et Psychoanalyse', in *La Psychanalyse* 3, 25).

[12] See R. Ingarden, *Der Streit um die Existenz der Welt.*

[13] *Coll. Works*, Vol. XV (ed. by Glockner), p. 450.

[14] See H. Wagner, *Philosophie und Reflexion*, p. 36: "Ohne Vorausgehen des Außersichseins ist dem Bewußtsein kein Beisichsein möglich."

[15] R. Ingarden, 'Über den transzendentalen Idealismus bei E. Husserl', p. 190.

[16] *Logical Investigations* II/2, p. 236.

[17] *Ha.* I, p. 32.

[18] 'L'idée fichtéenne de la doctrine de la science et le projet husserlien', p. 180ff.

[19] Similarly in J.-P. Sartre, *L'être et le néant*, e.g. p. 17: "...qu'il n'est pas de conscience qui ne soit *position* d'un objet transcendant..." – "...toute conscience connaissante ne peut être connaissance que de son objet." (*Ibid.*, p. 18).

[20] Here Husserl's concept of intentionality is essentially different from that of Brentano.

[21] *Coll. Works* XI (ed. by Glockner), p. 415 (emphasis by author).

[22] "...la conscience ou le moi *sont* ouverture, ordination *à* l'autre, *négation du repos en et sur soi-même*, et donc, en quelque mesure tout au moins, *négativité*. Que la conscience, en d'autres termes, n'est pas une intériorité pure, mais qu'elle est à comprendre comme *sortie de soi* et ex-sistence." ('L'idée phénoménologique d'intentionnalité', p. 119).

[23] *Ideas* I, *Ha.* III, p. 92.

[24] Totally different from Kant (see KRV, ...'Widerlegung des Idealismus').

[25] L. Eley, *Die Krise des Apriori*, p. 10.

[26] L. Eley, *Die Krise des Apriori*, p. 13.

[27] G. Brand, *Welt, Ich und Zeit*, p. 28.

[28] *Ha.* **VI**, p. 429.

[29] Cf. N. Hartmann's critical notes, in *Grundzüge einer Metaphysik der Erkenntnis*, p. 167ff. (in general: p. 164ff.): "...fraglich ist... die... Einseitigkeit, die Inhalte der Anschauung zu isolieren und als Fürsichsein zu betrachten."

[30] Ms. K III 6, p. 386ff.

[31] *Ha.* **VIII**, p. 480.

[32] *Ibid.*

[33] This then would mean the return to the starting point, however in a mediated, transcendentally reflected mode.

[34] *Ha.* **VIII**, p. 480.

[35] Hegel, *Logik* (ed. by Lasson, *Coll. Works* III), p. 30.

[36] *Ibid.*

[37] Hegel, *Logik* I (ed. by Glockner, *Coll. Works* IV), p. 60. See also the introduction to *Phänomenologie des Geistes*.

[38] *Ha.* **VIII**, p. 182.

[39] Cf. N. Hartmann, 'Das Problem des Apriorismus in der Platonischen Philosophie', in *Kleinere Schriften*, Vol. II, Berlin, 1957, p. 48ff.

[40] It leads to truth since truth is (transcendentally) dependent upon the subject.

[41] *Ha.* **VIII**, p. 281. See also *Cartesian Meditations*, *Ha.* **I**, p. 39; *Ha.* **VIII**, p. 282: "Universale transzendentale Erkenntnis, in sich beschließend universale Welterkenntnis" (emphasis in original).

[42] *Ha.* **VIII**, p. 281.

[43] *Ibid.*, p. 292.

[44] See Husserl's attempt to solve this problem: *Cartesian Meditations*, *Ha.* **I**, §63, p. 178.

[45] This holds true, without doubt, for Husserl's concept in *Ideas* and *Cartesian Meditations*.

[46] See espec. *Ha.* **VIII**, p. 480.

[47] The same problem arises in Hegel's philosophy.

[48] A. de Waelhens, 'Phénoménologie Husserlienne et...', p. 15.

[49] For a more detailed explanation, see my paper on 'Die Subjekt-Objekt-Dialektik in der transzendentalen Phänomenologie. Das Seinsproblem zwischen Idealismus und Realismus'.

[50] See T. W. Adorno, *Zur Metakritik der Erkenntnistheorie*, p. 242; cf. Husserl, *Cartesian Meditations*, p. 35ff.

[51] Cf. L. Eley, *Die Krise des Apriori*, p. 9: "Die Transzendentalität wird... von der Subjektivität her verstanden...."

[52] See above.

[53] Cf. E. Coreth, *Metaphysik*. p. 37ff.

[54] Cf. E. Coreth, *Metaphysik*. p. 158.

[55] "...Wahrheit soll immanente Übereinstimmung sein; Evidenz rekurriert nicht bis auf die Wirklichkeit, sondern nur bis auf einen 'gebenden Akt'. Alles bleibt innerhalb des geschlossenen Kreises des 'Bewußtseins von etwas'. Daß das Bewußtsein mit diesem 'Etwas' ein Ansichseiendes meint, wird willkürlich vom Phänomen ausgeschlossen." (N. Hartmann, *Grundzüge einer Metaphysik der Erkenntnis*, p. 166).

[56] *Ha.* **VIII**, p. 179.

[57] Cf. *Formale und transzendentale Logik*, §99, p. 221: "eine Welt, Seiendes überhaupt jeder erdenklichen Artung, kommt nicht 'θύραθεν' in mein Ego, in mein Bewußtseinsleben

hinein. Alles Außen ist, was es ist, in diesem Innen und hat sein *wahres Sein* aus den Selbst-gebungen und Bewährungen innerhalb dieses Innen...."

[58] See *Ideas* I (*Ha*. **III**, p. 93ff.): "Realität, sowohl Realität des einzeln genommenen Dinges als auch Realität der ganzen Welt, entbehrt wesensmäßig... der Selbständigkeit. Es ist nicht in sich etwas Absolutes..., es hat gar kein 'absolutes Wesen', es hat die Wesen-heit von etwas, das prinzipiell *nur* Intentionales, *nur* Bewußtes, ... Erscheinendes ist."

[59] Cognition thus would be mere selfrecognition in the end.

[60] Cf. *Ideas* III, *Ha*. **V**, p. 113ff.

[61] Cf. E. Fink, *Welt und Geschichte*, p. 150ff.

[62] *Ha*. **VIII**, p. 480 (emphasis by author).

[63] *Ibid.*

[64] *Ibid.*

[65] *Phénoménologie de la Perception*, p. XV.

BIBLIOGRAPHY

Adorno, T. W., *Zur Metakritik der Erkenntnistheorie (Studien über Husserl und die phäno-menologischen Antinomien)*, Stuttgart, 1956.

Brand, Gerd, *Welt, Ich und Zeit (Nach unveröffentlichten Manuskripten Edmund Husserls)*, The Hague, 1955.

Broekman, Jan M., *Phänomenologie und Egologie (Faktisches und transzendentales ego bei Edmund Husserl)*, The Hague, 1963 (*Phaenomenologica* **XII**).

Coreth, Emerich, *Metaphysik. Eine methodisch-systematische Grundlegung*, Innsbruck/Vienna/Munich, ²1964.

Eley, Lothar, *Die Krise des Apriori in der transzendentalen Phänomenologie Edmund Hus-serls*, The Hague, 1962 (*Phaenomenologica* **X**).

Fink, Eugen, *Sein, Wahrheit, Welt (Vor-Fragen zum Problem des Phänomen-Begriffs)*, The Hague, 1958 (*Phaenomenologica* **I**).

Fink, Eugen, 'Welt und Geschichte', in *Husserl et la Pensée Moderne* (ed. by H. L. van Breda), The Hague, 1959 (*Phaenomenologica* **II**).

Hartmann, Nicolai, *Grundzüge einer Metaphysik der Erkenntnis*, Berlin, ⁴1949.

Hartmann, Nicolai, *Kleinere Schriften*, Vol. II, Berlin, 1957.

Hegel, G. W. F., *Phänomenologie des Geistes* (ed. by J. Hoffmeister), Hamburg, 1952.

Heidegger, Martin, *Nietzsche*, Vol. II, Pfullingen, 1961.

Hoeres, W., 'Zur Dialektik der Reflexion bei Husserl', in *Salzburger Jahrbuch für Philos. u. Psychol.*, **II**, 1958.

Husserl, Edmund, *Husserliana (Ha.)* Vol. I, III, IV, VI, VIII, IX. The Hague, 1950ff.

Husserl, Edmund, *Logische Untersuchungen*. Vol. II/1/2, Halle a.d.S., ³1922.

Husserl, Edmund, *Formale und transzendentale Logik (Versuch einer Kritik der logischen Vernunft)*, Halle a.d.S., 1929.

Hyppolite, Jean, 'L'idée fichtéenne de la doctrine de la science et le projet husserlien', in *Husserl et la Pensée Moderne* (*Phaenomenologica* **II**), The Hague, 1959.

Hyppolite, Jean, 'Phénoménologie de Hegel et Psychanalyse', in *La Psychanalyse* 3, p. 17ff.

Ingarden, Roman, *Der Streit um die Existenz der Welt*. Vol. I, Tübingen, 1964; Vol. II/1/2, Tübingen, 1965.

Ingarden, Roman, 'Über den transzendentalen Idealismus bei E. Husserl', in *Husserl et la Pensée Moderne*.

Landgrebe, Ludwig, *Phänomenologie und Metaphysik*, Hamburg, 1949.

Leibniz, G. W., *Philosophische Schriften*, Vol. VI, Leipzig, 1932.
Lotz, J. B., *Sein und Existenz (Kritische Studien in systematischer Absicht)*, Freiburg/Basel/ Vienna, 1965.
Merleau-Ponty, Maurice, *Phénoménologie de la Perception*. Paris, 1945.
Merleau-Ponty, Maurice, 'Le Philosophe et son Ombre', in *Edmund Husserl (1859–1959)*, The Hague, 1959 (*Phaenomenologica* **IV**).
Sartre, Jean-Paul, *L'Être et le Néant. Essai d'ontologie phénoménologique*, Paris, [18]1949.
Waelhens, Alphonse de, 'L'idée phénoménologique d'intentionnalité', in *Husserl et la Pensée Moderne*.
Waelhens, A. de, 'Réflexions sur une problématique husserlienne de l'inconscient, Husserl et Hegel', in *Edmund Husserl (1859–1959)*.
Waelhens, A. de, *Existence et Signification*, Louvain/Paris, 1958.
Waelhens, A. de, 'Phénoménologie Husserlienne et Phénoménologie Hégélienne', in *Existence et Signification*.
Wagner, Hans, *Philosophie und Reflexion*, Munich/Basel, 1959.

DISCUSSION

Special Contribution to the Debate

R. M. ZANER

Passivity and Activity of Consciousness in Husserl

I have been asked to preface the discussion with some brief comments concerning the theme, 'the activity and passivity of consciousness in the work of Edmund Husserl'. After the presentation of these comments, I should like to call on Professor Tymieniecka, in view of the fact that Professor Barral's presentation has given a good deal of attention to the former's own thinking on the topic. After that, the floor will be open to general discussion.

I certainly do not need to insist here, to this gathering, how important is the distinction between activity and passivity to Husserl's thought. As you know, the German expressions, *Aktivität* and *Passivität*, cannot be rendered into English very easily, and translations certainly help to complicate many discussions. As, for instance, the still lively debates centering around the Husserlian notions of 'ego' – for which the present distinction is quite crucial. The same is the case as regards such basic themes as 'synthesis' and more particularly 'constitution'.

To help advance the discussion today, permit me to quote Alfred Schutz, for whom the problematic of constitution, and the distinction between activity and passivity of consciousness, are central – merely as a way of setting up how some of the complications get introduced to the discussion, thanks to a considerable extent to the linguistic problem of attempting to give adequate renderings of these terms. In his criticism of Husserl's transcendental theory of intersubjectivity, Schutz states:

At the beginning of phenomenology, constitution meant clarification of the sense-structure of conscious life, inquiry into sediments in respect of their history, tracing back all *cogitata* to intentional operations of the ongoing conscious life. These discoveries of phenomenology are of lasting value; their validity has, up to now, been unaffected by any critique, and they are of the greatest importance for the foundation of the positive sciences, especially those of the social world. For it remains true that whatever is exhibited under the reduction retains its validity after return to the natural attitude of the *life-world*. But unobtrusively, and almost unaware, it seems to me, the idea of constitution has changed from a clarification of sense-structure, from an explication of the sense of being, into the foundation of the structure of being; it has changed from explication into creation (*Kreation*). The dis-

Tymieniecka (ed.), Analecta Husserliana, Vol. III, 199–226. *All Rights Reserved.*
Copyright © 1974 by D. Reidel Publishing Company, Dordrecht-Holland.

closure of conscious life becomes a substitute for something of which phenomenology in principle is incapable, viz., for establishing an ontology on the basis of the processes of subjective life.[1]

Schutz is obviously not alone in pointing to the difficulties in understanding what Husserl meant by 'constitution' – witness Merleau-Ponty, for instance. The fundamental effort of Husserl to develop an autonomous discipline of philosophical criticism (i.e., transcendental phenomenology) has its *locus* in his attempt to explicate and clarify the full stratification of mental life (or, consciousness). This effort is particularly notable with respect to the problem of judgment in *Formal and Transcendental Logic*, as well as perceptual theory (both in *Ideen*, I, and in *Erfahrung und Urteil*).

As is known, the concepts of intentionality, passive synthesis and constitution form the most general or universal descriptive characteristics of mental life. Furthermore, none of these notions are for Husserl comprehensible apart from his central distinction between *Ich-akte* (*ego-activities*) and what might best be called pre-predicative or operative consciousness (which Cairns translates as 'automatic' processes of consciousness) – i.e., the realm of *Passivität*. Since one is here focusing on *Erlebnisse*, I have sometimes opted for the expression, 'consciousings', in order to emphasize that Husserl can in no way be accused of reifying conscious*ness* (*Bewusstsein*); the distinction between ego-acts and *passiven Erlebnisse* (activities, and pre-predicative processes of awareness) cuts across all 'experiences'. 'Acts' are, in the terms of *Ideas*, those processes (*Erlebnisse*) of consciousness which are descriptively characterizable as having an '*ego*-quality', that is, there is an *ego* busied with the objects of the various awarenesses in question. Distinct from these ego-activities, or *ego*-acts there is the region of 'non-acts', as he sometimes calls them, namely those *Erlebnisse* in which there is no describable characteristic or property of an '*ego*' kind; there is no '*ego*-busiedness' with the objects of the awarenesses in question; they go on without there being any *ego*-presence 'living-in' them directed to their respective objects.

The concepts of synthesis and constitution are understandable only in terms of this distinction, between *ego*-acts (or, simply, acts) and process of awareness – which is, fundamentally, that between activity and operative or automatic 'passivity'. However, further analysis shows Husserl that the latter is distinguishable into two general regions: secondary passivity or the secondarily operative, and the primarily operative. The former includes all those processes which have become sedimented, habituated, typified; those which either were at one time activities and now have become sedimented, or those which in principle could become acts, those in which there could be an 'ego-busiedness'. The primarily operative awarenesses are those to whose objectivities there can be no ego-directedness; those in which the ego cannot in principle 'live' and be 'busied' with the objectivities thereof.

Acts stand out from, and presuppose, the context of pre-predicative operative awareness (both secondary and primary), and benefit from the synthetic deliverances of these ongoing processes. Thanks to their functioning, there is always-already available to the concrete human ego a set, indeed a veritable world, an already constituted complex, of objects – conceived as sense-unities – of all types. The so-called logos of the esthetic world, of which Husserl spoke in Formal and Transcendental Logic, is found precisely in the operative performings (Leistungen) going on in the region of 'passivity', not only those going on in inner-time consciousness, but those that form the core of the crucial concept of sedimentation. Constitution must be understood in terms of the passive synthesis of inner-time consciousness, the effectuating of sense-bestowals going on in the stream of consciousness (Erlebnisstrom): the synthesis of identification, differentiation, and associative sense-transfer. It does not in any sense signify 'creation' – so far as I can tell, despite Schutz's contrary judgment, this is so for all of Husserl's works – but rather intentive synthesis. Thanks to these, there is always-already available to the concrete human ego, not only an entirely already-formed world of objects, experienced through a complex of typifications, but other subjects as well. It is fundamentally at the passive, operative levels, in Husserl's view, that these issues first find their locus.

Thus, the operative associative sense-transfer, for instance, as developed particularly in his Cartesian Meditations, is in no sense an active inferring by analogy to the other human subjectivity, but is rather an operatively, automatically ongoing series of synthetic processes of awareness, thanks to whose functioning certain items or affairs acquire, or have 'already' acquired, the sense: 'other human persons', cultural artifacts, and so on. To be sure, there is no settled view as yet on such issues as the constitution of others (witness Schutz's view as opposed to that I suggested), 'ego', and other issues. The clarification of them depends in part on getting clear on this crucial distinction between the activity and passivity of consciousness. Perhaps the discussions of today's papers will further help to clarify this distinction, and thereby aid us in understanding not only Husserl's work, but that of other philosophers as well who have addressed these problems. At this point, then, having only sketched the bare outlines of the Husserlian approach here, I will call on Professor Tymieniecka to comment on Professor Barral's paper.

Tymieniecka: First I would like to thank Professor Sancipriano who has shown in a penetrating way some aspects of the empirical soul with which I am in perfect agreement. Then, Dr Barral has opened such an extensive range of questions I cannot possibly give her but a few sketchy and seemingly disparate answers to her challenge. What seems to me important in our debate about the *a priori* and the passivity and activity of consciousness is that the question of passivity and activity of consciousness in its function of constituting the world at various levels, starting by kinesthesis and sensitive perception, is intimately related to the question: "What are the *regulations* which are guiding these processes?" Husserl encounters this question first in his analysis of outer perception concerning the continuity in the anticipation–and–fulfillment structure of the specific act in the perceptual process. In the serial progress of perception there is a unity of the same identical object that is anticipated, and which is being slowly constructed from one instance to another. The question arises then: "What is the regulative principle of this sense continuity, the point of reference of this whole process?" Then the same question recurs with respect to the passive synthesis, wherein it entails the problem of the '*a priori*'. But the pivotal issue is: how to understand 'passive' and 'active'? What does '*a priori*' mean? As a rule we employ the *a priori* either in the metaphysical sense, as Dr Barral has said, according to the manner of the Platonic ideas governing the structure of concrete beings, or in the epistemological sense of the Kantian dichotomy between *a posteriori* and *a priori*; but what is the meaning of the *a priori* in phenomenology? Should we even introduce it at all? My third and final point. I have indicated in my lecture some possible meanings of 'passive' and 'active' in Husserl and have shown the consequences of his definitions. Contrasting, like Dr Murphy, Husserl with Kant in this respect, we see that there is a 'receptivity' of the basic genesis in Kant, which seems to correspond to what Husserl calls 'passivity'. Obviously they both have in mind the same phenomenon but with nuances of difference. For Husserl this fundamental flow of experience through which we perceive, remember, or imagine is 'passive' in the sense, that it is a *spontaneous unfolding*. If by 'voluntary' we mean an act of deliberation and design, it is 'involuntary'. Indeed, it is not a result of deliberation but of 'routine' or 'spontaneity'. As such, I call it an 'automatized' procedure. We flow along with it, we cannot stop it at will without stopping to live. When we lose consciousness it is disrupted, or at least its main constitutive dimension is torn apart. Does this passive synthesis demand an effort from our part or is it effortless? In Husserl's view we simply accept it, we naturally consent, adhere to it, we do not even wonder that there could be something curious about it; thus in this sense it is completely effortless. On the contrary an active mode of operation would be firstly, voluntary, in some relation to the will; secondly, deliberate, and thirdly, it would be dependent upon effort. These clarifications seem to me extremely important, because the question of the *a priori* investigated in its farthest ramifications appears to be related, as I have shown it, to the way in which consciousness works. And here we are faced again with the question: 'How is this routine-flow through which we exist, which we do not choose, which we just accept – regulated?' At this juncture we encounter the problem of the principles of regulation of the passive synthesis or constitution. Let me call your attention to another point of my criticism of Husserl. Although, throughout all his works there are explicit indications that the principles of reference are ideas, especially in the constitution of the body Husserl speaks of ideas as *Leitfaden* of the constitution directing the constitutive process in the passive synthesis. But nowhere in his works has Husserl clarified sufficiently the notion of the idea or that of the *eidos*; consequently, there is an abysmal gap between this painstakingly explained process of constitution and the ideas which are supposed to be its points of reference. We are left completely in the dark as to *what* function they do exactly accomplish, and *how* they do accomplish it. We could make a reference to Husserl's 'free variation', as Claesges has done in his analysis of the constitution of a special object. In this essay Claesges refers to ideas as being a sort of model for the 'free variation' in the process of constitution as Husserl's takes them to be for the 'eidetic insight' or 'ideation' with which Husserl has been greatly concerned in his later years.[2] In fact, in the routine of the perceptual process there is the further question of continuity and error. For instance, in looking at a chair, one could have a glimpse of the chair and immediately thereafter of the table standing next to it, and then again back to the chair. How would we construct the one and the same object and not the other? How would we bring our perceptual series into a unifying unit and help to correct our error? How do we go back with another glance to continue the perceptual series and help to correct the error? Husserl attributes the correction to 'imagination'. 'Imagination', as understood by Husserl, belongs to the constitutive system; the role of its 'free play' is to present all the variants of individual possibilities which fall under one idea; imagination according to him by presenting the variations of all the individual possibilities accounts, first, for the obvious flexibility of the perceptual process, for error and correction; it offers them as points of reference. But here again there is a further point of my criticism of Husserl: if this 'free variation of imagination' which supposedly introduces ideas as the prototypes is the integral part of the constitutive process we are still left in the dark as to how ideas may accomplish this role. Husserl leaves unexplained both: the role as well as the meaning of this function. Are ideas transcendent to the variation or are they immanent? This and similar relevant issues have never been resolved by Husserl. Gurwitsch's theory of the 'field of consciousness' maintains that whatever regulations of the constitutive process there would be, they are immanent within the field of consciousness; they are intrinsic and are motivated by the mechanisms of the conscious field itself. Neither Merleau-Ponty nor Gurwitsch accept *a priori* principles. As you realize, this approach closes us completely within the transcendental field, which means, that whatever there is constituted by the original intentional consciousness, motivated by itself, it has its regulative principles within itself. It is against this assumption, and against the Husserlian inadequacy in treating these problems that I have directed my own analysis of ideas and of their role. In my exploration of ideas I am indebted to Ingarden for this fundamental distinction, to which Dr Barral has already referred, between *idea qua idea*, namely idea as a special entity, and its *content*. The fact that the idea possesses a content which exemplifies the rational structure of an individual being is instrumental in solving the twofold problem Dr Barral has mentioned, namely the metaphysical problem of the Platonic *met-exis*: How can individual beings have a rational structure, that is, be self-centered agents of *actio* and *patio*? Plato in spite of his unsurpassed treatment of this question never really answered it satisfactorily. In my attempt to resolve the riddle: what is there in the nature of the idea that such a correspondence is possible? – I answer – thus far with Ingarden, that this is due to the correspondence between the content of an idea and the corresponding concrete being. Idea, in itself, as strictly idea could not participate in anything and nothing could participate in it. But parting with Ingarden, I see the nature of the content not as an ideal structure but as an *ideal functional system*. (Cf. *Why is there Something rather than Nothing?*) We could never think that there is a causal relation between, let us say, the universal pattern of becoming and the becoming itself, because, as we know, the causal relationship holds only within the homogeneous realm of the physical world and not beyond, so there must be another type of relationship. (I have tried to show in my work on Leibniz that there are at least four types of continuity explored by him; unfortunately, the fact that usually they are not clearly distinguished vitiates the understanding of Leibniz.) It is because of the fact that ideas have in their content a functional pattern isomorphic to the structure of the empirically perceivable individual things that they can fulfill the metaphysical function of being the points of reference for the rational structure of individual entities. That is the metaphysical aspect of it. But evidently the problem is not resolved by examining it only within the metaphysical perspective. The next and inevitable step must be taken to explore first, its relation to human consciousness, and secondly, to see what is it that fulfills the regulative role of the conscious constitution of things and beings as we know them. My distinguished colleague, Dr Półtawski, pointed out that Ingarden never escaped the dilemma he himself has created; for there remain two juxtaposed levels in Ingarden's interpretation of the issue which can never meet. In the essay, which Dr Barral has so searchingly analyzed, I have shown that ideas in fact *can* and *do* fulfill the role of points of reference in the constitutive process as well – but at the price of having had to part with my master altogether. Ingarden was utterly dismayed when he read my book, *Why is there Something rather than Nothing?* and saw that I analyse the ideas differently from him. But ideas, which, after all, we obtain by variation of individual cases 'mirror' in their content the individual's inward *mechanism* rather than their own *static skeleton* in such a way that it really 'participates' in the functioning of consciousness as well as in the functioning of nature. In fact, in this second study analyzed by Dr Barral, I proposed that the content of the idea – due to its special *operative* nature – enters into the functioning of the constitutive processes. (We really would not know how a process could refer to an ideal structure. We would be caught in the Platonic dichotomy of two realms and could never escape from it.) In other terms, this *functional conception* of the content of ideas shows that they have two operational values: one of them relates to the ontological inquiry, the other to the epistemological inquiry. It seems to answer the concise question of phenomenology: 'How can ideas at the same time be points of reference for the ontological structure of concrete beings and for the functional operations of the constitutive consciousness?' The second point seems to agree with the analysis of 'creative context' in which I develop a *functional theory of consciousness*. Emphasizing the 'routine *rationale*' of the world, Husserl, and Ingarden after him, conceive consciousness as an intentional system of relations pervading the psychic life, or as a distinctive realm of a rational, all-encompassing faculty. Whereas from the point of view of 'creative originality' as well as that of disintegration of the intentional pattern of relations in psychosis, which I discussed in the first volume of *Analecta*, I propose, that 'consciousness' is broader than 'intentionality', and man's *functional system* at all levels of man's manifestation is more adequate an approach than 'faculties'. It is, so to speak, the basic operational mode which can be orchestrated in different ways; it can be distinguished at least in its two major orchestrations; first, the *constitutive function* with its *a priori* of ideas; it is limited because it does not allow either novelty or originality. Secondly, there is the *creative function* where we are no longer restricted by the constitutive system with its ideas as points of reference and where *imagination*, emerging not as a faculty, but as a result of the *functional orchestration of the creative context*, can play the role of generating ever new possibilities.

Barral: I will be very brief. First, I want to thank Dr Tymieniecka for explaining what I was unable to do, namely, specifying what ideas are. However, I am still dissatisfied because I am not quite sure that it was clearly stated what is the origin of ideas. Suppose I am going back to perception at its genetic starting point, to the perception of anything as it organizes itself at its incipient stage. Maybe ideas are a product of my rationality and that is why they become a directive for perception at the incipient stage of the constitutive genesis. The other question I wanted to ask after your lecture was: what do you mean by 'orchestration' and 'structuration' of 'creative act'? I would like you to explain how this creative act is unfolding; what are its directions? If you use the word 'orchestration' and 'structuration' it must have some directions. I was rather imagining creativity to be a pure projection, absolutely new, absolutely free.

Tymieniecka: Your two questions give me opportunity to sketch my thought a little further; for details I refer you again to my brief treatise in cosmology *Why is there Something rather than Nothing?*, and other works. Dr Barral has pointed out an important question: 'Are the regulative principles of constitution *immanent* or *transcendent* with respect to the constitutive system?' Assuming that ideas are immanent to the constitutive field, this routine of perception would be established automatically by the mechanism of the whole consciousness as such without being directed by any special agent. *Mutatis mutandis*, the case can be likened to a law in biology which maintains that there is a code organising the organic operations without assuming special 'principles'. This code is supposed to account for the purposeful cooperation of various components of such a sort as biologists accept, for instance, in embryology. There always remains, however, the question as to what motivates the selection of such definite types to be constituted according to the given code and no other. It seems then that the recourse to an internal code does not solve the problem of the origin of the types in the world and the immanent principle interpretation has to be abandoned. Of course Bergson would say that Nature is unspoiled by rationality; that the mind rends it in order to construct its forms; but there are very good reasons why Bergson's explanation, as Ingarden has pointed out, is unsatisfactory. If they are immanent we are closed within the one and unique transcendental system of the world, and we have to submit to Husserl, as he said in *Ideas I*, that even to ask after another world than the one we live in is meaningless; it amounts to a total submission to a 'transcendental determinism' from which there is no way out. If, on the other hand, ideas – which I consider to be *a priori* and to be the principles of the constitution – are transcendent to the flow of experience and to the constitutive system as such, then we are faced with two questions: firstly, if they are transcendent what is their metaphysical status; secondly, if they are transcendent how can they fulfill their role? I treat these questions is I mentioned before, by referring to what I call, in opposition to Ingarden, their 'functional' content, through which they 'enter into' the operational system of consciousness. Regarding their metaphysical status I start from the premise in my cosmological work that, in fact, we cannot begin a conclusive analysis of the world in terms of the *life-world* or of the modes of the world's appearance in consciousness; rather we must commence from the viewpoint of empirical experience, on the one hand, and ontology on the other, what I summarily call 'the constitutive system of the Universe', of which the transcendental *life-world* would be just a human dimension. Both: the *architectonic scheme* of the Universe and the *life-world* have the same point of reference: ideas as prototypes of the constitution of the Universe and of the functional mechanics of consciousness.

Straus: I have a few rather simplistic questions. The title of Dr Barral's paper reads: 'Problems of Continuity in the Perceptual Process'. What does the singular 'the' mean? I see some thirty, forty people here; are there thirty, forty processes? And if there are thirty, forty processes, how can these thirty, forty people hear the same talk? Let me go one step further: is consciousness singular, or is it a universal thing? But if there is a plurality of consciousness, if there are many consciousnesses, how are these consciousnesses related to each other? How can we see something together? And when we come back here into this hall again is not the act of seeing renewed, but at the same time the walls seen are 'recognized', as 'the same one'; while the words we have been hearing during this meeting are gone. How can we account for anything of the common world if we start from either one consciousness, *the* consciousness, or if we accept a plurality of them? Where are they 'located'? 'Where' is my consciousness? What do we do with that term 'consciousness' in the plural? How do we transcend the confines of an individual, who hears, but what he hears is common to all in the audience? We speak about an audience, which is supposed to hear one and the same speaker seen from here and seen from there; furthermore, there remains a complete enigma, how can these visible things produce sounds? How can these sounds mean something so that we understand it? Yet every child can do that. When the child asks his mother: "What is this?" – he is convinced that the visible mother will hear him; this visible thing produces sounds which are audible by the child, that both are related to a common visible, and that the mother will tell him the 'name' of that thing which is a 'common name'. I am not sure really whether I should assume that all this is related to the question of reality, or that we really have to give an account of reality. Maybe we should start from the assumption which Husserl makes, we should bracket the thesis of reality. Is the thesis of reality a thesis for a consciousness? If that would be so, where is that consciousness 'located' that I see something over there? Or is that thing seen over there also 'within' the consciousness? I know I am asking rather heretical questions, maybe even silly questions, but I would like to have some answers.

Zaner: Is there one and the same person who in one and the same continuous process wants to take up that one and the same challenge?

Barral: Dr Straus just touched upon the main problem of phenomenology and intersubjectivity. I usually tell my students that the reason why they know that there is a tree 'over there' is because the tree is here, 'within'. I use the word 'within' to circumvent a further dilemma for, indeed, we cannot 'locate' a conscious act. But there is a serious problem – I don't think it can be explained – how is it possible to communicate with the Other as Other? I am speaking with someone and it seems that this someone understands what I say, although the only assurance I have that this someone understands me is my own subjective experience of it, and I never really know if the Other understood the way *I* think he understood. Merleau-Ponty would say that I can have an assurance of this intersubjective understanding when it refers to an intersubjective object; an object which is perceived, recognized, pointed out by more than one subject or when several subjects concur in the perception of one and the same object. Referring to Dr Tymieniecka I am not sure that I could think of another world, because whatever world I could think of, I would always think of it in terms of what I am and of my consciousness as it is. I do not see how I could imagine another world. If I create another world it is still in terms of my experience of this one.

Tymieniecka: I agree that we cannot think of another world than ours in precise terms, but functionality of consciousness shows us that it is virtually possible to do so in general contours.

Morin: The question you raise points out the limits of what I would call 'reflexive analysis'. In fact, just how far can we use reflexive analysis? What usually happens here, and it seems to me you tend to fall in this trap, is the assumption of the 'child in the state of nature'; we find this fallacy also in some of the language theories. Dr Barral talked about an 'unspoiled vision'. What can it mean for a child to have 'an unspoiled vision'? You are saying that it is possible for me to go back to an 'unspoiled vision'. I take vision or perception as some kind of a bodily activity. Would 'unspoiled' bodily activity mean that we just have not learned to see or have not learned to listen? In other words, a state prior to the acquisition of habitual activity? But there are difficulties in employing this unfortunate model of the child in a state of nature. A better way to approach the problem would be to treat the child as a kind of an object of inquiry through his overt behavior, or you can follow the child's experience through a reflexive analysis of imitation.

The second point I want to raise is about the importance of tactile perception. So far we have only talked about visual experience, seeing and so on. Now, if we took other sense experience – let us say touch – tactual perception – many aspects of experience are revealed that are completely absent in its visual version. By touch, for example, I can surround the object, I can close my hand, I have control of it – I can close my hand on it. I can stop at any moment and the information that I have of this object is already enriching at that time, but I also know that I will discover more if I keep on going. The notion in tactual experience is that there is place for experimentation; it is immediately evident the part that I play in it. Maybe tactual analysis or reflection on tactual perception would go a long way toward retrieving the 'originary' experience.

Barral: In the first place I should have made a few things clearer; I was referring to the very 'beginning' of a child's perception, when as we naively say he is 'just beginning to see the world'. I don't believe that there is a 'pure state of nature'; a child is born into a situation, a social community situation which is a kind of prepared place, but there must be a moment when the child is beginning to see, to hear, to touch, a moment where he first gets in contact with the world, that would be the 'unspoiled'. By 'unspoiled' I meant also that he has not yet been 'conditioned' in any way; so if it were possible to catch the first moment of the child's contact with the world that would be the 'unspoiled'. And yet it is already set up by the parents, by the situation of the family, by the time and place where he lives. As to your other point of 'tactual perception' I would have to refer you to Merleau-Ponty where he expounds his theory on the 'receiving' of the world. For him the whole body is in contact with the world and therefore the perception of the world by the body, I would say, is part of that background. It is against that background that everything appears. It is due to this totality of my body which is in the world and which has continual contact with the world, in which I can receive certain objects more specifically, like if I close my hand, that I will have the full perception. So when I have a full tactual perception of an object it would be a second moment in the process of perception. Referring to this specific point in Merleau-Ponty he says: "that wherein all our deliberate perception takes place, our real perception takes place." In other words, I have the sensation already but when I deliberately pick up an object it is then that I make it stand out in the world. There is a whole field of study open in this matter.

Ver Eecke: I would like to express my admiration for the attempt by Dr Barral to indicate the difficulties that are present in the phenomenon of perception. But I have the impression that in trying to do so she has deformed the view upon perception itself in two ways. First, she calls 'perception' both the perception of the chair and the perception of my inner will. Originally perception means visual perception of an outside object, and only by far analogy can we say that I perceive my will. Therefore, I would like to say on the point of view of the bodily dimension of perception she seems to diminish that impact. On the other hand, perception has a spiritual, in the Hegelian sense, dimension which is indicated in phenomenology by such terms as 'ideas' or *a prioris*. Now I believe that she underestimates again the spiritual dimension in perception, and that is indicated by the fact that she wants to go back to psychology and learn from psychology something that she doesn't believe she has to transform again in a philosophical problem. I would say if you want to go back to the origins it would not be to those of the child's experiences because the child's perception will be formed by the spiritual dimensions of the parents as well as the child's body. Therefore, if you want to study the spiritual dimension of perception you might go back or forward in the history of philosophy starting from Husserl. If you want to go forward you might go to the Structuralists, if you want to go backward you might consult Hegel, indicating that at every point in history perception is co-determined by the spirit of the people. When, for example, in the time of the Renaissance, men are looking to phenomena outside in the world they look with the following *a priori*: if I think enough I will understand the laws of the objects, and that is the basis of Galileo and Pascal. If you want to go a step further in Hegel you find that he analyses the Romantic reformer as someone who believes that he finds in his own heart the justification for how the laws of the world should be. And if you go the next step, that is, to morality in Hegel, the man of the moral vision believes that he finds in himself the moral laws by which the world will be improved. It means that spiritual ideas which are shared by a community are directly present in the perception of the world. So I would say that the phenomenon of perception involves these two dimensions: the bodily dimension and the spiritual dimension. In trying to show the difficulty of perception you cannot diminish either the bodily dimension or the spiritual dimension.

Barral: I thank you for your observation. I was only trying to say that phenomenologically this cannot be done. You spoke about the spiritual dimension and the different understanding of perception of the physical world. Certainly all the examples you gave are quite true but do we find them all in the phenomenological analysis of perception? I think that we go already beyond. Dr Tymieniecka's paper made me do a great deal of soul searching. Before I read the paper I was convinced that everything could be found phenomenologically; now I am not quite so sure. I don't know whether I could find everything that you describe phenomenologically, but I am not diminishing the role of the spiritual. But you refer then to Hegel's phenomenology not to the phenomenology of Husserl; it is an altogether different matter.

Eng: I think we are getting into a clarification of what phenomenology meant for Husserl. It seems to me that for Husserl the *a priori* is an absolute intuition. In 1913 Husserl cannot be accused of being a transcendental idealist. He begins with a correlation of self and world. The following quotation might well be expressive of his concern at the time: "Wherever we turn, every existent that is valid for me and for every conceivable subject as truly existing in reality, is correlative – and with an intrinsic necessity – with particular universals as an index of their systematic manifolds." And again: "The first breakthrough of this universal *a priori* of correlation between experiential object and modes of givenness (which occurred during the elaboration of my *Logical Investgations* approximately in the year 1898) shook me so deeply that ever since then all of my life work has been dominated by the task of a systematic working out of this correlational *a priori*". I would say with regard to 'activity' and 'passivity' that they are moments in this circular movement. *A priori* is an '*Urphänomenon*' for Husserl, and you cannot go behind the *Urphänomenon*.

Thompson: I would like to go back to something Dr Ver Eecke was saying previously. He suggested that if we looked at some historical situations we would see that the structuring of our perception is relative to the historical background of the spirit of the time. Let me suggest another line of approach, which is to look at the cross-cultural situation, it is then that we come across others who, in Dr Straus's words, don't have a common world with us; they are coming from another culture. They may not see certain social realities which we can see, and vice-versa. There are even some material objects which exist for them that we don't seem to be able to see. Perhaps the most profitable place to investigate perception would be then to study how an anthropologist enters a new culture. The first thing the anthropologist has to do is to 'see' reality the way it is seen within this culture. This is in effect entering a new scheme of perception altogether. This means also entering a new *a priori* (presuming that the anthropologist arrives with at least a partially different *a priori* for his perception than does the native of the culture.) So here we are meeting empirically with a situation where a phenomenologist could not investigate the clash of two *a prioris*, two types of regulative ideas. Margaret Mead particularly took the opportunity to reflect upon the activity she was performing as an anthropologist and came to problems which, to my mind, are extremely similar to the problems we have discussed today concerning the origin of the regulative ideas which govern the genetic constitution within a particular culture. Her solution, by the way, is to look to biology. I doubt if anybody here would accept such a crude naturalism. I would suggest, however, that perhaps there is the point where we could try and find out, as I might say, empirically, or at least phenomenologically, how an anthropologist actually does come to see reality, to understand reality in the way in which somebody with a different cultural *a priori* is able to do.

Półtawski: I want to add something to the discussion between Dr Barral and Dr Tymieniecka. In the first place there is the problem of the *Leitfaden*, the thing as the *Leitfaden* of the constitution. Husserl speaks in *Ideas* III about the thing or the original concept as *Leitfaden* of the constitution. He states that the original concept is not a generic concept at all, it is not a generalization, it is the 'form of the unified experience', then it is finally related to the forms of experience which are primordial. Secondly, a few remarks about the conception of cognition in Ingarden and in Husserl: I think that Ingarden stresses the fact that cognition is in effect 'creative'. There is the issue of 'creativity' of the intentional objects by means of which we aim at 'creating' works of art, etc. I don't think he worked out the problem quite clearly and at length, but there are some statements which unequivocally state that cognition as such is *creative* for him and at variance with Husserl. My next remark has to do with the problem of the state of ideas – where are they to be found? Well, I think that there are nowhere – there is the functional conception which, I think, Professor Tymieniecka is trying to develop. However, I believe, it would be a more pertinent question to ask: 'Where are they from?' I have in mind here psycho-therapy and the concept of the categorial and uncategorial attitude. When the patient cannot adopt the categorial attitude there occurs some modification of the way he perceives the world and when he adopts it he can, so to speak, construct ideas; construct them from something which is not in his experience all the time. It seems then that there are no ideas in experience. There is some structure of *what* we experience and we can, by adopting the categorial attitude, construct ideas from it. I would not be quite sure whether it is correct to say that for Husserl there is only one world, as Dr Tymieniecka did say. I think that there are for Husserl still very many possibilities of experience, which we don't have actually taken into account and in these possibilities there may be constituted other worlds not quite the same as the world we are experiencing.

Finally, I would like to ask Professor Sancipriano: "How do you understand the relation of affectivity to the various levels of consciousness?"

Sancipriano (translated by professor Attilio Salerno): I wish to preface my answer by the statement that my investigation has a synthetic character. This effort to synthesize, which has been preceded by analytical endeavors, especially by my work *Husserl's Logic,* which is subtitled: *The Genealogy of Logic and the Intentional Dynamics,* led me to see that the problem of understanding and of reciprocal understanding has also an emotive component. My intention is not to perform *eidetic* reduction but to live and reproduce 'affectivity'; to this effect, I think that Husserl and Bergson lend themselves admirably to an analysis.

Corello: We have not considered so far *Gestalt* psychology and the organizational principles involved, and yet it forms a background to several points of the discussions. It had a great impact on Merleau-Ponty, on Sartre, and, of course, on Gurwitsch, and it involves Goldstein's work. If we consider it in the sense that the whole organization of our perception divides into foreground and background, into parts and wholes, into elements and aggregates (some of these are under our deliberate control) then on this view we can work on the type of consciousness as a mode of attention, and varying the attention to the perceptual world we can sometimes reverse foreground and background, break totalities into elements and so that sometimes these transformations can even reverse the object, transform the object, or make present objects disappear. I think that in this consideration the whole notion of ideas can be reworked. I don't believe that there are such things as ideas, but there are schematic outlines of objects. In other words, there are typical modes of abstracting from perceptual objects and these typical modes of abstraction vary from child to adult, vary from culture to culture. They vary according to the lines of action perceived. I think this must be at least mentioned because it is essential to the study of the problem of perceptual organization and the modes of attention directed to the intersubjective world and the perceptual world.

Martin: I was very interested in Professor Tymieniecka's comments about functions, and I gather she is using the word 'function' in the usual, more or less, mathematical sense of a one-many relation. First I thought, perhaps one could make sense of the suggestion that if you take a kind of functional account of consciousness or of ideas you somehow can therewith do away with those ideas or at least you make them simply functions of something else. But you know that will not quite work on any notion of function that I can think of, because when you have a functional relation you must have some perfectly good objects on the one hand and some perfectly good objects on the other; your functional relationship is then something between them and you cannot make any sense of a functional analysis without both present in some very substantial way. Now, may I just ask one very brief question: How do you see the problem of possible worlds? Are you keeping them somehow as objects within intentional consciousness, or do you somehow destroy them all by positing them only in the immanent realm?

Tymieniecka: I believe I am using the term 'function' in two senses, very closely connected, one more general and comprising the other specific sense. Firstly, with respect to consciousness, the term 'function' means purposive organization of a set of operations as instances of activity; secondly, with respect to ideas as '*a priori* principles', 'function' means, as you said, a specific type of relation, namely, between the ideas as a perfectly determined type of objects exhibiting a rational pattern of all the possible variants of a certain type of individual concrete beings in their possible mode of constitution as being and becoming and these objects themselves. Both sides are perfectly definable. But the 'function' of ideas with respect to concrete beings is to install the regulative points of references for their actual constitutive frame and its progress. Your second question refers to the contrasting of the *constitutive* structuring function of the world with the *creative* function along the lines of voluntary/involuntary, effortless/effort and so on. I have distinguished precisely between two different 'functional orchestrations' of conscious operations such that, while the constitutive system, which as I propose, unrolls 'passively' and 'anonymously' refers to the ideas as the 'regulative *a priori*' and, consequently, doesn't allow for originality and novelty and closes upon itself, we would be stuck like Ingarden and Husserl with one possible world. On the other hand, the discovery of the *creative orchestration* of conscious operations, which – as I propose – works according to *self-generating possibilities* and unfolds a special type of functionality, referring to personal effort, deliberation, choice, etc., allows for other possible functional systems of consciousness to develop or enter into action which would bring different worlds from ours. I agree with Dr Barral that we cannot now with our present constitutive bonds and our creative capacities conceive what they would be like – Leibniz claimed that the worlds should have a spatio-temporal axis – yet it shows the virtuality of our consciousness to have other functional orchestrations, which would account for such worlds. In other terms, my view is an upshot of contrasting the *constitutive* with the *creative* orchestrations of consciousness, and its final outcome is to give another formulation of the possibility of different worlds than is Leibniz's. Leibniz, as you know, saw this possibility in ideas stored in the mind of God whereas my analysis refers the possibility of other worlds to the 'creative imagination' as it unfolds from within the creative orchestration of our operations generating the possible forms.

To Dr Półtawski I would answer that although Husserl's and Ingarden's transcendental constitution account for the evolution of experience, thus for possible experience which is not available as yet to unfold, and the *horizon* of the constitutive genesis plays the role of accounting for this mutability and change. The *horizon*, however, circumscribes the possible expansiveness and this with reference to the *a priori* structures or ideas, prescribing to the constitution its structural directives as such and no others beyond which it cannot reach. Thus in this renovating or enriching mutability we never reach the *novelty of types* and remain within the same pattern of order.

Dr Półtawski is right that Ingarden does not draw the distinction between constitutive and creative operations and confuses the two in the approach to the work of art. This, as I have pointed out to him in my essay, 'The Undivine Comedy, Structure versus the Creative Vision', published in the *Festschrift* for his seventieth birthday, makes his analysis of esthetics deficient. In *Eros et Logos* I attempt to establish the other side of the necessary approach besides Ingarden's intrinsic structural analysis of the 'creative experience' in contradistinction to the 'cognitive (or constitutive) experience'.

Barral: I have to answer so many questions that I don't know where to begin. First, I take up the objection: that consciousness is not consciousness 'of' something but that it is *I* who see and feel, etc. I agree with you that *I* am the one who is doing it, I as a subject, I as a person am conscious, but if I am under sedation, under anesthesia, then my consciousness ceases. I don't have any consciousness; consciousness is not a thing, I am the being who is conscious. Then I would like to correct perhaps a misnomer: I did not mean to say that we find all the answers to the enigma of consciousness in psychology. As a matter of fact, I left out some phrases in my paper which give references to the effect that what goes on in consciousness remains a problem because we don't really know if what psychology tells us is correct. The same way I am not sure that an anthropologist going into a different culture would be able to see the things as the natives see them. We can make the effort to devise a method of communication but we cannot really be absolutely certain how adequately it would report the facts. Our communication with others remains a problem.

Concerning the question: whether I fail to distinguish between the certitude of *knowing* something and the *thing* that I know, I must reply that I have no intention to 'put them all together' as you claim. But, on the other hand, I don't agree with Descartes either that I should or can doubt everything; in fact I am convinced that Descartes himself could not have put out of his mind all that he knew, or put everything in doubt, because it is impossible. Therefore, I do make a distinction between the two questions: 'Is something there', and 'Am I certain of it?' These are two different questions.

The last point was the issue of *Gestalt*. I agree that there is plenty to be done in this field, still a substantial investigation to be accomplished in this matter. Perhaps the idea of the *Gestalt* that gives us the complete undifferentiated thing is the closet possible manner to grasp the nature of something because we don't analyze yet, we see it all at once; as soon as I talk about consciousness I split this *Gestalt* and then try to put it back together by saying we are conscious beings, whence the question of inter-communications arises.

Mohanty: I find myself in a great deal of sympathy with the problem Dr Köchler finds in many idealistic interpreters of Husserl. Let me first give an expression to my own feelings about this whole issue of interpretation, and then give some justification of my feelings.

I feel that many of Husserl's writings – particularly the later writings – have an idealistic overtone. He uses idealistic language, deriving it from his once copious readings of Fichte. He was one of the German philosophers to whom Husserl owed much. And it was also through Fichte that Husserl came to Kant. However, I have also the feeling that something very essential about Husserl is lost, we miss something very distinctive of his way of doing philosophy, if we interpret his manner of thinking idealistically. Thus, I find myself in complete agreement with all sorts of phenomenological positions which would take us beyond the trenches of idealism and realism. The pressing question is: How? The belief to which I am trying to give an expression is: in a certain very important sense, and this is the distinctive feature of phenomenology, it places us in a position that actually is beyond idealism and realism. Because realism-idealism are, after all, wedded to classical philosophy, to metaphysical theories, whereas phenomenology as phenomenology is not metaphysics. This is the first point why phenomenology goes beyond idealism and realism. Furthermore, there is a distinctly phenomenological conception which helps us in this, and that is the concept of intentionality. Once consciousness is formulated as intentional, and once we seriously and consistently adhere to this notion, then the classical theses of realism-idealism either have to be excluded or seriously modified. So, to employ the same terms again: both realism and idealism might be misleading.

Let me elaborate on both of these contentions, taking the second one first.

It seems to me that the whole point of our intentionality thesis is this: Consciousness is not something in which a thing can have a habitation. It is being 'directed to' worlds and objects. So that the classical idealist thesis that there is something in consciousness, or even the classical realism-idealism problematic whether the world is *in* consciousness or *outside* of it cannot be formulated or stated. Because we cannot anymore regard consciousness as a container in which something is or something is not. Since consciousness is 'being directed toward' the classical realism-idealism thesis becomes irrelevant.

Why not then realism? It seems, at any rate, that in a certain sense the intentionality of consciousness points toward realism; for consciousness is directed toward something therefore there must be something toward which it is directed. However, it also seems to me that the thesis of intentionality does not – by itself – entail realism. Because as it has been pointed out by those linguistic philosophers and logicians who have been concerned with intentionality, the fact that there is an intentional act directed toward an object does not logically imply the existence of that object. The fact that I believe that I am thinking of a dragon does not imply the existence of the dragon. The fact that I believe in proposition p does not imply that p is either true or false. Thus, there is the fact of 'existence in dependence' with regard to the intentionality thesis. A certain 'truth-value' indifference.

It is with considerable hesitation that I would be inclined toward a realistic interpretation of intentionality. Consequently, from the fact that consciousness is 'directed toward' an object nothing follows regarding the possible existence or non-existence of that object. It seems to me that the intentionality thesis by itself – if rigidly applied – places me in the position which is beyond the classical alternatives of realism-idealism. However, I would further modify this by saying that it has a closer proximity to realism than to idealism.

Let me take up now my other contention, namely that the phenomenological method is incompatible with an idealistic interpretation in Husserl's writings. As I like to understand Husserl he is doing something like this: there is a world which is pre-given, the world is given; what are we supposed to do with it, we as philosophers? There are things, objects, and events, and all kinds of entities which we encounter with different attitudes. I as a mathematician, for instance, encounter numbers, in perception I encounter the world which is already there with different kinds of properties. Some of these properties are real, some imaginary, fictitious, possible, probable, or actual; I as a philosopher try to clarify the meaning of these entities as I encounter them. I find myself in considerable agreement with the way in which may of the linguistic philosophers pursue this issue and the phenomenologists agreeing with them.

Permit me to pose this parallel and also demarcate its limitations. There is a common ground for such an agreement. First there is a common denial to commit oneself to a factual inquiry. I mean it is not an empirical inquiry of the world, rather, what the world means to me. However, there are many philosophers who maintain that in order to clarify meaning we have to clarify the meaning of the words employed in the search, and in order the clarify the meaning of the words we have to see how they are used in language. However, there is a very different kind of clarification for Husserl which I think is more ultimate, much more basic. That means to clarify the meaning of predicates like, real, imaginary, probable, possible, and so forth, The most important and basic inquiry is not the use of those words but to go back to consciousness and to see *how* these words first acquired their meaning through the appropriate acts of consciousness. Thus, phenomenology is engaged really in the search for the clarification of meaning by going back to those acts of consciousness in which these predicates become meaningful and not merely empty concepts. So that there is nothing committed to the ideality of the world, but what is being asserted about the meaning of the predicate 'real', insofar as it can be made clear intuitively to the collecting consciousness. If this is the phenomenological endeavor then I think it bypasses in a very important way the metaphysical thesis of realism-idealism.

As I understand Dr Köchler's paper it also intends to show the way to pass beyond idealism-realism by going back to an existentialist thesis of Husserl. But, I suppose, if we correctly understand Husserl, we find it there. Now, it is undeniable that Husserl used metaphysical language, he did use idealistic language, but then we have to understand that his use of philosophical language was greatly influenced by the German philosophical jargon. Consequently, it is also possible to understand him in the manner that he was not committed to a metaphysical thesis but was trying to clarify the basic predicates, concepts, which we apply to things and to the world. And in this clarification of the meaning of these concepts he was returning to the intuitive evidence of reflective consciousness.

Póltawski: Originally I wanted to ask Dr Köchler how does he understand the term 'realism'? But after Dr Mohanty's rejoinder I have changed my find, and would rather comment on some of issues he has raised. Professor Mohanty said that phenomenology, as Husserl understands it, is beyond 'idealism-realism' controversy, and that it is not metaphysics. In a sense I agree, in another sense I do not. I agree that it is not metaphysics – in the sense that metaphysics aims at the understanding of the essence of the world and of man-in-the-world – rather that it is a method. If you say that by practising it we don't try to understand what the world and man really is, I disagree. Let us connect this with the problem of intentionality and mention the work done by the analytic philosophers. I think that you are quite right in saying that analytic philosophy is close to phenomenology as Husserl understood it, but why is it so close? You say that there is the old Cartesian, you didn't use this word, 'conundrum' of idealism-realism, and I agree with you that we are at present, in a way, beyond the realism-idealism issue, because it is nonsense to ask whether something is *in* consciousness or *outside* of it. But if you say that intentionality, in Husserl's understanding and, perhaps, also in the understanding of the analytic philosophers, is instrumental in overcoming this

'conundrum' I again have to disagree, because intentionality, as understood by Husserl, remains within the idealism-realism issue; it is an
abstraction of the actual human 'consciousness'. Then if you say intentionality does not give us any assurance that the object *is* because it lends
itself strictly to analysis, you are speaking only about one object before
you; but if one raises the question whether 'I am in the world or *not*', then
one cannot say: 'I don't know, I am not too sure'. I think that if one
wants to understand the situation of man inside the world intentionality
should be understood not in the pure intellectualistic way of Husserl but
as a side of man's not only being but *acting* in the world as well as being
acted upon; in a way it would be a realistic position. If this position would
be interpreted as going beyond the realism-idealism issue, it would be in
a special way, because, in fact, I think it is *realistic* in its basic attitude.

Matczak: I am very glad that Dr Köchler compared Husserl with
Hegel; I think it is an illuminating approach and sheds light on Husserl's
approach to philosophy. Although I may not agree with his comparison,
still I would like to address a question to Dr Köchler. First of all for
Hegel the idea of complete reality has its ultimate basis within the Absolute and the Absolute is 'active', its very essence, and this is Hegel's
starting point, is becoming: being becomes non-being. In fact, the weak
point of Hegel is his analysis of becoming at the speculative level of the
concepts of being-non-being. But for Husserl, as Dr Köchler presented it,
the basis of reality would be ultimately consciousness; intentional consciousness which by its very essence is a cognitive factor: if it is a cognitive
factor would then this be an 'active' factor? It seems to me that cognition
by its very essence is awareness of something, with the amphasis falling
upon its object. Its awareness is 'dispassionate'. Consciousness is active
if we consider it from the point of view of 'affection'. But 'affecting'
relates more to pleasure and displeasure than to cognition. In other words,
I don't see *activity* in cognitive consciousness. Yes, there is activity in
perception but still this activity is due to the fact that perception is connected with affection and in concrete it is richer than an intentional
correlation 'act-object' in its modalities. It is due to these other aspects
of perception that, as it was pointed out previously, it is a factor of
spontaneity, of creativity, and of action. In Husserl I don't find these
factors. This singling out of the one, cognitive-logical side of perception
is in Husserl presupposed from the start, and consequently we do not

have what phenomenology claims 'the absolute starting point'; phenomenology is not presuppositionless. Dr Mohanty mentioned, among other things, that consciousness is not a 'container'. I would agree with him, but why? I will give an example of the problem if we move to another question which he brought out too, namely, that ideas would be neutral with respect to existence. This is, I think, the basic question for phenomenology. To make, like Husserl does in the *epoché*, an abstraction of the existential status of ideas seems to me provisory. Do ideas themselves have no existence? We cannot dodge the issue by distinguishing, like Ingarden does, between the essence, as the content of an idea and the idea itself. If there is an essence then this essence has necessarily an existential status. In whatever sense we interpret it, whether as an appearence or as an ideal or real existence the existential status of essence has – sooner or later – to be clarified. The whole problem for Husserl is how to come from this existentially neutral approach to a more adequate, a more complete one, integrating the results of his empirical reflection from which the *epoché* takes its starting point. In this perspective the whole problem of realism-idealism in Husserl's position is extremely difficult. I think we do not have a key in Husserl's philosophy to such a distinction.

Meyn: After Professor Mohanty has spoken so eloquently to Dr. Köchler's paper there remains merely to make a methodological plea for interpreting Husserl. But first, though, I should like to mention that an unjustified presupposition of Dr Köchler's interpretation seemed to be that Husserl is to be seen in the old Cartesian inside/outside framework. It might be useful to look at the Appendix to the *Logical Investigations*, 2nd edition, where Husserl explicitly rejects that framework.

The methodological point I want to make is simply this: Husserl has a very difficult style and he tries to express his thoughts in a terminology that doesn't very easily lend itself to what he wants to say. Instead of reading him literally we have to try to understand what it is he tries to express. Sepcifically, I would like to know how Dr Köchler understands the claim he made in the early part of his paper (I don't know if I have it *verbatim*) to the effect that every *a posteriori* claim is determined by and can be deduced from a subjective *a priori* one. Allegedly this is something Husserl said. Now, surely, in the ordinary sense it would not be *a posteriori* any more. Anything that can be deduced from something that is *a priori* is itself *a priori*, and, surely, Husserl as a logician was aware of this. We cannot give any example in Husserl of an *a posteriori* claim being deduced from an *a priori*. Then the thesis as stated is meaningless and we should rather try to find a different interpretation. We will come back to Husserl's belief that what we start out with – what is given – is the ordinary world. Thus we will come back to the empirical starting point of our investigations, It is at this level of investigations that Husserl's examinations, and phenomenological ones in general, diverse from empirical investigations. But the starting point is certainly not inconsistent with that of the empiricist or the scientist.

Póltawski: Dr Köchler shows us how in Husserl's thought there is a reduction to absolute consciousness – this consciousness is an absolute being and this excludes the possibility of dialectic. Yet, he makes the comparison with Hegel and wants to plead for a certain kind of 'immanent dialectic'. In conjunction with this he also mentions that the *a priori* includes all the requirements of experience, that is to say, that the *a priori* cannot be affected by the content of experience. I would like to note in this connection that his presentation of Husserl's notion of consciousness is somewhat one-sided; it neglects the genetic side of Husserl's analysis. It is basically a static analysis which has taken its guideline from *Cartesian Mediations* and *Ideas* I. This starting point, however, falls prey to a danger that Husserl expressly warns against. Thus, my question is somewhat difficult to answer, in as much as I say, that we seem to take in this connection the static analysis as governing Husserl's thought. And yet, perhaps, it is the genetic analysis which is the field within which dialectical analysis would be appropriate. I would agree with Dr Mohanty's suggestion that the clue to this is the problem of meaning. We see, especially in Husserl's analysis of the passive synthesis, that the genetic formation of sense occupied him a great deal. Thus it seems to me that when one seeks dialectic in Husserl's thought the clue to it would be a genetic analysis of meaning.

Barral: I will now call on Dr Köchler to reply to the many questions he has received.

Köchler: (translated by Dr Eng): Much depends on what one understands under 'phenomenology', how it is defined. One can take a methodological approach. But with respect to the issue of realism-idealism Dr Köchler feels that the central point of Husserl's philosophy lies in the turn toward absolute subjectivity which is productive of all that is constituted. With regard to Dr Matczak's distinction between the more 'contemplative' emphasis of Husserl contrasted with the logical emphasis of Hegel, Dr Köchler points out that in Hegel's stepwise development in the dialectical unfolding of the worldmind there is, at particular junctures, a return of the spirit to itself; the spirit is turning inward on itself, and in this return to itself there is also the 'meditative moment' latent or patent, so there seems to be a kind of homology or analogy between the way in which a 'contemplative' moment appears in Husserl and the way it appears in the return of the spirit to itself in Hegel. Dr Köchler also observes that Hegel is dealing, albeit in a different way, with the very same process of reflection which Husserl demonstrates. There is a number of passages in Hegel that have a remarkable significance when compared with certain passages in Husserl. Surprisingly these similarities have not sufficiently been noticed.[3] Of course Husserl rarely, maybe in only one or two places mentions Hegel explicitly. Perhaps that is significant.

With regard to Dr Meyn's objections to the statement that every empirical determination can be referred back to a subjective *a priori*, a number of quotations can be given. What, however, matters is the underlying transcendental attitude of Husserl which can be expressed in the following way: A world which is a world of entities of any particular sort does not come 'from without' into my *ego*, into my life of consciousness. Everything 'without' is what it is in relation to this inwardness, and has its 'true being' from the self-givings and validations within this inwardness. So everything that is developed is a constitution out of transcendental objectivity.

Tymieniecka: The question of Dr Meyn deserves more attention still from a different point of view. You asked whether in Husserl all empirical instances are supposed to be subjectively regulated, and it seems to me that this question leads, in fact, to the question of the nature of intentionality which has already been touched upon here, even though not completely answered. This is precisely – as I emphasized in my lecture – a point of great difficulty in Husserl, since Husserl, like Kant, giving radical priority to rationality, to the strictly rational type of ordering, interpreted intentionality from the very start as an agency rationally organizing according to the rules of logic. Consequently, Husserl claims indeed throughout his work that all instances of experience at all levels, starting with the empirical as far down as the kinesthetic operations, are regulated in the final analysis by *the rules* of subjectivity, notwithstanding the difficulties *how* this can be done. He himself, in his works following *Ideas* I, has been struggling to find the connection between intentionality *par excellence*, which is strictly rational, and the empirical levels of acts and operations. Later on, he found a subterfuge, in the notion of the 'soul'. Professor Kuypers very correctly pointed out at the conference at Waterloo that Husserl in his later writings, (*Ideas* II, *Husserliana* IV) considers this empirical complex of the soul as having 'a mediatory role' between body and spirit. He was compelled to admit a mediating link between the kinesthetic operations and between the strictly rational intentionality. In spite of his best efforts Husserl never really accomplished the task of unification, nor did Kant for the matter. We have to improve upon the masters in this respect. It remains our task to resolve whether, in view of the assumptions with which it is conceived, intentionality can be reinterpreted at all.

NOTES

[1] Alfred Schutz, 'The Problem of Transcendental Intersubjectivity in Husserl', in Alfred Schutz, *Collected Papers*, Vol. III, *Studies in Phenomenological Philosophy* (ed. by I. Schütz), with an introduction by Aron Gurwitsch, Martinus Nijhoff, The Hague, *Phaenomenologica* 22, 1966, pp. 83–84.
[2] Cf. V. Claesges, 'Intentionality and Transcendence: On the Constitution of Material Nature', *Analecta Husserliana* II, pp. 283–292.
[3] K. Kuypers in 'Husserl's Two Attitudes', *Analecta* II, emphasizes 'Hegel's metaphysical tradition' in Husserl's later investigation. [The Editor].

PART III

PHENOMENOLOGY AND NATURE

SERGE MORIN

SENSE-EXPERIENCE: A STEREOSCOPIC VIEW

The name of Maine de Biran figures rarely in historical accounts of phenomenology. Whenever he does appear, commentators will usually present him as being mainly preoccupied with the task of reconciling the two main philosophical schools of his period – rationalism and empiricism (sensationalism). Such a view of Maine de Biran misses his main contribution to philosophy, that of presenting us with a new direction – a new direction which makes Maine de Biran into a sort of nineteenth-century Merleau-Ponty. It is true that by no stretch of definition could he be turned into a full-fledged phenomenologist. Nevertheless, it is possible to show that there is no other philosopher in the nineteenth century who has been as important as M. de Biran for the spread of phenomenology. In other words, with M. de Biran we are dealing with both a sensationalist 'philosophe' and a budding phenomenologist. His phenomenological approach – his new direction – becomes evident mainly in his analysis of sense-experience.

Experience reveals an opposition between what is owing to me and what is not, or, as Jeanson would say, there is a duality of events, viz. those which I produce and am responsible for, and those which just happen to me. Biran, in turn, expresses this duality in terms of being aware of ourselves as *passive* and *active*. In every experience, Biran tells us, we can distinguish this duality: "Such is the nature of our organization, ... the intimate connection between our faculties of sensation and movement, that there is hardly any impression which does not result from their mutual concourse, and which is not consequently active in one respect and passive in another" (*I.H.*, p. 16). We can distinguish (though not separate) these two aspects of our experience, i.e., the voluntary and the involuntary, activity and the passivity, what we do or initiate, and what happens to us. I say that we can distinguish though not separate the two because one cannot be without the other (like colour and shape).[2] Our being active is always experienced, as it were, on a background of passivity or necessity, and *vice versa*. The two go in pairs though, for the purpose

Tymieniecka (ed.), Analecta Husserliana, Vol. III, 229–251. *All Rights Reserved.*
Copyright © 1974 by D. Reidel Publishing Company, Dordrecht-Holland.

of analysis and understanding, we must break the unity. Reflecting on
our experience we can focus our attention on either the active or passive
aspect of it. By stressing the one we purposely neglect the other or, to use
a more popular expression, we put it in brackets. Now, there are moments
of our existence where this bracketing or suspending of one of the element
is, as it were, already prepared. By this I mean that there are moments
where we feel more passive than active and vice-versa: *'on sent plus qu'on
n'agit'*. For example, when I experience a tickle, an itch, a cramp, or again
it may be a general feeling of *'malaise'* or well-being etc. These are things
which *happen* to me: they come and go without my consent, as it were.
I experience myself as *passive* in the face of these happenings. Some, like
itches and tickles, are easy to describe and to locate while others present
special difficulties, e.g. our moods, character, feeling of well-being, etc.
For the moment suffice it to say that they seem to invade or 'overcome'
me, and are present like a foul taste or odour. On other occasions, I am
active. Things do not just happen: they are initiated by me. I am responsible for them. Speaking, listening, watching are examples which
immediately come to mind. Thus, our waking life is, as it were, 'two-
sided': activity and passivity, voluntary and involuntary.

*S'il y a donc, comme on n'en saurait douter, une activité sensitive, je la distinguerai de l'acti-
vité motrice à laquelle je donnerai exclusivement ce nom, parce qu'elle se manifeste à mon
sens intime avec la plus grande clarté.*[3]

It may be claimed that Biran's choice of words here is unfortunate.
Evidently, he does not want to be caught in the old soul-body trap: a
different terminology seems inevitable. Nonetheless, his point remains
indisputable, i.e. reflexive analysis reveals that our waking life is 'two-
sided'. Similarly for Merleau-Ponty when he says that "Each time I experience a sensation, I feel that it concerns not my own being, the one
for which I am responsible and for which I make decisions, but another
self which has already sided with the world, which is already open to
certain of its aspects and synchronized with them".[4] In other words, I feel
the influence or presence of a life or activity which goes on in me but
without me, as Biran so often says. It is, as Merleau-Ponty says, the life
of my eyes, hands and ears.[5] In sum, it is the life of the organism. Reflexion reveals this 'other' aspect of life, i.e., life as something which is
already determined as opposed to life as a task or project for which I am
held responsible. This is not to say that there are two lives, but rather it

is to say that life reveals itself from two aspects. My life is both something which I received and something for which I am responsible. Reflexion makes explicit this ambiguity in that it reveals this interweaving of two experiences of life, i.e., as a task or project and as a problem which is, as it were, already solved. In a way we can say that it is the awareness of the body-organism and the body-subject. Likewise, there are evidently not two bodies but rather two experiences of the body. We feel our heartbeat, we see our wounds sealing themselves, etc., all this goes on in me but without me, and yet it is I who must eat, cure, protect, etc. Reflexion reveals this interweaving of two experiences of life: *"la vie de sentiment et la vie de mouvement"*, *"un aspect subi et un aspect voulu de mon existence,"* the voluntary and the involuntary. This, I believe, is what Biran is trying to get across. But this relation between the two experiences of meanings of my life (existence) is often but a vague feeling, a feeling which becomes stronger in times of ill health.[6] Merleau-Ponty noticed also that *"la conscience découvre d'autre part, en particulier dans la maladie, une résistance du corps propre"*.[7] It is Biran's intention to make more explicit this relation (passivity- activity, involuntary-voluntary), and this he does in his brilliant and bold analysis of sense-experience. In fact, one can say that with Biran we have the first attempt at a phenomenological description of what Merleau-Ponty calls *'le sentir'*.

Biran's analysis of sense-experience is another instance of what he calls *"réfléchir l'habitude"*, and here *"l'habitude"* means in effect the classical empirical view of sense-experience. In contrast with Biran's analysis, the classical empirical theory of sense-experience is awfully restricted to one sort of sense-experience, i.e., visual: the same applies today with the sense-data theories.

On peut voir dans les Dialogues d'Hylas et de Philonous *comment Berkeley fonde la meilleure partie de ses raisonnements contre l'existence des corps sur des exemples tirés du sens de la vue. On peut voir ensuite dans le* Traité de la nature humaine *comment le sceptique Hume, poussant les conséquences jusqu'au bout, s'est servi des mêmes armes pour détruire toute existence substantielle et réelle, jusqu'à celle du sujet pensant individuel. Ce dernier philosophe a raisonné conséquemment, et comme pourrait le faire un être intelligent réduit au sens de la vue, si tant est qu'un tel être pût penser et raisonner.*[8]

Add to this the fact that the classical empiricist analysis of visual experience was directly deduced from the reigning scientific explanations (mechanical, physiological, etc). Why were the other senses, if not totally neglected, at least relegated to an inferior role? Are most philosophers afraid of becoming blind more than anything else? It must be said that most empirical philosophers did deal with the other senses (second was the sense of touch), but when they do so venture out it is mainly to pay lip service to the other senses. And even then, their skimpy analysis is visually tinted. The consequences of such a bias are still very much with us today. It has become, as Biran would say, a force of habit, even a near cultural habit at times. First of all, as Biran points out:

... les termes appropriés uniquement aux perceptions de la vue, par exemple, ne peuvent s'appliquer que d'une manière tout à fait illusoire aux perceptions du toucher et de l'ouïe....[9]

Or, as Merleau-Ponty puts it:

L'étonnement du malade, ses hésitations dans le nouveau monde visuel où il entre, montre que le toucher n'est pas spatial comme *la vision.*[11]

In sum, to see is not to touch or to hear, etc. Furthermore, such notions as causation,[12] matter, substance, force, resistance, memory, imagination, etc., are all drastically affected by this visual explanation of experience. Hume, for example, gathered most of his arguments for causation from looking at them: he looked at the billiard balls and rightly affirmed that all one could see was that one ball moved whenever the other struck it. The correctness of his analysis is evident for the simple reason that there is nothing else to be seen but a succession of events. Similarly with Russell for whom matter consists of being a kind of construction out of actual or possible sense-data. Indeed, if one *looks* at a table from various points of view, what else does one see except a series of coloured shapes? What! No table? In sum, we can say with H. H. Price that most Empiricists have been, and continue to be, 'Visual-Philosophers'.[13] With Biran we are for the first time presented with a 'stereoscopic view of sense-experience' (Price's expression). Biran remedies the visual bias which he found in most theories of sense-experience by bringing to light how each sense addresses, as it were, the world in its own particular way, and how all the senses communicate among themselves and complement each other.

I wondered a moment ago why it is that most Empiricists are visual-philosophers. Aside from any psychological reason, their visual bias is quite understandable. First, their theory of sense-experience is a theory of sensation directly deduced from the physical and physiological theories of the time. The theory gave the appearance of being so simple and clear that it seemed self-evident. Second, the term 'sensation' stood for a general abstract idea, and consequently it allowed Empiricists and Sensationalists to take only one kind of outward sensation – usually visual – as the prototype and thus passing over the many crucial differences among the other types of sensations. Third, because their task is, in part at least, to explain the mechanism of experience, they did not need to multiply the various sorts of sensations. As the saying goes, "if you've seen one, you've seen them all". One sort of sensation seemed to be quite sufficient. They believed that what can be said of this sensation (visual) also applies to any other sensation. As Biran puts it:

Les sens forment ordinairement pour eux un chapitre assez court; ils prennent la sensation en général, ou toute manière (égale et uniforme par hypothèse) de sentir les impressions ou de percevoir passivement les résultats de l'action des objects, pour ce qu'elle est our paraît être en effet, lorsqu'on la considère du sein même des habitudes de tous les sens, ou dans certains modes prédominants qui se rattachent à l'exercice de l'un d'eux en particulier, et tel mode spécifique d'un sens individuel (comme celui de la vue par exemple), comme servant de type aux autres ou formant le seul caractère réel de l'idée abstraite complexe exprimée par le terme sensation; *on suppose l'analogie ou même l'identité de nature de tous les éléments qui se trouvent compris dans l'extension de ce terme, en se dispensant de la vérifier.*[14]

Fourth, with most visual-philosophers, because their theories were directly deduced from the physical and physiological sciences, sight was not treated as sight, i.e., it was subordinated, following the scientific explanations, to incoming light rays which impinged or touched on the sense organ in question. Thus Biran observes that all the senses were, in the final analysis, really reduced to 'touch'.[15]

Now Biran, as we saw, does not want to *explain* experience, i.e., to reconstruct it piece by piece, as one would reconstruct a house, and using some hypothetical material – sensation – as the building block.

L'entreprise de reconstruire, pour ainsi dire, l'entendement humain avec certains éléments artificiels hypothétiques (donnés souvent par les seules classifications de notre langage) ne serait peut-être guère moins téméraire que ne le serait celle de recomposer chimiquement un produit naturel du règne végétal ou minéral, altéré etrésous dans ses éléments par le creuset.
... Il faut le dire: indépendamment des difficultés extrêmes attachées à un projet, tel que l'avait conçu le célèbre auteur du Traité des sensations, *le point de vue où il se plaça, dès son début, devait en rendre l'exécution impossible ou tout à fait illusoire.*[16]

Rather, he wants to clarify experience: he wants to *catch* us ('*nous surprendre*') perceiving. In sum, he wants us to regain the ability to describe our actual perceptual experience. In so doing, he will go a long way in refuting the classical empiricist view of perceptual experience which consisted essentially in the mental inspection, by a disembodied mind, of impressions rolling in through passive senses (usually vision). On the contrary, Biran will show that perception is essentially active, or, more explicitly, that voluntary movement is a condition of perception.

Il faut prendre conscience de ce qu'il y a de volontaire dans la connaissance digne de ce nom pour dérouler les prétentions d'un sensualisme superficiel qui ferait du moi un simple divers d'impressions sensibles, un 'polypier d'images'; si tous les actes du Cogito sont des rayons d'un même moi, c'est parce qu'ils participent d'une unité d'effort, d'une discrète tension qui nous tient éveillés, alertes et accordés sur le monde. Le regard révèle le voir comme acte. Ici Maine de Biran est invincible.[17]

A purely passive being would not be able to know anything: he would not even suspect his own existence.

Si l'individu ne voulait pas ou n'était pas déterminé à commencer de se mouvoir, il ne connaîtrait rien. Si rien ne lui résistait, il ne connaîtrait rien non plus, il ne soupçonnerait aucune existence, il n'aurait pas même l'idée de la sienne propre.[18]

Perception is active: it is bodily activity. To perceive is to be 'attentive' to a certain degree, and to be attentive demands a certain effort.[19]

To see I have to open my eyes, guide them, give them a rest, etc. To hear I have to listen, turn my head and even cup my ears with my hand. To smell I have to close my mouth, accelerate or slow down my respiration, move my head about slowly, etc. To touch I have to reach out, grasp, explore, etc.[20]

Evidently these are not the experience of a passive, contemplative, disembodied being, but rather of a performer, an active embodied performer. Our perceptual experience is that of an embodied subject. For a passive, comtemplative, disembodied being, these experiences have to be denied.

In perception one does not passively wait for the world to come and 'impinge' itself on us. Rather one goes out and meets the world, one addresses oneself to it. One has to go out and gather, as it were, the 'material'. It is because philosophers have been too caught up in this 'material' that they have neglected or taken for granted the 'gathering' of this material.[21]

I am aware that in perception I am responsible for something. I am more or less aware of being responsible for the way I go out to meet the world, for the way I address myself to it. In sum, something depends on me, or is 'up to me' in the way the world appears. It is all up to me to ask questions, to close or open my eyes to some event, to listen or not to listen to a murmuring suspicious couple, to step forward or retreat, etc. I am aware that the 'gathering' or the 'exploring' is up to me, that it is, as we say, 'in our hands'.

Le mouvement commencé, s'il s'arrêtait à la première résistance (par exemple, si, lorsqu'un corps est posé sur ma main, ses doigts, en se fermant, s'arrêtaient au plus léger contact), l'individu saurait simplement qu'il existe un obstacle; mais non point si cet obstacle est absolument impénétrable, solide, dur ou mou, etc.[22]

It is up to me not only to go out to the world, but also to analyse it, to observe it, to scrutinize it. This will evidently depend a great deal on the motility, or lack of it, of my body and my sense-organs.

Par le mouvement seul nous ne connaîtrions guère que des masses diversement résistantes; la main décompose en quelque sorte ces masses, met à nu leurs éléments, distingue leurs propriétés, démêle leurs nuances; c'est le premier des instruments d'analyse, et tous ses avantages dépendent évidemment de sa construction, de la motilité supérieure de ses parties et de la nature même de sensibilité.[23]

Here Biran gives us the two sides of the picture, i.e., that in perception *all* is not up to me. To use Pierre Janet's example, it is I who gathers the apples into the basket but it is not up to me that there be apples nor that they be what they are.[24]
Or, as Ricoeur puts it:

J'ai la conscience obscure que quelque chose dépend de moi dans l'apparence du monde; mais ce n'est pas que l'objet soit tel que tel: tout mon pouvoir est de l'interroger, en me tournant ou me détournant de lui, en extrayant du fond tel ou tel caractère ou en l'y laissant glisser.[25]

There is in perception something active and something passive: a voluntary and an involuntary aspect. This involuntary aspect belongs, as it were, on the side of the organism: the sensibility of the organs, as Biran says, or the life of my eyes, ears, and hands, as Merleau-Ponty puts it. "Each time I experience a sensation," we heard Merleau-Ponty say, "I feel that it concerns not my own being, the one for which I am responsible but another self which has already sided with the world, which is already open to certain of its aspects and synchronized with them".

La sensibilité recueille à mesure les découvertes du mouvement, s'empare des nuances les plus délicates et se les approprie; elle saisit ce filet imperceptible, ces petites éminences, ces saillies, qui disparaissaient dans la résistance totale ou dans la rapidité de la course, et dessine exactement ce que l'organe moteur ne pourrait pour ainsi dire qu'ébaucher, si on le supposait calleux à l'extérieur.[26]

Thus, for Biran, our perceptions are owing as much to the motility of the organs as to their degree of sensibility.

Ces propriétés de la matière ne peuvent se manifester à lui qu'autant qu'il veut continuer le mouvement et c'est l'intensité de son effort qui en est la mesure ; presse-t-il l'obstacle de toutes ses forces, sans pouvoir fermer la main, il a un terme fixe qui lui fait connaître l'impénétrabilité, la dureté ; si l'obstacle cède plus ou moins facilement, il a la mesure de ses divers degrés de mollesse, de mobilité, etc. L'individu ne perçoit donc le premier rapport d'existence qu'autant qu'il commence à mouvoir, et les autres rapports successifs, qu'autant qu'il veut continuer le mouvement.[28]

But it is the active aspect of perception which traditionally has been neglected. It is this which gives the sense of touch such discriminatory power.

C'est ainsi, en effet, que l'aveugle géomètre doit la netteté et le nombre des perceptions qu'il se forme des modes de l'étendue figurée, autant à la délicatesse du sentiment des houppes nerveuses, qu'à l'agilité et à la flexibilité de ses doigts.[27]

Biran, with his analysis of our tactual perception is making explicit a very important feature of all perceptions. His analysis clearly brings to the surface the fact that when I reach out and touch something, I am aware that my actual experience of this object is not complete and that the object hides from me certain features, that it is up to me to proceed with the inquiry and to, as it were, 'close-in on the object'. I know that I will discover other features of this object if, for example, I tighten my hold or if I spread my fingers, etc. Or again, I know that if I walk a bit further I will see more and more of the object. In sum, I am aware of communicating with a real world, and not a fiction of the imagination. In the imaginary world, as Sartre has shown in his work, *L'imaginaire*, I do not observe or discover anything new about the imaged object. Our tactual experience is well suited to make this feature explicit.

This sense is also best suited to bring out the interweaving of two aspects of our existence, i.e., the voluntary or active aspect and the involuntary or passive (*'subi'*) aspect.

Dans les circonstances simultanées de l'exercice de ce sens, on peut faire abstraction tour à tour, tantôt de la motilité ou de l'effort dont l'organe du toucher actif est le principal instrument, tantôt de la sensibilité affective qui a son siège dans le même organe considéré comme celui d'un tact passif.[29]

What does he mean by *'sensibilité affective'*? What is this *'tact passif'*? There is nothing very mysterious in this strange terminology. What Biran is trying to say is that, in our exploration of the world we are aware of being, from a certain point of view, *passive*, and, from another point of view, *active*. I move my hands and fingers but the data received during this exploration (the given of experience) is not entirely owing to my initiative, but is owing to the life or the sensibility of the organs. Now the sense of touch enables us to focus on either of these two aspects. Reflecting on our tactual experience, we can become aware that, from one point of view, there is an aspect of my experience over which I have no control, and another aspect for which I am responsible. I perceive the role which is mine and the role which belongs to the organs. Now the sense of touch enables us to focus on either of these two aspects. Reflecting on '*le toucher*', Biran makes explicit the two-sidedness of our waking-experience: "*les événements*", as we heard Jeanson say, "*qui relèvent de deux ordres distincts*". Biran does not talk so much in terms of 'events' but of different lives interweaving, as it were: '*vie sensitive*' and '*vie personnelle*', '*le sentir*' and '*le mouvoir*'. In other words, Biran believes it possible, as he says, "*de faire la part du sentiment et celle du moi*" (*I.H.*, p. 16). He begins with the former.

> *Que l'on applique sur ma main un corps dont la surface soit hérissée d'aspérités, ou polie, d'une chaleur douce ou d'un froid piquant, etc., tant que le contact dure, j'éprouve une impression agréable ou douloureuse, qu'il n'est point en mon pouvoir d'augmenter, de diminuer ni de suspendre en aucune manière: voilà la part du sentiment; et quand même la faculté motrice serait paralysée, il s'exercerait de la même manière. C'est à des sensations de ce genre que le tact serait borné, s'il n'était pas doué de mobilité, et dans ce cas il serait bien inférieur à plusieurs autres parties du corps recouvertes par la peau, mais dont la sensibilité est bien plus délicate, plus exquise.*[30]

Do we not, for example, often say that "*Je sens bien plus que j'agis?*" When, for example, driving along and suddenly your car is filled with the foul odour of skunk, or when one feels a cramp in one's stomach, or a twitching eyelid, then, as Biran says, "*je dis que je sens, que je suis ou que je me sens passif*" (*I.H.*, p. 14). These are what one ought to call 'sensation', i.e. when one is affectively modified. Biran, was worried about the deep-rooted ambiguity of the concept of sensation. Here he is telling us how the concept ought to be used: notice also how his use of the concept of 'sensation' is truer to experience.

> *Lorsque le sentiment prédomine jusqu'à un certain point, le mouvement qui concourt avec lui est comme nul, puisque l'individu n'en a point conscience, et l'impression demeure passive. Je conserverai à toutes celles de ce genre le nom de sensation.*[31]

In another work he is more explicit still:

Lorsque l'affectibilité, *mise en jeu par la cause extérieure, est prédominante, la* matière *de la sensation prévaut sur l'acte* perceptif, *toujours inhérent au déploiement de l'effort. Cette prédominance est bien exprimée par le terme* sensation *qui convient alors à cette classe.*[32]

Biran refers, as he often does, to ordinary language for support, i.e., the way we ordinarily describe what we smell.

Ce sens, mis en jeu d'abord par l'instinct, demeure presque entièrement sous sa dépendance ; son immobilité absolue annonce combien il est passif, et on pourrait dire qu'il tient, parmi nos sens externes, le même rang que le polype ou l'huître, dans l'échelle de l'animalité ; ses fonctions, il est vrai, se rallient au mouvement de la respiration, mais ce mouvement premier est nécessaire, forcé, continu par sa nature, et par là-même presque insensible ; aussi les odeurs sont les sensations *par excellence comme l'indique le langage même ; ce sont celles qui se distinguent le moins : lorsque plusieurs se trouvent unies ensemble, elles se fondent dans une sensation unique, dont l'analyse nous est impossible, malgré* l'attention *volontaire que nous donnons au mélange ; remarquons que cette attention ne consiste que dans un mouvement d'inspiration uniforme, lente et prolongée. Ce sont là les bornes de notre* pouvoir *sur ces modifications.*[33]

Biran proceeds very cautiously in his clarification of the concept of 'sensation': he constantly tries not to remove himself from direct experience.[34] It is this effort which enabled him to distinguish 'sensation' from what he calls 'intuition'.

L'organe peut être tellement constitué, et l'agent externe auquel il est soumis, d'une telle nature, que les impressions reçues soient très peu affectives d'une part, pendant que l'effort est comme inaperçu dans son déploiement peu intense, d'autre part. Il y a là un certain rapport entre les deux forces, qui ne permet guère d'assigner quelle est celle à qui appartient l'initiative ou la prédominance. La nullité d'affection directe exclut le caractère sensitif, la faiblesse de l'action motrice exclut presque le caractère aperceptif, l'individu ne sent *ni n'agit et pourtant le phénomène de la* représentation *s'accomplit, il y a un* objet *extérieur ou intérieur passivement* perçu. *C'est ici que l'idée de* sensation *paraît exister par elle-même et venir toute faite du dehors ; ce phénomène mérite bien d'être analysé dans ses circonstances et conditions particulières ; nous le distinguerons sous le nom de perception simple ou objective* (intuition).[35]

This distinction is an extremely interesting one, as I shall soon show. But first, it is necessary to continue with Biran in his clarification of the concept of 'sensation'. The next two distinctions have to do with a greater and greater activity or participation on the part of the subject. After intuition and sensation, there is what he calls, 'perception':

Le moi *s'unit d'une manière bien plus intime à une autre espèce de modes auxquels il participe ou concourt par un effort voulu. La possibilité de ce concours actif suppose que l'organe qui reçoit l'impression fait partie du sens de l'effort, ou rentre dans le domaine de la puissance motrice.... J'appelerai* perception *toute impression non affective à laquelle le* moi *participe par son action, consécutive à celle d'un objet extérieur. Le système perceptif embrassera donc tous les phénomènes qui naissent de l'action des objets sensibles, combinée avec celle d'une volonté subordonnée, encore aux impressions qui occasionnent ou motivent son premier dé-*

ploiement. Ce système comprendra aussi les phénomènes de l'ordre sensitif précédent, en tant que le moi *qui y était spectateur passif pourra y rendre une part active.*[36]

What Biran is after here is the following situation: something is placed in my hand. It is this impression of 'something is in my hand' which then draws my attention. I then begin to observe this something; I scrutinize it, I close my hand around it, I turn it around, etc. In sum, I become active, and not simply a passive spectator.

Observons bien ici que l'influence de l'attention ne consiste point, comme on l'a dit souvent, à rendre l'impression plus vive. On peut s'assurer dans tous les cas, qu'il n'y a point d'action directe de la volonté sur les organes sensitifs, dont les dispositions propres et bien indépendantes de tout exercice de la volonté, donnent à toutes les impressions reçues ces caractères si variables de vivacité ou de langueur, d'énergie ou de faiblesse. Tout le pouvoir de l'attention consiste donc à fixer les organes mobiles à volonté, comme l'ouïe, le toucher, sur l'objet présent, à les détourner de toutes les autres causes d'impressions, et à rendre ainsi l'intuition non pas plus vive que ne le comporte l'état naturel de la sensibilité de l'organe combiné avec la manière d'agir de l'objet, mais relativement plus claire, plus nette, par une véritable concentration de la faculté représentative.[37]

It is this active participation which ordinary language aims at when it distinguished between 'seeing-looking', 'hearing-listening', 'smelling-sniffing', etc.

Biran will yet distinguish 'perception' from 'aperception'.

J'appellerai aperception toute impression où le moi *peut se reconnaître comme cause productive, en se distinguant de l'effet sensible que son action détermine. C'est ici que s'applique parfaitement la définition de Leibnitz:* Aperceptio est perceptio cum reflexione conjuncta. *En effet, tant que la volonté est subordonnée à une impression étrangère, le* moi *peut ignorer la part qu'il y prend, cette part active se confondant avec celle de l'objet ou de la cause extérieure qui prédomine. Aussi l'attention commandée, la plus active en apparence, exclut-elle la réflexion ou l'aperception interne de l'action ou du sujet qui l'exerce. Mais lorsque la volonté a l'initiative sur les impressions et se commande elle-même, le* moi *ne peut méconnaître sa propre causalité. Il a l'aperception immédiate de la cause qui est lui, en même temps que l'intuition de l'effet, ou du résultat sensible, qui vient de lui ou de son effort.*[38]

It is here that Biran is most ingenious. With the previous three distinctions (sensation, intuition and perception) we witnessed a gradual increase in the initiative or activity of the conscious subject. With 'sensation' we have, what Ricoeur calls, a passive attention: here I am absorbed, fascinated or caught up by the object (here the subject's participation is hardly noticeable). With 'perception' we meet a subject who is definitely aware of his active role, of his contribution, but even here, this activity or participation of the subject can become neglected or taken for granted.

The reason for this is that we are usually much too pre-occupied by what is in the world to pay attention to ourselves in it – our way of addressing the world. We are so busy describing what we see, hear and touch that we neglect or fail to notice our participation. Gradually the activity of the subject is taken for granted. Its 'history', as it happens with any habitual activity, is forgotten. Our 'looking', 'listening', 'sniffing', etc., are activities which have forgotten their history, their origin – we do them *as if* we had done them all our lives and without ever having learned them. As Biran says, *"c'est comme une pente où l'on glisse sans s'en apercevoir, sans y songer"* (*I.H.*, p. 7). Where once our perceptual activity was, as it were, problematic, it is now a quasi-automatic activity. We are now in a position similar to that of a person who has learned how to skate, i.e., we can now move on to better things.

Our attention is focused on things and events in front of us. Our participation (looking, touching, listening...) forgets itself to the advantage of the things or events in front of us. But, as Biran is trying to say in the above passage (see p. 239) this is not the end of the story. Not only does our activity forget itself, but it is transferred onto the object. It is somehow the object which has the *active* role and the subject is but the passive receptor of sense-data: Biran's passive spectator. Perception then has become like sensation: it has become, or so it seems, the simple reception of impressions.

> ... *plus les mouvements deviennent faciles, plus la perception qui en dépend acquiert de précision et de clarté, plus aussi l'individu méconnaît la part qu'il y prend, plus son activité s'enveloppe: le sujet et le terme de l'action, l'effort et la résistance se trouvant enfin presque identifiés, tout semble revêtir le caractère passif de la sensation.*[39]

Now Biran has shown that it is the natural developement of an embodied consciousness to misunderstand (*'méconnaître'*) the active role which the subject plays in perception because perceiving is a bodily activity, and as such it tends, as Ricoeur points out, to forget its history or its origins. Perception gradually becomes more and more like 'sensation'. Biran is somewhat surprised by this phenomenon but failed to exploit it fully. In fact, this phenomenon seemed to count against his thesis that without a sustrained effort I would know nothing.

> *L'individu ne perçoit donc le premier rapport d'existence qu'autant qu'il commence à se mouvoir, et les autres rapports successifs, qu'autant qu'il veut continuer le mouvement. Mais, si nous supposons que la résistance diminue progressivement au point de devenir insensible, le*

dernier terme de l'effort décroissant sera la limite, et pour ainsi dire l'évanouissement de toute
perception, de toute connaissance.[40]

One could then ask himself if Biran is not asking too much from this couple 'effort-resistance', and by so doing if he has not missed the essential of *'la connaissance'*? It may be because he has stressed too much the analysis of our tactual experience and not enough our visual one, that he failed to exploit the fact that knowledge is more like *'voir'* or *'entendre'* than *'regarder'* or *'écouter'*.[41]

Yet, as I have mentioned, Biran is surprised by what he has called the phenomenon of 'intuition' (see above, p. 238). In this sense he was very close to recognizing himself the limits of his theory of effort. Sight exemplifies his phenomenon of intuition: here the passive impressions are hardly affective, and effort is nearly nil and goes unnoticed. It is here that simple or objective perception (intuition) reaches its maximum.

La nullité d'affection directe exclut le caractère sensitif, la faiblesse de l'action motrice exclut
presque le caractère aperceptif, l'individu ne sent ni n'agit et pourtant le phénomène de la
représentation s'accomplit....[42]

Biran wants to show, via these distinctions, that the subject's participation in perception gradually increases. In the first three (sensation, intuition and perception) the subject's role was either that of a passive spectator or, with time and habit setting in, the subject came to take his activity for granted or transferred it to the object. But which 'aperception' this neglect or transfer is impossible for here the subject creates, as it were, the intuitions or sensations.

Mais lorsque la volonté a l'initiative sur les impressions et se commande elle-même, le moi
ne peut méconnaître sa propre causalité. Il a l'aperception immédiate de la cause qui est lui,
en même temps que l'intuition de l'effet, ou du résultat sensible, qui vient de lui ou de son
effort.[43]

Where will we find such a situation? Biran comes out with a brilliant answer: *"dans le sense de l'ouïe unie à la voix"* (*O.I.*, 2, p. 228).

Biran found in all the various sense-experience an active and a passive aspect: what we put in and what we receive, as a Kantian might say. Furthermore, he showed that the discriminatory power of the sense is owing to their mobility. Although none of our external senses are completely without it, some are evidently at a disadvantage when compared with, say, the *'tact actif'*. Now, though each sense is, as it were, opened to its own world, they nonetheless communicate with each other. We do not see as if we had never touched, nor do we touch as if we had never seen; we do not talk as if we had never heard; everything tastes different if one's nose is not functioning properly, etc.

Le sens du goût est celui qui paraît d'abord avoir le plus de rapport avec le toucher; les saveurs ne sont en effet que le tact propre de la langue et du palais…. Différentes saveurs peuvent bien se comparer aux sensations tactiles de froid, de chaud, de doux, de rude, de piquant; aussi ces deux genres de modifications ont-ils plusieurs noms communs dans nos langues.[44]

Or to put it more bluntly and with a grain of salt, we do not see the beauty of a woman as if we had never touched a woman or smelled her perfume. Thus, Biran is making explicit the great amount of inter-weaving communication which goes on between the various senses. We *see* that something is soft, we hear that something is hard metal, etc.[45]

Along a similar line, it is Biran's keen observation which enabled him to notice that the more passive sense acquire their discriminatory power by their close association with more versatile organs, e.g., the eye, though it has little capacity for movement, acquires its discriminatory power by association with the other senses but mainly with the more versatile organs of touch (*I.H.*, pp. 23–24). Also the sense of smell, which at first appears to be the most passive, acquires activity by communicating with the respiratory system and with a more active sense – taste.

L'organe du goût (qui est en même temps celui de la parole), est doué d'une très grande mobilité; l'effort qui a lieu dans la mastication ou la pression des lèvres, des dents, du palais contre les corps solides, suffirait sans doute pour nous donner les idées plus ou moins confuses de la résistance, et de quelques-uns de ses modes; plusieurs espèces d'animaux ont, comme on sait, leur tact dans la bouche et le museau.[46]

Thus Biran goes a long way to show the essential unity of what the Empiricists have called, sense-experience. By returning to lived experi-ence, Biran has re-united what his predecessors have separated.

Here the reader will notice that I have said little about the sense of hearing, how it communicates with the other senses, from what organ does it acquire its mobility, etc. I have kept Biran's analysis of this sense for the last because of its importance and originality.

In order to hear well one has to listen. And, what is it to listen?

… sinon déployer une action sur les muscles destinés à communiquer divers degrés de tension à la membrane du tympan, etc.? Il est vrai qu'ici l'effort est devenu imperceptible, que le jeu et l'appareil du mouvement, étant tout à fait internes, ne se manifestent point comme termes de la volonté; que l'oreille étant dans l'homme extérieurement immobile, ouverte à toutes les im-

pressions, sans moyen direct de s'y soustraire ou de les modérer, paraît être un organe d'autant plus passif, que sa sensibilité est plus prédominante.[47]

But, Biran tells us, nature has re-established the missing equilibrium by associating, in a very intimate manner, this sense with the voice-organs. The following passage is a gem as far as reflective analysis is concerned.

Lorsque nous percevons des sons (et nous les percevons toujours d'autant plus distinctement, qu'ils sont plus de rapport avec ceux que nous pouvons rendre, imiter ou articuler nous-mêmes), l'instrument vocal contracte donc des déterminations parallèles à celles de l'ouïe, et se monte, pour ainsi dire, au même ton: en entendant chanter ou parler, nous chantons, nous parlons, tout bas; c'est un instinct d'imitation encore plus marqué ici que dans aucun autre mouvement, il nous entraîne le plus souvent, sans que nous nous en apercevions.[48]

Thus, when we listen we are, as it were, our own echo: *"l'oreille se trouve comme frappée instantanément, et du son directe externe, et du son réfléchi intérieur"* (*I.H.*, p. 27). The interesting and important fact here, for Biran, is that the exterior sound is imitated or reflected; a kind of echo. What is especially of interest is that a person, according to Biran, independently of any exterior cause, can create or impress by his own activity, an essentially passive sense. The other senses must wait, as it were, for some exterior event.

Dans le regard le plus actif l'oeil ne s'éclaire pas lui-même, et si dans le toucher la résistance n'est aperçue que dans un effort que la volonté détermine, il n'en est pas moins vrai ... que la matière et la forme de la perception ne viennent pas de la même source; que le sujet ne s'oppose point à lui-même .... Mais dans l'exercice simultanée de l'ouie et de la voix, le mouvement, et le son qui en est le produit, émanent bien de la même source et s'adressent au même sujet qui les réfléchit, étant approprié également au double effet dont il est cause. Le son étranger qui vient effecter l'ouïe extérieure est vraiment distingué du son articulé que la volonté détermine, comme un mouvement libre, volontaire, ou accompagné d'effort, est distinct d'un mouvement contraint, dénué de tout sentiment de pouvoir. Dans le premier cas, c'est le moi qui est cause; dans l'autre, c'est une cause non-moi sous laquelle l'individu est, ou se sent passif, et nous retrouvons encore là, dans ce contraste, un double fondement à la croyance et au jugement d'extériorité.[50]

In these 'reflected' sounds a person is aware that the initiative is his and that the sounds which he hears (passive) are his doing. He is aware of both making sounds and hearing himself. I 'see' myself in the effect: I am aware of 'something done by me', as opposed to something just 'there'.

L'activité qui produit immédiatement les uns se réfléchit dans les autres; l'individu qui émet le son et s'écoute, a la perception redoublée de son activité. Dans la libre répétition des actes que sa volonté détermine, il a la conscience du pouvoir qui les exécute, il aperçoit la cause dans son effet, et l'effet dans la cause; il a le sentiment distinct des deux termes de ce rapport fondamental; en un mot il réfléchit.[51]

In the other senses, the subject could always become absorbed or distracted by the object, and eventually became unaware of his participation in perceiving. But here, with the sense of hearing united with the vocal organs, this neglect or unawareness is impossible. Even if the subject's attention was drawn by the object, this object is perceived as his own product.

L'exercice du sens de l'ouïe interne considéré dans son union ou sa correspondance intime avec l'organe vocal ... est le seul sens par lequel l'être sensible et moteur puisse se modifier lui-même sans le concours d'aucune cause étrangère à sa propre force motrice. Dans les autres sens, le vouloir et l'effort ont toujours un terme ou un objet extérieur, dont l'impression distrait ou absorbe le sentiment interne de la cause qui concourt à la produire. Ici, l'effort et le mouvement n'émanent profondément du sujet que pour s'y réfléchir sous les formes d'une perception sensible, que l'être actif adopte et reconnaît comme son ouvrage.[52]

This sort of experience correctly deserves to be set apart from the others, and this is why Biran calls it by a special name – *aperception* – for only here *"le moi ne peut méconnaître sa propre causalité. Il a l'aperception immédiate de la cause qui est lui, en même temps que l'intuition de l'effet, ou du résultat sensible, qui vient de lui ou de son effort"* (*O.I.*, p. 9).

Thus, also, Biran completes his clarification of the concept of sensation which had previously been used interchangeably to mean either (what he calls) sensation, intuition, perception and aperception: while not forgetting those occasions where it was used to mean a sort of physiological occurrence.

Biran, as we saw, does not want so much to *explain* experience, i.e., to reconstruct it piece by piece and using 'sensation' as the building block, but rather, as we saw through this essay, he wants to clarify experience. He wants to show how, through our activity or initiative, the world unfolds itself to our questions. He has shown that the traditional empiricist and sensationalist view of perceptual experience is far removed from actual everyday experience. His reflective analysis has made more than explicit the fact that perceiving is a bodily activity, and that the perceiving subject is an embodied subject.

NOTES

[1] Throughout this paper the following abbreviations are used to refer to Maine de Biran's works (in their order of appearance):

I.H.: *Influence de l'habitude sur la faculté de penser* (ed. by Pierre Tisserand), PUF, Paris, 1952.

P.J.: *Oeuvres de Maine de Biran* (ed. by Pierre Tisserand), Vol. 1, Alcan, Paris, 1920 – various fragments described by Tisserand as *'le premier journal'*.

O.I.: *Oeuvres inédites de Maine de Biran* (ed. by Ernest Naville), 3 vols., Dezobry, Paris, Magdelaine et cie, 1859.

M.D.P.: *Mémoire sur la décomposition de la pensée* (ed. by Pierre Tisserand), 2 vols., PUF, Paris, 1952.

M.P.O.: *Mémoire sur les perceptions obscures* (ed. by Pierre Tisserand), in the series *Classique de la Philosophie*, Colin, Paris, 1920.

N.E.A.: *Oeuvres de Maine de Biran*, Vol. 14, PUF, Paris, 1949 – various tests grouped by Henri Gouthier under the title *'Nouveaux Essais d'Anthropologie'*.

[2] *Et pourtant, pour moi qui meus mon corps et essaie de me surprendre dans l'acte même c'est tout un de vouloir, de pouvoir, de mouvoir et d'agir; l'ordre adressé au corps, la disposition de l'organe à répondre à l'ordre, la réponse effective sentie dans l'organe, l'action produite par moi, tout cela constitue une unique conscience pratique que non seulement je réfléchis difficilement, mais que je ne comprends qu'en la brisant. Le dualisme est la doctrine même de l'entendement.* (P. Ricoeur, *Le volontaire et l'involontaire*, p. 202.)

[3] "If therefore there is a *sensitive* activity, and there can be no doubt of it, I will distinguish it from the *motor* activity, to which I will apply that name exclusively, because it manifests itself to my intimate sense with the greatest clarity" (*I.H.* p. 14).

[4] M. Merleau-Ponty, *Phenomenology of Perception* (transl. by Colin Smith), Routledge and Kegan Paul, London, 1962, p. 216.

[5] *Ibid.*, p. 250.

[6] *P.J.*, p. 65.

[7] M. Merleau-Ponty, *La structure du comportement*, PUF, Paris (6th ed.), 1967, p. 204. "On the other hand, consciousness discovers, especially in the case of illness, a resistance of the *'corps propre'*."

[8] "One can see in the *Dialogues Between Hylas and Philonous* how Berkeley based the greatest part of his arguments against the existence of bodies on examples drawn from the sense of sight. One can also see in the *Treatise of Human Nature* how Hume, a skeptic, pushed the consequences to the end, used the same weapons to destroy any real and substantial existence, including the individual thinker's own existence. He argued in a consequential manner, as might do an intelligent being reduced to only the sense of sight, provided such a being could think and reason" (*O.I.*, 2, pp. 103–104).

[9] The words applying to visual perceptions alone, for example, can apply only in the most illusory manner to tactile and auditory perceptions (*O.I.*, 1, p. 87).

[10] And a few lines later, Merleau-Ponty adds:
"Cela prouve que chaque organe des sens interroge l'objet à sa manière, qu'il est agent d'un certain type de synthèse, mais à moins de réserver par définition nominale le mot espace pour désigner la synthèse visuelle, on ne peut pas refuser au toucher la spatialité au sens de saisie des coexistences" (*La phénoménologie de la perception*, p. 258).

[11] "The sick person's astonishment, his qualms when faced with the new visual world he has entered, show that the sense of touch is not spatial in the same manner *as* sight is" (*La phénoménologie de la perception*, p. 257).

[12] See, A. Michotte, *La perception de la causalité*, Paris, Editions *Erasme*, 1954.

[13] H. H. Price, 'Touch and Organic Sensation', *P.A.S.* **44**, 1943–44.

[14] "For them, senses usually occupy a short enough chapter; they take sensation in general, or any manner (equal and uniform in the hypothesis) of feeling impressions or passively perceiving the results of action on things, for what it is, or seems to be in fact, when considered from the very heart of the habits of all senses or considered in some predominant mode attached to the use of one of them only, and one specific mode of one individual sense, as being typical of all the others or as being the one real characteristic of the complex abstract idea expressed by the word *sensation*; the analogy, or even the identity in the nature of all the elements comprised within the extension of the word, is assumed without any verification" (*M.D.P.*, 2, p. 94).

[15] *M.P.O.*, p. 7.

[16] "To undertake rebuilding, so to speak, human understanding from certain hypothetical artificial elements (often made available merely through the classifications in our language) would be nearly as presumptuous as trying to restore chemically some natural product, either vegetal or mineral, altered and reduced to its components in the crucible.

... One must say it: notwithstanding the extreme difficulties attached to the project as the celebrated author of the *Traité des sensations* had conceived of it, the point of view he adopted made its completion impossible, or completely illusory from the start" (*M.D.P.*, 2, p. 12).

[17] "One must become aware of all that is voluntary in knowledge worthy of the name so as to cast off all the claims of some superficial sensualism according to which the self would be a mere diversity of sensible impressions, some 'polyp of images'; if all the actions of the Cogito are rays of a single self, it is because they partake of a unity of effort, an unobtrusive tension which keeps us awake, alert and attuned to the world. To look reveals sight as an action. Maine de Biran stands invincible on this point" (Paul Ricoeur, *Le volontaire et l'involontaire*, p. 318).

[18] "If the individual did not *will* or was not determined to start moving, he would know nothing. If no resistance was encountered, he would know nothing again, he would not suspect any existence, would not even be aware of his own" (*I.H.*, p. 18).

[19] Cf. *O.I.*, 2 p. 84.

[20] When I speak of 'guiding my eyes' or of 'reaching out for a pen', etc., it then becomes possible to bring out the following objection: it seems intelligible to ask "What do you guide your eyes with?" and "How do you guide your eyes?" and "How do you cause your movements?" and "How do you cause your reaching for the pen?"..., "by some secret shoving and lifting of your arm?" The apparent mistake here is to think of the body as an instrument. My action is the cause of the hammer's action on the nail, but what is the cause of my action? Some prior action of mine? Biran is strongly opposed to such an instrumental view of the body. Such a view would mean that my body becomes, at one point, the term of my action, and this it is not. I never experience movement (my movement) as an *intermediary* between me and the world. We often characterize the body by saying that it is the instrument of my action on the world, but such a characterization can be as misleading as speaking of images as if they were something that I see in my mind's eye. A moment's reflexion will reveal the mistake of this characterization. First, if the body were an intermediary then there would be times when the body would cause me problems, i.e. moving myself would indeed be problematic. But this evidently goes against our experience. Children are not aware of their body as an ensemble of means which they must activate in order to accomplish such and such a thing. Our actions accomplish themselves without our having to resort to our body as a mean. Our movements have no 'instruments'

which would enable us to execute them. We move easily or with great difficulty but never *with* a body as an 'instrument'. I do not have to resort to my body for the simple reason that I am my body. The body is not an instrument, because what we call an instrument is always at the service of some other thing which uses it. This other thing which would use the body – the disembodied subject or mind – would have to know this body as being something different and apart. The theme of my thought would then have to be this instrument, and not the purpose of the action which it wants to accomplish, which is absurd. Such a view is deduced from our knowledge of the body *qua* an organism – the biological, physiological body.

[21] *See O.I.*, 2, pp. 83–85.

[22] "If any movement in progress stopped as it met the least resistance (for example, if when something is put in his hand, his fingers, as they closed, stopped at the very first contact) a person would know only that there is an obstacle; not, however, whether that obstacle is absolutely impenetrable, solid, hard or soft, etc." (*I.H.*, p. 18).

[23] "Through movement alone we would know only masses of varying resistance; the hand breaks down these masses, reveals their elements, distinguishes their properties, sorts out their nuances; it is the first among the instruments of analysis, and all its advantages obviously depend on its construction, on the superior motility of its parts and on the very nature of sensibility" (*I.H.*, p. 20).

[24] *See* Pierre Janet, *L'intelligence avant le langage*, Paris, Flammarion, 1936, pp. 7–25.

[25] "I am dimly aware that something depends on me in the way the world looks; yet it is not that the object be as it is; my only ability is to ask questions by turning toward it, or away from it, extracting from its depths this or that characteristic, or letting it lie there" (Paul Ricoeur, *Le volontaire et l'involontaire*, p. 314).

[26] Sensibility gathers progressively the discoveries made through movement, grasps the most delicate nuances and makes them its own; it perceives that imperceptible crack, those minute peaks and those ripples which disappeared with total resistance or in the speed of a movement, and makes an exact drawing of what the motor organ could only, as one might say, outline roughly, if one supposed its outside were coarse (*I.H.*, p. 20).

[27] "Thus, actually, a blind geometer owes the clarity and the number of the perceptions he has of the modes of imagined areas as much to the delicacy of feeling in the nerve clusters as to the agility and flexibility of his fingers" (*I.H.*, p. 20).

[28] "Those properties of matter can become manifest to him only inasmuch as he *wants* to continue a movement, and the intensity of his effort becomes the measure of his will; if he presses the object with all his strength and yet cannot close his fingers, he meets with a set limit which conveys to him hardness, impenetrability; if the obstacle gives way more or less easily, he can measure its various degrees of softness, mobility, etc. The individual thus perceives the first relation of existence only as he starts moving and the other successive relations only inasmuch as he is willing to continue the movement" (*I.H.*, p. 18).

[29] "In the simultaneous circumstances surrounding the exercise of that sense, one can put aside now the motility or the effort of which the active tactile organ is the main instrument, now the affective sensibility located in the same organ when considered as a passive tactile instrument" (*N.E.A.*, p. 153).

[30] "If one were to put in my hand some object whose surface were either spiky or polished, either softly warm or so cold that it stings, etc., as long as the contact lasted, I would experience either an agreeable or disagreeable sensation which I could not either increase, diminish, or still in any way: that is the part of feeling; and even if the motor faculty were paralyzed, it would function all the same. Touch would convey only that type of sensations

if it were not coupled with mobility, and it would then be much inferior to many other parts of the body covered with skin yet where sensibility is much more delicate, more exquisite" (*I.H.*, p. 16–17).

[31] "When feeling is predominant to a certain extent, the concurring movement is somehow nullified, since the individual is not aware of it, and the impression remains passive. I shall apply to all of these the name *sensation*" (*I.H.*, p. 16).

[32] "When affectibility, actuated by an external cause, predominates, the *matter* involved in the sensation overcomes the act of *perception*, always inherent in the expenditure of effort. That predominance is aptly expressed by *sensation* which then applies to that class" (*M.D.P.*, 2, p. 911).

[33] "That sense, first brought into action through instinct remains almost totally dependent upon it; its complete immobility shows its passiveness, and one could say that, among our external sense, it ranks with the polyp or the oyster on the animality scale; true, its functions are linked with the movement of breathing, yet that movement is by nature necessary, imposed, continuous, and therefore almost insensible; thus odors are the epitome of sensations, as language itself indicates [in French]; they are the least distinguishable; when many are united together they melt into one sensation, which we are unable to analyse, despite the voluntary attention we give the blend; notice that such attention is only a uniform inhaling, slow and prolonged. These are the limits of our *power* over these modifications" (*I.H.*, p. 29).

[34] Biran is not absolutely correct in saying that (a) odours are the sensations which give us the most difficulty in distinguishing, (b) that they are the prototypes of sensations, i.e. passive *'par excellence'*. The following passage certainly refutes this view.

"It's the smell – once you step in that door you are never free of it. It is the smell of the death-house. I've been in five of the joints, and it's the same in all of them.

Sometimes in dreams, when you are out and holed up someplace, because you know they are looking for you – in your dreams you wake up and smell it.... I mean, in your dreams, you smell it and wake up.

It was the odor of men.

It was the smell of men who lived without women, and without the hidden ways of women in cleanliness. It was the smell of old soap along seams; of old soap and detergent left in the corners and cracks of the floors; the rancid, musty smell of old buildings only partially masked and clouded by the smell of lye.

It was the smell of clean bodies and of clean clothes; of dirty bodies and of dirty clothes – for there were those who had forgotten cleanliness, and the occasional one who had not known cleanliness, and some who no longer knew there was cleanliness.

It was the stale, acrid smell of old tobacco, of leaves and paper, of burning string and charred cloth... sometimes of burning hair, as a man for lack of something better to do, sat and burned the hair on his arms and legs.

It was the suffocating odor of several hundred men sleeping with all windows closed to keep out the cold.

It was the odor of several hundred men living in the staleness of the little air that seeped up through the ventilators; the little air that seeped through the debris of years – the debris of prison cells stuffed into the ventilators over the years for as many reasons as there had been occupants.

It was the smell of food eaten by several hundred men in several hundred cells three times a day.

And to temper the smell of men were the odors of the shops: the smell of canvas, of machines, of coal, of sawdust, oil and grease; the odors of the farm: of cows and horses, of pigs, of manure, of milk, ensilage and fodder; the smell of the cannery – the clean smells of work and honest toil.

All these were mingled yet again in the odors of the other inhabitants of the building, the small fourfooted beasts that lived in the vents and conduits.

It was the accumulated odor of fifty years or more, and of the hundreds of men, who whether they lived within the five-by-seven foot cells, or walked the corridors by day and night as guards, sweated with the sweat of anxiety and fear". (Unpublished paper of the late Maurice J. O'Connor, M.D., C.M., *The Convict: A Study of Imprisonment.*'

[35] "The organ can be built in such a manner, and the external agent to which it is subjected of such a nature that on one hand the received impressions bear little affect while on the other hand the effort is somehow not perceived as it is so weak. There exists some relation between the two forces which does not allow us to decide which holds the initiative or which is predominant. The absence of any direct affect excludes the sensitive aspect, the weakness of motor action almost excludes the apperceptive aspect, the individual neither feels, nor acts and yet representation occurs, some external·or internal object is passively perceived. At this point the idea of sensation seems to exist by itself and to come from the outside ready made; this phenomenon deserves to be analysed in its own special conditions and circumstances; we shall distinguish it as simple or objective perception (*intuition*)" (*M.D.P.*, 2, p. 9).

[36] "The self unites in a much more intimate manner with another type of modalities in which he participates or with which he concurs through a voluntary effort. For that active collaboration to be possible, it must be supposed that the organ receiving the impression is part of the sense of effort, or is part of the motor activity. I shall name *perception* any non-affective impression in which the self participates through its own action, following that of an external object. Thus the perception phenomenon shall include all phenomena born through the action of sensible objects, combined with the action of a will still submitted to the impressions that occasion or motivate its initial application. This system will also include the phenomena of the preceeding sensitive category, insofar as the self, which was then a passive observer, can now play an active role" (*O.I.*, 2, p. 7–8).

[37] "One must note at this point that the role of attention is not, as has often been said, to sharpen impression. One can ascertain in all cases that there is no direct action of the will upon the sense organs, whose readiness, totally independent from any action of the will, imparts to any impression received those diverse characteristics of liveliness or languor, of vigor or weakness. The role of attention is thus only to hold as it wishes the mobile organs, such as hearing and touch, upon the attendant object, to divert them from any other source of impressions, and in so doing to make intuition not sharper than it is in the natural combination of the organ's sensibility and the object's mode of action, but somewhat clearer, more definite through the genuine concentration of the representation faculty" (*O.I.*, 2, p. 88).

[38] "I shall call apperception any impression where the self can recognize itself as the cause, distinguishing itself from the sensible effect determined by its action. Leibnitz' definition applies very well in this case: *Aperceptio est perceptio cum reflectione conjuncta.* Indeed, as long as the will is subordinated to an outside impression, the self can ignore its part in it since its own action blends with the action of the object or the predominant external cause. Thus willed attention, which appears to be the most active, excludes reflection or the inner apperception of the action or its author. Yet when the will is in command of impressions and of itself, the self cannot escape seeing its own causality and has the immediate apperception of itself as a cause, as well as an intuition of the effect, or the sensible result of its action or effort" (*O.I.*, 2, p. 9).

[39] "... The easier the movements become, the more precise and clear is the perception acquired through them, and the less aware is the individual of his part in the process, the more involved his activity becomes; the subject and the end of the action, effort and resistance, become at last like one, and everything takes on the passive appearance of sensation" (*I.H.*, pp. 198–199).

[40] "The individual thus perceives the first relation of existence only as he starts moving, and the other successive relations only inasmuch as he is willing to continue the movement. If however we suppose that resistance decreases until it becomes insignificant the last step of that decreasing effort will be the limit and the disappearance, so to speak, of any knowledge" (*I.H.*, p. 18).

[41] "That initiative in exploring establishes the distinction between voluntary and passive attention; in the latter one is totally absorbed in the object, occupied, caught, even fascinated by it; yet that initiative does not produce the essential part of perception which is really to see, to hear, and which is linked with the presence of the object. One sees here the limitation of principle in a psychology of effort; there occurs a moment where acting gives way to knowing and serves it, where effort turns into an opening to the world, an interrogative ingenuity; doing becomes the tool of seeing, yet only to render it more docile, more readily available" (Paul Ricoeur, *Le volontaire et l'involontaire*, p. 314).

[42] "The absence of any direct affection excludes the sensitive aspect, the weakness of any motor activity almost excludes the apperceptive aspect, the individual neither *feels*, nor *acts*, and yet *representation* takes place" (*M.D.P.*, 2, p. 9).

[43] "Yet the self cannot ignore its own causality when the will directs impressions and commands itself. The self has an immediate apperception of itself as a cause at the same time as the intuition of an effect, or a sensible result which he or his own effort produced" (*O.I.*, 2, p. 9).

[44] "Taste is the sense that at first glance appears to have the closest relation to touch; flavors are indeed but the tongue and palate's own sense of touch; (...). Various flavors can easily be compared to the tactile sensations of cold, hot, smooth, rough, sharp; as a consequence those two types of modifications share many words in our languages" (*I.H.*, p. 27).

[45] "Comme de longs échos qui de loin se confondent,
Dans une ténébreuse et profonde unité,
Vaste comme la nuit et comme la clarté,
Les parfums, les couleurs et les sons se répondent"
(Baudelaire, *Les fleurs du mal*, présenté par Jean-Paul Sartre, Gallimard, Paris, 1964, p. 21).

[46] "The organ of taste (which also serves speech) is extremely mobile; the effort made during chewing, or the pressing of the lips, of the teeth or of the palate against solid objects could probably be sufficient to give us vague ideas of resistance and some of its modes; many species of animals, as is well known, experience touch with their mouth and nose" (*I.H.*, p. 28).

[47] "... if not to act upon the muscles which communicate various degrees of tension to the eardrum, etc.? Effort, it is true, has now become almost undetectable; movement and the moving parts, since they are completely internal, do not appear to be subjected to the will; it is true that the ear is immobile on the outside, is open to any and all impressions, has no immediate way of stopping or attenuating them, appears all the more passive because of its great sensitivity" (*I.H.*, pp. 25–26).

[48] "When we hear sounds (which we always perceive more easily if they are more closely related to sounds we can reproduce, imitate or articulate ourselves), the oral instrument acquires patterns parallel to those of hearing and reaches, so to speak, the same pitch:

when we hear speaking or singing, we speak or sing mentally; this imitative instinct is stronger here than in any other movement, and it carries us away unwittingly the most often" (*I.H.*, pp. 26–27).

[49] Biran is evidently on to something but it doesn't seem to be quite what he says. It seems to be fundamental in learning a language – one can't really follow a spoken language till one can speak it more or less. One has to mimic to learn. In a way Biran is suggesting we do this in listening to people talking our own language, and this seems a bit odd. What Merleau-Ponty says is nearer the mark: *"On pense d'après un tel".* Yet, I suppose what Biran is after are cases where we do mimic, where we feel that we are, as it were, holding back our mimicry (it is as if the sounds we hear go directly to the tip of our tongue). When for example, I listen closely to what I am told: here I start, at one point during the conversation, anticipating what will be said, and this anticipation is more than just a guess. We often see others listening and moving their lips, eye-brows, etc., in imitation of the speaker; in fact we need not even see the speaker as many a telephone conversation has revealed. The imitation is not just a repetition of the others's movements but an imitation which, as it were, anticipates what is said. Here it seems to make sense to say that we are our own echo.

[50] "In the most active stare the eye will never illuminate itself, and even if, as we touch, resistance is only felt through an effort determined by the will, it remains... that the matter and the form of perception do not come from the same source; that the subject never opposes itself.... But in the simultaneous exercise of speech and hearing, the movement, and the sound it produces, issue from the same source and are directed toward the same subject which reflects them since that subject is equally receptive to the double effect of which he is the cause. The foreign sound which impresses external hearing is in fact distinct from the articulate sound determined by the will, in the same manner that any movement that is free, voluntary or accompanied by some effort is distinguished from an imposed movement, devoid of any idea of intention. In the first case the self is a cause; in the second, we have a *non-self* cause under which the individual is, or feels himself to be, passive and we find again, in this contrast, a twofold basis to the belief and the judgement of exteriority" (*O.I.*, 2, p. 230).

[51] "The activity which immediately produces the one is reflected in the other; the individual who makes sounds and listens to himself has a double perception of his own activity. As he freely repeats the acts determined by his will, he is aware of the ability which executes them, he sees the cause in its effect and the effect in its cause; he is distinctly aware of the two elements of that basic relation; in a word, he is reflecting" (*O.I.*, 2, p. 232).

[52] "The exercise of internal hearing seen in its intimate union or relationship to the voice organ... is the only sense through which the sensible and motor being can modify itself without the help of any cause external to its own motor ability. In the other senses, the will and the effort always have as their end some external object whose impression either distracts or absorbs the internal feeling of the cause which serves to produce that feeling. In this case effort and movement come from deep within the subject only to be reflected in it as a sensible perception, which the active being adopts and recognizes as his own work" (*O.I.*, 2, pp. 232–233).

[The author is indebted to Prof. Hugues Roy and Patricia Clarke-Roy for translations of quotations from French into English.]

WILFRIED VER EECKE

FREEDOM, SELF-REFLECTION AND INTER-SUBJECTIVITY OR PSYCHOANALYSIS AND THE LIMITS OF THE PHENOMENOLOGICAL METHOD

I. INTRODUCTION

For Husserl, philosophy has an important task: it makes man self-aware through reflection and thus contributes to the humanization of our personal life and to that of history.[1]

Phenomenology is for Husserl a unique way of doing philosophy, because it does away with some great obstacles to self-awareness. Phenomenology refuses to accept the hidden objectivist presuppositions of rationalism and empiricism, whereby one believes that there is a world independent of the subject. Through his *noetico-noematic* analysis, Husserl shows that there are no *noematic* aspects of the objects without a corresponding *noetic* act of the subject. Becoming aware of this all-important dimension of the subject is, for Husserl, equated with reaching the transcendental dimension of philosophical reflection.[2]

It would, however, be mistaken to equate the *noetic* acts of the subject with pure intellectual acts. Phenomenology has stressed again and again that the *noetic* acts or the human intentionality is a bodily intentionality.[3]

The bodily dimension of intentionality has several consequences. In so far as the world I live in is constructed through my bodily intentionalities, I cannot hope to live in a universal world. My world is my particular world.

Phenomenological reflection taught us also that although I have, through my reflection, a privileged access to my world, my bodily involvement in the world makes my world in principle available to the other just as his bodiliness makes him accessible to me. The uniqueness and the inter-subjectivity of my world are both related to my body.[4] The work of Sartre relates my bodily intentionality both to my freedom and to my finitude. Sartre sees the awareness of my *noetic* or intentional acts as the guarantee for my freedom, and my bodiliness as the limit of my freedom imposed on me by other people.[5]

Phenomenology, from Husserl to Merleau-Ponty and Sartre, seems

Tymieniecka (ed.), Analecta Husserliana, Vol. III, 252–270. All Rights Reserved.
Copyright © 1974 by D. Reidel Publishing Company, Dordrecht-Holland.

therefore to inter-relate closely the concepts of intentionality, bodiliness, intersubjectivity and freedom. The method achieved to arrive at these conclusions is that of the reduction, i.e., suspension of the involvement in the every day world whereby my world can be taken apart in *noematic* and *noetic* aspects. This phenomenological method is also sometimes called a reflection on the pre-reflectively given world.

In this paper I would like to argue the following points:

(a) The analysis of the complex interrelationship between the body and intersubjectivity brings to light a new problematic aspect of human freedom.

(b) The phenomenological method (being a personal reflection on the experiences I recall) is only of preliminary use for the problem we have in mind.

II. MY BODY AND MY CHARACTER: FREUD

Phenomenology is an adult activity. It is a reflection on a fully constituted world, on a body that I experience to be mine and on others who are part of my world but separate from myself. Phenomenology, however, often proceeds as if there is no problem with the origin of my body-relations, with the beginning of my world-constitution and the genesis of my intersubjective relations.

Psychoanalysis and child psychology have given us extensive descriptions and deep insights into the prehistory of the adult relationships that exist with the body, the world and other people. This information is invaluable for philosophical reflection because as adults we have no access to this important prehistory, through ordinary memory or through reflection. We will use the writings of Freud, Erikson and Lacan to describe this prehistory.

In 1908, Freud published a short article of seven pages on 'Character and Anal Eroticism' which aroused astonishment and indignation. Freud communicated two observations he had made and tried in a somewhat unsatisfactory way to interrelate theoretically these two observations.

The first observation was that a number of people regularly have a combination of three characteristics; those people are orderly, parsimonious and obstinate.[6]

The second observation was that Freud had learned from this kind of patient that they normally had needed a longer time for toilet-training and that accidents still had occurred in later childhood.

Freud tries in his short article to explain that it is reasonable to see a logical connection between these two observations. He has at his disposal the following theoretical framework:

(a) Human sexuality is present from infancy. It develops and uses successively different body zones as sources for erotic pleasure. The anal zone is one of them.

(b) There are a number of mechanisms such as repression, sublimation, and reaction-formations which are useful to explain human behavior. The insights of the natural sciences are often less useful than those of folklore, poetry, proverbs, etc., for explaining human behavior.

Within this framework Freud proceeds as follows. He first states that the time required for toilet-training by these persons indicates that they must have derived a *subsidiary pleasure* from defecating. He then points to the fact that in growing up in our society anal eroticism must be repressed. Orderliness is then explained as a reaction formation. The relation between obstinacy and anal eroticism is explained by referring to the obstinacy of children during toilet-training to the practice of breaking obstinacy in children by spanking them, and finally, very surprisingly, to an ancient linguistic expression wherein an invitation to a caress of the anal zone is used to express stubborn defiance. Even a passage of Goethe's *Götz von Berlichingen* is mentioned as an argument.[7]

The relation between stinginess and anal eroticism is explained by referring to the fact that long-standing cases of habitual constipation in neurotics can be cured by an interest in money. Freud also refers to the common usage in speech, which calls a stingy person 'dirty' or 'filthy'. Freud proceeds then by showing that the connection between dirt, feces and money is an ancient insight of mankind expressed from antiquity in myths, fairy tales, superstition and dreams.

Freud himself seems somewhat unsatisfied with the explanation of the connection between money and feces because he adds another possible explanation of a totally different kind. It is an argument based on continuity and contiguity. It happens, says Freud, that anal eroticism starts being repressed at the age when children are taught the importance of money.

Freud ends his article by mentioning that he has noticed a similar relationship between character complexes and excitations of particular body zones in the people who have an intense burning ambition. They usually suffered earlier from enuresis.

This article astonishes us in many ways. We are surprised at the proclaimed factual relation between interest in a body zone and a typical character. We are uneasy because Freud does not coherently explain the factual relations he discovered. And most of all we are disturbed by the fact that Freud first reduces the problem of the relation between anal eroticism and obstinacy or stinginess to a purely physical problem. He says that those persons must have derived a *subsidiary pleasure* from defecating whereas he in fact looks for an explanation in purely cultural phenomena: Proverbs, language-usages, myths, folklore, poetry.

The above reactions arise from the fact that Freud seems to be operating with a dualistic concept of man: There is his body and there is his character. Freud thinks about the young child in terms of physical stimuli and physical pleasures. He knows, however, of the cultural creations of man. Freud very hesitantly combines the physical and the cultural in man without having any philosophy to justify it.[8] This has as a further consequence that Freud must factually discover connections between the physical and the cultural dimensions of man. He cannot predict them on theoretical grounds. Thus the discoveries are as much a surprise for Freud as for the reader. (If one can accept at least that the hesitant style of many of Freud's writings is not purely pedagogical....)

Freud's shortcomings allow us to appreciate the progress made by Erikson.[9] Erikson not only describes the relations between one erotogenic body zone and one character-complex, but also describes these relations for a whole series, adding to Freud's anal body-zone his analysis of the oral and the phallic-zone. Whereas Freud restricts himself to an odd character-complex, Erikson claims in general that a person's character is co-determined by the way he went through the different infantile body-problems. Furthermore, Erikson expands his analysis to include non-Western civilizations, especially American Indian tribes. Erikson also claims that not only the attitude to the different body-zones but also the sequence of attitudes is important. And finally Erikson indicates that there is nothing magic about a particular body-zone (except that that zone may be the dominant one in that period), but that the whole body is relating to the world and other people as to that particular body-part. Erikson thus starts from the presupposition that my character is rooted in my bodily relations and that those, in turn, were developed through the intersubjective relations I had in my childhood. Let us therefore look more closely at Erikson's analysis.

III. MY BODILY CHARACTER AND INTERSUBJECTIVITY: ERIKSON

Erikson does not separate the body and the mind as Freud does. His concepts of body-zones, which relate in a particular mode to the world, and therefore develop a charactermodality, serve to mold the unity that phenomenology expresses by defining man as a bodily being in the world.[10] Erikson develops, as we have seen, a complete theory of the bodily relations from birth to old age.

The early periods relate closely to the developmental periods postulated by Freud. The later ones are elaborations by Erikson. The series of periods are the oral, anal, phallic, latency, adolescence, young adulthood, maturity, old age periods.

For our purpose it is enough to analyse the early periods. In each of them Erikson believes, just as does Freud, that one body-zone is more important than others. This body-zone allows the child to develop a particular relation with other people and with the world. At each period the child, then, makes special demands upon the other, and has special fears and weaknesses that adults can exploit. In this way adults mold definitely the future character of the child.[11] Furthermore, Erikson believes that the total body has a relation to the world similar to that of the dominant body-zone.

Within this framework Erikson describes the first years of the life of the child as that period where the mouth is the body-part with which he relates to the outside world.[12] Relating with the mouth to the outside world makes the child totally dependent upon that outside world. The child must be given – he must be given enough and at appropriate times. The initiative lies with the parents. The child can only wait or ask to receive. The parents can choose to adopt two extreme positions: they can decide to feed the child on schedule – as has been the case at some times in the western civilization – or they can decide to feed the child on demand. This latter attitude was chosen by the Sioux-Indians.[13] When a strict schedule is imposed, the child might often cry in hunger before it is time to be fed, or might have to be awakened in order to be fed. This attitude, therefore, forces the child to experience the needs of his body as a demanding force, and he will experience the world as a place where he is not certain to find gratification for his needs. This attitude might have the advantage of preparing the child for a well organized adult life. On the other hand, the decision to feed the child on demand requires a societal attitude that allows the mother to give exclusive attention to the baby day and night. With the Sioux-Indians this social attitude includes prohibition of intercourse with the mother for the entire period of breast-feeding. This period sometimes lasts three to five years.[14]

This decision made by the parents and the society to feed the child on demand has, as a consequence, that the child experiences the world as a place of abundance. This experience makes it possible for the Sioux to become generous as adults, because even in poverty they experience the world as a place of abundance.

But it is not only the mouth which must be taken care of by the parents. The whole body of the child needs to be given appropriate stimuli: the child needs to be softly and warmly spoken to, it needs skin-contact and, somewhat later, it must be subjected to moving object and colors. At all these levels the child depends upon the other.[15]

This extreme dependency upon the other for his bodily well-being and his bodily development forces the child to a psychological experience of the world which can be trust or mistrust, and which will dominate his future character.[16] The knowledge that the character of the child is involved in the way the adults respond to the bodily needs of the new-born is not alien to peoples, as for example, the Sioux-Indians say: "A small child [is] not supposed to cry in helpless frustration".[17]

Around the age of six months, the child's muscles have already developed enough for him to use somewhat actively his hands, but – more importantly – he can also use his jawbones for biting purposes. At the same time, the teething-process starts. The child discovers its body as a hostile and alien dimension of himself and will often in anger or despair bite on any object arriving in his mouth.[18]

At this point the Sioux-Indians change drastically their permissive attitude towards their children. As soon as he bites, they 'thump' the baby's head and let him fly into a wild rage. Sioux mothers comment now that: "crying will make him strong",[19]

On top of that, the Sioux strap up to his neck their angry baby on a cradle board. This prevents him from expressing his rage in bodily movements of his limbs, and forces the child to internalize his aggressivity. This enforces upon him those sado-masochistic characteristics for which the Sioux is famous.

The next moment in the child development is the anal phase. The child has sufficiently developed his body and his muscular system such that he can be subjected to toilet-training, whereby he is asked to hold on and to let go at particular times and in particular places. He is already sufficiently developed to assert himself against other children and to snatch away or to hold on stubbornly to his toys.[20]

Again the parents are free to choose different attitudes towards this problem. They can be extremely lenient and allow the child to master toilet-training by imitation of older children, which was the case with the Sioux-Indians or they might impose a strict learning process upon the child. The imposition of an early and strict training method runs the risk of requesting an achievement from the child that he cannot yet perform. The child might use this training-experience to affirm in stubbornness his opposition to the parents, who in turn might show their displeasure with his behavior. Doubt and shame might then alternate with stubbornness.[21]

A similar choice is possible at the phallic phase where the parents can look with amusement upon the sexual explorations of the children, or where the parents under societal pressure can try to repress sexual curiosity as something bad.[22] The former attitude is taken by the Sioux-Indians; the latter was often common in our civilization. Similarly the child might at that age be allowed to develop some activity and initiative even at the expense of the parents' furniture and property, or the child may be anxiously confined to his proper place.

The bodily development of the child thus poses at every period a new challenge to the parents, who will encourage or enforce a particular way of bodily behavior. This attitude will in turn encourage or prevent a particular character.

Erikson shows very well how the Sioux-Indians consistently approach the child-rearing methods in such a way that the child will experience trust in the world and in himself, while making him capable of sado-masochistic hardship. The frustrations imposed on the Sioux-Child are then later used as cultural virtues in hunting, warring or religious worship. The large degree of leniency granted him in his early childhood becomes the foundation upon which the proverbial Sioux generosity is built. In a hunting society where survival depends on the willingness to share of the best and luckiest hunters, this virtue was as necessary for the Sioux-economy and the Sioux-society as was the readiness to endure hardship.[23]

On the other hand, the child-rearing methods of the Western civilization, where training methods and feeding schedules are imposed on the young child, prepare him for an economy and a society based on the clock. The awareness of bodily needs and the experience that gratification does not always come prepare him psychologically for an attitude necessary for our society: saving.

Erikson, like other cultural anthropologists such as Benedict, Kardiner and Mead, discovered that societies are built upon, and require for their survival, a particular set of virtues and vices. These virtues and vices are perpetuated in the societies through child-rearing methods.

If this analysis is correct then we cannot avoid the conclusion that the belief that as an adult I adhere freely to a set of moral values and strive freely for a series of virtues is an illusion. The bodily relations developed during childhood, and imposed upon us through child-rearing methods, make us almost bodily inclined to a particular life style, a particular set of virtues and moral obligations. The fact that we have no recall of our child-rearing, or that we are not aware that our relation to our body was influenced by our relationship with other peoples, makes us naively ready to accept the notion that freedom is not at stake in the choice we make of our moral obligations.

Even worse, phenomenology is a method based on a rigorous analysis of my personal experience in the world. The limits of active memory are therefore also the limits of the phenomenological method in the orthodox way as developed by Husserl.[24]

An analysis of the child-rearing practices and their consequences on character formation poses therefore a deep challenge to the phenomenological method. This is almost ironic, since the challenge as we have outlined here, was only possible through the fact that Erikson used one of the main conclusions of the phenomenological movement as an instrument for analyzing the consequences of child-rearing practices. (We are referring to the refusal of both phenomenology and Erikson to accept the body-mind dualism. Erikson is here in direct opposition to Freud.)

Thus we have encountered two problems that are closely interrelated. The affirmation by the phenomenological tradition of the importance of my body allows us to accept philosophically the analysis of Erikson. This analysis tells us that the virtues I choose can be traced back to the bodily relations I have been trained to develop in my child-rearing. Thus arises the philosophical problem whether or not I am free in the choice of my moral virtues.

The second problem concerns the phenomenological method itself. Erikson's analysis allows us to discover that the choice of moral virtues is so deeply engrained in the relations we have with our bodies, with the bodily styles that we maintain, that this choice seems to be made for us rather than by us. The bodily origin of this choice takes this choice out of the reach of the self-reflection practiced in phenomenology. The threat to freedom posed by child-rearing cannot be perceived by the phenomenological method. We can therefore also not hope, as Husserl expressed it, that the phenomenological reflection will make us free, as it cannot even discover the threat to freedom. The question thus arises whether the discovery of the bodily dimension by phenomenology is not at the same time the discovery of a limit in principle to that philosophical method.

IV. GROUNDS FOR THE LIMITATION OF THE DETERMINISTIC INFLUENCE OF THE OTHER UPON ME

The problem we are faced with is the following: is a society capable, through its child-rearing methods, of successfully imposing those virtues and vices required for its survival, and does that represent a limitation in principle to personal freedom?

When we formulate the problem of freedom in this way, we are almost confronted with a choice between appreciating the efficiency of a society which succeeds in adequately preparing its members for their adult tasks, and being afraid, in the name of personal freedom, of this efficiency.

Let us therefore look into the possibility that there are principal limits to the efficiency of the socialization process.

The first indication of such a limit is the identity-crisis during adolescence. The identity-crisis as a period is described by Erikson as a moratorium, in which a young person tries to understand his society, tries to understand himself and tries to project a role for himself in the society to which he would like to be loyal – a role that he can see as useful in the society and that he can exercise in conformity with the moral standards he has adopted.[25] This period is called a crisis because the young person finds himself caught from all angles. He might dislike the moral compromises required to be successful in any role in the society, he might resent the mechanization of all roles in contemporary society, he might be frightened by the minimal influence that any role has upon the society – but the aspect we are now concerned with is that he might feel inadequately prepared to face the jump into a social role with a permanent commitment.

This feeling of being inadequate for any future role normally leads to a crisis. The adolescent then recaptures in imagination how he came to be what he now is through a series of interactions with other persons, particularly with his parents. He might not be able to recall the early child-rearing methods, but he has a wealth of information available. He knows the virtues that guide his parents and that they imposed on their children; he might have seen how they behaved towards younger brothers or sisters; he might have heard comments made by his parents to their friends. In short, the adolescent in this crisis has almost all the available information upon which Erikson has built his theory. It is then not surprising that a young adolescent finding himself to be inadequately prepared for any role in the society tries to mediate this deficiency in the easiest, but often, the only immediately available solution: i.e., an imaginary effort in which he projects himself as becoming a totally different person if his parents had been different or if they had acted differently with him. The adolescent therefore creates in imagination an alternative child-parent relationship in the name of a better and more efficient preparation for his entrance into the society.

A second moment in which a form of liberation from the child-rearing practices occurs is the experience of love or courtship. As Erikson notices, a lot of adolescent love consists in talk.[26] The young persons want to find out who they are. In that dialogue about themselves, they discover effective differences in personality that can be attributed to different parent-child relations. At the same time love is a relationship of good will, or – in psychoanalytic terms – it is a relationship whereby one mutually projects in the other capacities which are not there, but because they are projected in them, are made possible. As is sometimes said: love allows a person to blossom. We might add, however, that it is very possible that the liberation occurring in courtship is but a liberation within the main direction of the societal virtues, and that it only confirms the complete grasp that the society has upon the individual. It nevertheless indicates that child-rearing practices are not totally determinant: within limits changes in character are possible.

A third experience which reveals the limits of child-rearing is that of psychoanalytic therapy. Psychoanalysis claims that in talking we can become aware of our long forgotten childhood relations with other persons, and that with the help of transference we can liberate ourselves from undesirable character-traits which resulted from our childhood-relationships.[27] Again, no miracles are possible and the history of the 'wolfman' is there as an example to show that the liberation through psychoanalysis is but a partial one.[28]

Finally as a parent we have the capability to develop an alternative child-rearing method for our own children. The possibility of introducing an alternative or a slightly changed child-rearing practice depends on the young parents having liberated themselves partially from their own childhood experiences in their identity-crisis. Otherwise the saying: "What was good for us as children, is good enough for our children now," will become practice. Still another condition for effective change in child-rearing methods is that the parents have thought out alternative attitudes towards their children, and alternative attitudes towards the reactions of their children to their new methods. Otherwise one falls back easily upon the patterns that one remembers from one's own childhood.

Freedom from child-rearing is therefore very limited. The freedom from its influence is greatest in an inter-generational context. If we cannot make ourselves free from undesirable childhood influence, we can at least rear our own children differently.

Thus, ironically, the freedom from the child-rearing methods discovered in our analysis is based on their partial inefficiency. As an adolescent with an identity-crisis, as a psychoanalytic patient, as a person in love, or as a young parent, liberation from childhood is motivated by the inefficiency of the child-rearing methods. We have now to analyze whether these inefficiencies could be remedied. If they can, we have lost the grounds for claiming a possible liberation from the deterministic influence of child-rearing.

V. PHILOSOPHICAL FOUNDATIONS FOR THE NON-DETERMINISTIC CHARACTER OF CHILD-REARING PRACTICES

Contemporary psychoanalytic theory and philosophers of the unconscious have already begun to accept some of the basic assumptions of Freud. All men, even the mentally ill or perverted, had the potential to develop normally. Expressed differently: perversion or mental illness is a developmental deficiency. Man goes through a series of stages and periods. If he does not go successfully through them he will not reach 'normal' life.[29]

Freud furthermore believes that in the early periods of his life man is confronted with some dramatic and traumatic experiences. Freud expressed the childhood tragedies in his theory of the Oedipus- and castration-complexes or in that of the duality of human drives. This latter view leads Freud to postulate the opposition between Eros and Thanatos, between the love- and death-instincts.[30]

Contemporary psychoanalytic theoreticians such as Lacan and his school, and philosophers of the unconscious influenced by Lacan, have represented child-development as a series of complexes, which force the child to accept what it never wanted and will always struggle against accepting: i.e., that in order to be man one must accept being only one among many.[31]

This theory is expressed in a remarkable article by Lacan in the *Encyclopédie Française*.[32] There Lacan analyzes childhood up to roughly four or five years. Yet, he does it with a different framework than Erikson. Where Erikson indicates how the body-experiences imposed on the child prepare him for the virtues necessary in his society, Lacan shows that the child encounters three times a radical frustration of his deepest needs and desires. A satisfactory incorporation of these frustrations is then seen as a precondition for avoiding mental illness.

The third frustration is the well-known Oedipus-complex. Because the two others are of a similar nature Lacan calls them also complexes.

The first complex is called the weaning-complex.[33] It includes not only the weaning itself but all the discomforts that the child experiences with feeding in his first year. Those discomforts include stomach-trouble, moments of hunger, etc. The child remedies these discomforts by psychologically refusing to accept them. Physically, however, they are present. The non-acceptance of the weaning is expressed in the imagination and phantasies of the child where the mother is not just imagined as present, but where the mother (or in psychoanalytic language: her breast) is incorporated by the child's imagination. This illusionary incorporation is sometimes physically supported by sucking on the thumb or some other object.

The further development of the child requires that the imaginary presence of the mother can serve as a substitute for the perceived absence of the mother. The effort of imagination must therefore serve twice as a form of illusionary solution to a frustrating experience.

The second complex is called the complex of intrusion.[34] Lacan identifies the height of this complex with the jealousy of siblings. Sibling-rivalry, however, is accompanied by the discovery of the sameness of the others. This discovery is greeted with jubilation. It is to account for this jubilation that Lacan postulates the mirror stage, in the development of the child. At first, the child discovers his body in parts: his hands, his feet, his legs. In the mirror, or in the recognition of the sibling as being the same, the child discovers the totality of his body. This gives the narcissistic tendencies of the child a real object. In a second moment, however, the child has to discover the alienating dimension of having a body. This dimension is a permanent threat for man as Sartre's analysis of the look indicates.[35] The shame and fear for the look of the other finds its reversal in the attempt of jealousy to destroy the otherness of our own ego in the other. The tragedy of Cain, the tragedy of human jealousy, and the discovery of our own body are the essence of what Lacan calls the complex of intrusion.[36]

The third complex is called the Oedipus-complex, and coincides with the discovery of the bodily sex-differences and the typical role of the father.[37] The discovery of the sex-differentiation means for the child that it cannot be everything at the same time, and that it therefore has to come to grips with the human problem of finitude.[38] The dramatic dimension of this experience is underlined in the somewhat unhappy choice of label given by psychoanalysis to this experience: castration.

The awareness of the role of the father teaches the child that the task of both the mother and the father is not solely to take care of his wishes and desires. He has to accept that he cannot count on exclusive attention, not even from his parents. One could translate this in existential language by saying that the child discovers that he is alone in the world. In psychoanalytic language, that is expressed by saying that the child has to give up or has to repress his libidinal ties with the mother.

This Lacanian version of psychoanalytic theory thus stresses that the child in his development encounters deep frustrations and is required to renounce some of man's deepest desires. Erikson talks about the fact that the child in each period of his development encounters typical fears and anxieties, which were then exploited by the society to mold the character of the child. Erikson stresses the successful use made by society of those anxieties. This led us to ask if man had any freedom in the choices of the virtues that he would adhere to in moral life.

Lacan underlines the fact that in his development the child is forced to give up the protective and reassuring presence of the mother (complex of weaning), the ideal of self-possession (complex of intrusion and mirror-stage) and the hope of being the unique object of attention from the parents (Oedipus-complex).

These forced renunciations and these imposed frustrations can, I believe, indicate why the rewards promised and given by the society for the successful acquisition of the virtues and skills imposed on the child during the child-rearing, are always experienced as insufficient. The gratification received for compliance with the cultural virtues is thus experienced against a background of disillusion about the price paid for this socialization.

To become man is to undergo these frustrations. The image presented by the Lacanian version of psychoanalytic theory is close to a saying of Hegel: that man is a sick animal and that his sickness, his frustrations are his humanness.

This sickness, these necessary frustrations, guarantee that no man will coincide completely with the virtues and skills imposed on him during child-rearing. This vision of man forms thus an ontological safeguard against any fully deterministic influences from a society. Any reward by a society can only be experienced in a horizon of disappointment.

This leads us to the uncomfortable conclusion that freedom from the influences of child-rearing practices can only be accepted by affirming ontologically the sickness of man, or to use Sartrian terms, the absurdity of the human condition.

VI. CONCLUDING REMARKS

We have presented a form of possible determinations of human morality which questioned the possibility of human freedom. In order to discover that dimension, we had to make use of documentary evidence presented by Erikson. This evidence made us aware that intersubjectivity is not just co-constitutive of our subjectivity, but that intersubjective experiences in childhood are quasi-determinative of our lifestyle. We also discovered, with the help of Lacan's theory, a way to point to a philosophical foundation for claiming that the quasi-determination could not be total. This philosophical foundation brought us in contact with the nature of human needs and desires. In these concluding remarks we want to put our previous analyses in a proper perspective.

1. *The Limits of the Phenomenological Method*

To discover the deep quasi-deterministic influence of the other on myself we had to study the child-rearing practices as Erikson analyzed them with the help of psychoanalytic theory. Self-reflection is twice insufficient for the discovery of this level of human determination. Self-reflection is insufficient because I cannot recall personally the data necessary for recognizing this deep form of determination. Furthermore, the deficiency of my memory is not something that can be remedied by exercises. Pure memory-training will not make available to me my childhood-experiences because between the facts of my childhood and my adult memory stands not simple forgetfulness but active forgetting brought about by repression. In the debate between philosophers of the unconscious, repression has indeed been credited with being a barrier in principle to everyday memory and thus also with being a limit in principle to the use of the phenomenological reflection for the conquest of the unconscious and its influence.[39] If I do not undergo psychoanalytic treatment, whereby through transference I am overcoming repression, then the only way by which I can understand the level of determination about which we have been talking is by studying the impact that other adults have had on other human beings and accept this determination for myself on the basis of analogy.

However, this method highlights the second limitation of the phenomenological method for our problem. The study of the influence of the others on another human being does not allow me to apply the usual *noetico-noematic* analysis. It forces me to complete that technique with the use of a theory (here, the psychoanalytic theory), which allows me to interrogate the data. The validity of this theory depends on the facts used to support it, but these facts can only be detected by the theory. The circularity goes even further because the theory is sometimes capable of creating the necessary facts for its support. See, e.g., the fact that patients with Jungian therapists have dreams about churches and patients with Freudian therapists have dreams with sexual themes. We are here not any more confronted with a phenomenological problem but with a hermeneutic one.

However, we should not accept so readily the limitations of the phenomenological method. There are two lines of argumentation available to question the validity of the limitations put upon the phenomenological method.

The first argument is that repression does not succeed in creating total forgetfulness. As Freud says, man is always somehow aware of the repressed and the unconscious.[40] He is struggling against accepting fully what he is aware of: his forgetfulness is thus but partial forgetfulness. Even in our analysis we have used the partial awareness of the influence of the child-rearing methods as an indication of the limited freedom from this form of determination. We have pointed to the fact that in the identity crisis in adolescence, man creates in phantasy an alternative childhood that would have made him different. In this identity crisis, the veil of forgetfulness is thus partially lifted. This memory, however, does not provide us with the richness of material given to us by Erikson. Furthermore, the use of those forgotten childhood-experiences is – strangely enough – often limited to phantasy. In the identity crisis, man does not seem to be able to make fully conscious use of his forgotten childhood. Repression is overcome in the identity crisis, but not enough to allow the full efficiency of the phenomenological method, which rests on the full control by my consciousness of my memory.

The second argument for not accepting the limitations in principle of the phenomenological method is that Husserl himself has in at least one important work extended the phenomenological method to cover a domain that needs – besides information available through memory – information only available through a learning process. We are thinking of his *Crisis of European Sciences.*

We have, however, to recognize that *Crisis* is an extension of the phenomenological method. Furthermore, Husserl is making explicit use of the way he personnally experiences the crisis of the European sciences. Husserl is, therefore, analyzing as much his own experience as the objective status of the European sciences.[41] The *noetico-noematic* analysis is therefore still useful. Matters are somewhat different for an analysis of the influence of child-rearing methods. The depth of these influences is not revealed to me by a *noetico-noematic* analysis, but by the study of objective information and the belief in the validity of psychoanalytic theory.

With the study of the constitution of the subjectivity through intersubjective influences we therefore come clearly to a problem for the phenomenological method. That this method is, however, not completely alien to those problems can be seen from the fact that we first needed a phenomenological theory of the body before the full philosophical importance of psycho-analytic theory could be formulated.

2. *Freedom and Human Desires*

The concrete problem we have analyzed in our paper is the threat to human freedom posed by the apparent efficiency of child-rearing methods. Husserl proposed self-reflection as an instrument for achieving freedom. Our study indicated that self-reflection is useless to discover the quasi-determinism of the child-rearing practices. We had to study directly the intersubjective relations involved in child-rearing. When we then looked for a possible liberation from this kind of determination we did not turn to reflection, we turned to man's experience of his desires. These frustrations are first imposed during child-rearing. The rebellion against these imposed frustrations makes man capable of taking distance from the results of his child-rearing, even – if as we said – man might not be capable of effectively undoing the practical results of his child-rearing. It nevertheless allows him some effective freedom in the choice of his parental attitudes when he will bring up the next human generation.

On the other hand, his rebellion against the frustrations imposed on his desires cannot be too successful because the result of not containing the limitlessness of human desires is madness or perversion. And on this path man will not encounter freedom either.

If we want to situate human freedom we therefore have to situate it in the rebellion of man's desires[42] against the successful imposed limitations during childhood. This rebellion must, however, be restrained. Man can only afford the rebellion against his childhood-training if he is capable of replacing the social limitations of his desires by personally imposed limitations.[43] Otherwise, madness is around the corner. Therefore – contrary to some of the deepest aspirations of Husserl – phenomenology and reflection cannot deliver freedom. Freedom is a battle won upon desires. A more reconciliatory formula may be that reflection is not a sufficient condition for freedom; it is only an auxiliary instrument. Husserl says that one of the important virtues of the philosopher is humility, but I am not certain that Husserl had that kind of humility in mind.

NOTES

[1] Husserl, *Husserliana* I, pp. 272, 273. Yamamoto, p. 131. Husserl, *Husserliana* I, p. 39.

[2] Husserl, *Husserliana* I, p. 272.

[3] Merleau-Ponty referring to Husserl calls this bodily intentionality: 'operative intentionality (fungierende Intentionalität)' in *Phenomenology of Perception*, p. xviii.

[4] Merleau-Ponty, *Phenomenology of Perception*, pp. xii–xiii.

[5] Sartre, *Being and Nothingness, passim.*

[6] S. Freud, *G.W.*, VII, p. 203; *S.E.*, IX, p. 169.

[7] *Ibid., G.W.*, VII, p. 207; *S.E.*, IX, p. 173.

[8] The first time that Freud did this consciously and methodologically was in another small but important article: 'Some Points for a Comparative Study of Organic and Hysterical Motor Paralyses.' S. Freud, *G.W.*, I, pp. 39–55; *S.E.*, I, pp. 160–172.

[9] Erikson, *Childhood and Society*, particularly pp. 72–97 and pp. 247–274.

[10] Erikson, *Childhood and Society*, pp. 72–97.

[11] *Ibid.*, p. 79.

[12] *Ibid.*, pp. 72–80.

[13] The Sioux case is described under the title: 'Hunters Across the Prairie', *ibid.*, pp. 114–165. One finds an excellent summary on pp. 153–156.

[14] *Ibid.*, p. 135.

[15] *Ibid.*, p. 72.

[16] *Ibid.*, pp. 247–251.

[17] *Ibid.*, p. 135.

[18] *Ibid.*, pp. 76–80.

[19] *Ibid.*, pp. 136–137.

[20] *Ibid.*, pp. 80–85.

[21] *Ibid.*, pp. 251–254.

[22] *Ibid.*, pp. 85–92.

[23] *Ibid.*, pp. 154–155.

[24] This can be seen concretely in Husserl's analysis of duration and succession in time. Husserl, *Husserliana* **X**, 18, pp. 42–45.

[25] Erikson, 'Identity and the Life Cycle', e.g., pp. 113, 122, 149; 'Identity: Youth and Crisis', e.g., pp. 15–19; 'Insight and Responsibility', e.g., p. 90; 'Childhood and Society', e.g., pp. 261–263.

[26] Erikson, *Childhood and Society*, p. 262.

[27] The strongest claim that psychoanalysis is limited to talk is made by J. Lacan in an article translated and commented by J. Wilden, 'The Language of the Self'.

[28] Gardiner Muriel (ed.), *Wolf-Man: His Memoirs: With the Case of the Wolf-Man* by Sigmund Freud.

[29] Freud stated this for the first time systematically in 'Three Essays on the Theory of Sexuality', *G.W.*, V; *S.E.*, VII.

[30] Freud, 'Beyond the Pleasure Principle', *G.W.*, XIII; *S.E.*, XVIII.

[31] E.g., A. de Waelhens, *La Psychose*, Chapters II–III.

[32] Lacan, 'Le complexe, facteur concret de la psychologie familiale', *Encyclopédie Française*.

[33] *Ibid.*, p. 8.40–6/8.

[34] *Ibid.*, p. 8.40–8/11.

[35] An explicition of the relation between Sartre's analysis of the look and the mirror-stage can be found in our article 'The Look, The Body and The Other', or 'Sartre and Psychoanalysis'.

[36] Vergote has elaborated on the tragedy of Cain in 'Ethik und Tiefenpsychologie'.

[37] Lacan, *ibid.*, p. 8.40–11/16.

[38] An interesting connection can be made with Hegel's analysis of desire in *Phenomenology*, pp. 217–227. For an elaboration of this comparison see our article, 'Hegel and Freud'.

[39] For a good survey of the debate see Ricoeur, *De l'interprétation. Essay sur Freud*, pp. 366–406.

[40] Freud, *G.W.*, I, p. 305; *S.E.*, II, p. 299.

[41] *Husserliana* **VI**, e.g., first 20 pages.

[42] See a similar line of argumentation for the problem of revolution in W. Ver Eecke, *Law, Morality and Society: Reflections on Violence*.

[43] The idea that freedom in the full sense of the word is connected with a struggle against the desires is not a discovery of psychoanalysis. It is an insight that can be found in the writings of many religious thinkers. See, e.g., *The Spiritual Exercises of St. Ignatius*: The Expression used is struggle against inordinate attachment, pp. 21, 23, 149–157; K. Rahner, *Theological Investigations*, Vol. II, p. 2; 'Freedom in the Church', Vol. III, p. 4; 'Reflections on the Theology of Renunciation', p. 5; 'The Passion and Asceticism'. The importance of the concepts of renunciation and frustration for a psychoanalytically inspired anthropology can be seen in Huber, Piron, Vergote, pp. 214ff.

DISCUSSION

Eng: I take it that our discussions are placed within the context of phenomenology and not of psychoanalysis, although whatever may touch upon psychoanalysis will be examined phenomenologically. The analysis we have been presented with in the previous lecture would be an analysis in terms of a 'hyletic' and not 'pure' phenomenology, to use Husserl's distinction, because a 'pure' phenomenological analysis proceeds by the subject's personal authentic experience. A psychoanalyst's report is 'hyletic' phenomenology, reconstituting the experience of someone else. In psychoanalysis the most purely phenomenological activity was Freud's self-analysis. That is the closest we come to phenomenology in psychoanalysis. Erikson at first departed from Freud's theory in the direction of a broader notion of the transcendental subject, nearer to the *life-world*, by introducing his notion of the stages of life in which the various stages embody the development. Different stages were correlated with the development of various character qualities. Initially he developed the notion of the achievement of identity in adolescence, which meant the formation of a *unified self concept* out of the variety of self concepts that had emerged during the young person's earlier development. In effect only to the initial stages of life that he applies Freud's libido theory, which is tied to empirical, natural bodily instincts, but from the middle stage on it is no longer linked to any empirical bodily zone. For example, in his biographical study of Ghandi the discovery of Ghandi's identity is no longer understood as the level of the achievement of adolescent character. There emerges a level of life-*Gestalt* which becomes the subject Ghandi. Identity is becoming more and more referential to the transcendental subject, in Erikson's view, it becomes more and more comprehensive.

There is no point in comparing phenomenology with psychoanalysis simply because psychoanalysis begins as a reflection on other people – on the Other; it does not start from *my* direct experience, it is instead a re-constructed experience. If I as a psychotherapist want to talk about

Tymieniecka (ed.), Analecta Husserliana, Vol. III, 271–278. *All Rights Reserved.*
Copyright © 1974 by D. Reidel Publishing Company, Dordrecht-Holland.

my experience of the other then I might enter into the sphere of phenomenology, but as soon as I step out of that sphere and begin to make statements about the life of the other in a therapeutic relationship, I am doing something different and in this instance the problem of freedom becomes paradoxical. I think many of us feel that psychological and psychiatric theories leave us with very little room for freedom. In psychoanalysis it is the therapist who is the agent of change, it is he who suggests the keys to the transformation and the reconstitution of the life of the patient according to his patterns, whereas, it seems to me, the very soul and essence of phenomenology is that I discover my self, my own transformation, I am enhanced by the discovery of the self transformations in my own experience, for instance, when past debris of my life become transformed into guiding exemplars for my future. There is a transformation. Freud touched on this when he introduced the notion of the *super ego* as being formed out of the *id* residues. But he never discussed that. 'Transformation' and 'sublimation' are words he uses only in his very early works; he discarded them later on. Freud becomes increasingly mythopoetic in his thought, but, in the accounts of the analytic process this is missing. These mythopoetic exchanges between the analyst and the patient are left out of the theory, but it is precisely here, where the phenomenological element exists, in this transformation of our experience, whereby the facts of past history become the guiding exemplars of my existence; the past, so to speak, is retrieved in the future, it is no longer a sheer necessity but a guide toward the realization of my own freedom.

Mohanty: I think part of the problem which has been raised in Dr Ver Eecke's paper is the relation between phenomenology and the natural sciences. That means, the same kind of limitation could be raised to phenomenology at a very different level, let us say, by biochemistry. Appealing to biochemistry one possibly would come up with a deterministic thesis, leaving even less room for freedom, whatever may be meant by freedom. I think that any natural science – any theory, and psychoanalysis is a theory, like biochemistry – would presuppose a certain conceptual framework, would make use of concepts, which phenomenology should precisely be in a position to clarify. This was Husserl's intention when he spoke of a 'presuppositionless philosophy' that would offer a ground for the sciences. I will take only one or two examples from this paper's discussion. The psychoanalytic theory makes use of concepts

like imagination, repression, rivalry, and, of course, there is the concept
of causality, and so on. Furthermore, the psychiatric theories have been
constructed not only on the ground of these notions, but also making use
of basic logical rules. If the phenomenologist cannot accept the finality
of the psychoanalytic theories, that means neither the rejection of their
content nor of the language they use, it is merely the demand to clarify
their basic assumptions. For example, taking the concept of imagination
psychoanalysts alone cannot clarify it, consequently psychoanalysis or,
for that matter, any natural scientist instead of limiting phenomenology,
enhance and expand it, because they depend on it for the clarification of
the assumptions and concepts which are used but left unclarified within
the scientific framework. Phenomenology alone can elucidate the meaning of the ground of scientific activity.[1]

Ver Eecke: It is difficult to answer immediately all the points raised,
but I have discovered, in the way you have explained your point of view,
certain elements with which I can agree, and others which point to a
different direction than the one I take. I believe that the role of phenomenology was very well placed by Dr. Mohanty, and was indeed used to
show the inadequacy of the natural sciences. However, I do not believe
that the same applies to psychoanalysis except, as you say, in a subjective
way, namely, to clarify concepts psychoanalysis employs but doesn't
care, or might not be able, to clarify. Thus, I accept that phenomenology
can have a clarifying function but I maintain that there is a radical
difference in the relation between phenomenology and natural sciences,
on the one hand, and phenomenology and psychoanalysis on the other
hand. What phenomenology is doing in relation to the natural sciences
(i.e., to show that the human subject is a crucial factor even in the activity
of natural sciences) cannot be claimed for psychoanalysis.

Let me indicate some complications concerning the parallelism you
were drawing. In my opinion psychoanalytic theories have become
philosophically a problem because phenomenology is not capable to deal
with man properly. Although it understands man as a bodily subject,
yet its method ultimately alienates phenomenology from the concrete
natural man, whereas psychoanalysis provides information concerning
that dimension which phenomenology has discovered as alien to itself.
Secondly, the theory provided about the bodily subject by psychoanalysis
was discovered, as Dr Eng already pointed out, by a phenomenological

method used by Freud against his own training that he tried at every point to erase. I merely wanted to show in my paper that it was by rejecting the naturalistic method that Freud came up with possible phenomenological insights which, nevertheless, he could not utilize. Thirdly, supposing that I grant you your parallelism between psychoanalysis and the natural sciences, as a philosopher you have to take into account the fact that we live in a culture which believes in psychoanalysis. Therefore, psychoanalysis and phenomenology have to engage in a dialogue as was done by Freud himself. I do not believe that the phenomenological method is a method to attack psychoanalysis, but rather to provide psychoanalysis with a philosophical basis to challenge philosophy.

Girndt: I want to make a historical remark to Professor Morin. Contemporary anthropology has developed a similar theory to what Fichte first discovered and proposed, namely, that reality can only be found by presupposing a model of 'reaching out', 'moving toward'. Although the idea of visualizing is predominant in Fichte, the theory of reality is based on the premiss that we have a striving *ego* which completes reality toward an idea, and only on this basis can we have an experience of reality; on this basis can we have a perception. It is the theory of resistance which was here, I think, developed for the first time.

My second comment is directed to Professor Ver Eecke. I have the feeling that you missed the point altogether. Moreover, the replies that have been made to your exposition have not been radical enough. First, I do not think that one can treat this theme adequately be leaving out Schutz and G. Herbert Mead, who are actually key figures in this debate on self consciousness and the relationship to the Other. I also feel that you miss the point on the phenomenological method as well. It cannot be denied that Freud had naturalistic presuppositions and a naturalistic background. I may remind you that on such ideas like mechanism, determinism, one can clearly show that Freud consciously or unconsiously bases his theory on the basic premises which were introduced by Feuerbach, namely, the 'reduction theory' and the 'projection theory'. We find the same in Durkheim and practically in all modern anthropologic approaches, and this is the exact opposite of what Husserl had in mind. What you leave out is transcendental consciousness which in principle can never be the object of any science; in leaving it out you certainly run into the problems of freedom. Besides, you define freedom only as a leg of necessity, which is certainly a totally insufficient idea of freedom. What you are doing is exactly eliminating the main point of the phenomenological method and then try to approach the phenomenological method from the premiss that is the very subject matter of investigation of the phenomenological method. Nor do you make any reference to the reply that Sartre has given to Freud, because what you presuppose as a last resort is that a natural anonymous agent, *id*, 'drive' or whatever, by genetic development eventually comes to consciousness, to judgment, and carries all the aspects that make a uniquely human dimension. You presuppose this explanation as a last resort with naturalists like Feuerbach, you believe that we can give a genetic account of the dimension of consciousness which in principle can never be an object of an objectifying approach.

Meyn: I have the feeling that phenomenology is sometimes taken to be a *panacea* for all sorts of things. I was surprised to learn in Dr Ver Eecke's presentation that the phenomenological method is supposed to be used to discover causal relationships. I was under the impression that Husserl was looking for logical relationships. He asserted, time and again, that he was not interested in explaining causal connections, but rather in understanding the logical relationship of horizons and frameworks. Now I think the reason for this identification of causal with logical explanations is that the notion of reflection you function with seems to be the notion of introspection, but certainly the Husserlian notion of reflection is not that of introspection. If we take the discovery of logical relationships as a paramount task of the phenomenological enterprise, under this interpretation the question of freedom is: whether there are logical frameworks that are eternal and that we can in some way discover – in which case there is no freedom to make these frameworks – or whether in some sense these logical frameworks that we discover by phenomenological analyses are in a sense made by us and created by us, and hence can be changed. I would have rather liked to see, since the topic of the paper had to do with self-reflection and intersubjectivity, how, and in what sense self-reflection can disclose the necessity of intersubjectivity.

Półtawski: I think we are involved in a very interesting discussion and that it touches on several crucial points: the problem of the situation of science, phenomenology, and even metaphysics. Answers and solutions to these problems will partially depend, I think, by what we understand by 'phenomenology'. When we say that we can or cannot use, say, psychoanalysis to enable us to learn something about phenomenology, the answer will be conditioned by our definition of phenomenology. Then there is the problem of Freud, as he was a historical man expounding his theories, discovering the subconscious. It is true, as you said, that Freud worked on the assumptions which are not compatible with phenomenological approach and which are not compatible with freedom, for that matter. But does that mean that if we pursue phenomenology and accept freedom then we have to reject the subconscious in man? I don't think so. And concerning the problem of what phenomenology is, I think, that Husserl was still very much in the old vein of science as it was understood in the nineteenth century. You may say that phenomenology, and this is I think partly the Husserlian approach, is just speculations about what is absolutely given and what is not absolutely given, and the result is some theory which is beyond phenomenology. The idea of this conception of phenomenology would be in fact of the old brand. But I don't think that is the real contribution of Husserl's phenomenology to the conception of philosophy and our knowledge. There is a problem whether it is true that this striving to grasp the given is everything; that description of givenness is all we can and have to do while doing philosophy. There is even the problem of theoretical metaphysics which perhaps has to supply what is immediately given in Husserl's sense. Of course, these things must fit together, we must build up the whole picture of the world the best we can, and I think this is Husserl's idea of the different *a priori* regions which supplement each other. I think that we must give up the idea that we can only start and limit ourselves to describing what is actually and immediately given. We must build an adequate picture of reality and man in it, and then it is rather ontology which we must try to reinforce. Ontology would have to fit in with what is immediately given, but there is no immediate givenness without the framework of the *a priori* you are speaking about. Then I think that science, and Freud, and whatever we try to find out about reality has a right to get into the picture we are trying to make.

Ver Eecke: I am willing to say that I sinned by omission; I believe that Shütz and Mead might be interesting. Sartre's objection against psychoanalysis is just a first attempt to digest psychoanalysis, as Merleau-

Ponty's rejection of psychoanalysis is a first attempt to digest psycho-analysis, which Merleau-Ponty himself withdrew later on. Concerning Feuerbach I believe that you touch on something the consequences of which, I believe, you do not draw. Freud has basically conflicting pre-suppositions, he has presuppositions from Feuerbach concerning culture to which you have drawn attention. He had presuppositions concerning man also. But independent of one's presuppositions that are present in one's cultural period we are capable of seeing the real as it is given, and Freud used to describe and to analyze what *he* experienced, often against the theoretical presuppositions. Therefore, I thought that in my analysis of Freud I showed, very clearly, the discrepancies that were at work. Concerning the notion of freedom, I am completely in agreement that any analysis of it is not complete. There is a threat to freedom coming from our desires and it is only the threat that I have analyzed, not the concept of freedom. I ended by saying that freedom should be located in the struggle against the desires – that means, that freedom is not in reflection but it is in a task, a form of containment of the desires. Instead of accepting the absurdity of life, which is the other extreme of the deter-minism of psychoanalysis, one should see it as a task of constraint not by society but by oneself, the limitlessness of one's desires.

In reference to Dr Meyn, I believe I did not confuse logical relations and causal relationships. I was not interested in going from there to limit phenomenology; I am limiting phenomenology from the point of mem-ory. My objections setting limits to the possibilities of phenomenology are those that memory has encountered, at least, if you can believe psychoanalysis' principal limits, and therefore, to use the words of Dr Półtawski, phenomenology can only work with the absolutely given, whereas psychoanalysis draws all the attention to things that are not absolutely given, which we first have to mediate before they become available for philosophical reflection. I did appreciate very much the way you tried to situate the problem.

Meyn: I am not sure that I see the relevance of memory as you want it – and it seems that to trace back your memory or my memory – would again bring in the causal aspect. Answering Dr Półtawski's remarks about the 'immediately given', I want to mention that Husserl, to my knowledge, did not talk about the 'immediately given'. He said in fact that the mistake that philosophers have made before, namely Berkeley et al., was to distinguish between the 'immediately given, and the object behind it and he felt this created undue difficulty. Secondly, Husserl didn't draw the distinction between anything 'absolutely given' and something 'merely given', he took something as the given and put it into brackets. That means he takes it as provisionally given and wants to analyze its structure. But he never, and this is what the effort of my earlier remarks were, he never can be taken as defending a pure foundation interpretation.

Matczak: I really think that phenomenology and psychoanalysis are two absolutely, diametrically opposed views. Psychoanalysis functions with certain naturalistic presuppositions which are basically those of the nineteenth century – I think they are untenable – wherein we refer to the object of psychological investigation as something which is objectifiable, which has characteristics that can be outlined, which can be seen in a causal-motivational sequence, which has tendencies, a line of development pre-supposing time and space, a world, and so on. Whereas it is exactly the aim of phenomenology in the Kantian tradition to base reality on something else, namely on a transcendental consciousness. And transcendental consciousness is not a possible object of any scientific investigation which would objectify it as it is wont to do in the natural sciences. I do not see how you reconcile a metaphysical naturalistic presupposition on the one hand, and a transcendental investigation, as started with Kant and then pursued further on by Husserl, on the other hand. I do not see that there is any possible compromise between the two. That does not mean, as you very rightly stated, that we have to deny the fact of psychoanalysis and make sense of it from a phenomenological point of view. That would be a new science worth to develop, but not with the premises which Erikson, Freud, and all the others employ, because we don't know what a transcendental term is.

NOTE

[1] Cf. G. Lautéri-Laure, 'Imaginaire et Psychiatrie', *Evolution Psychiatrique* I (1968), for an interesting study of imagination both from the psychiatric and phenomenological point of view. [The Editor].

ERLING ENG

CONSTITUTION AND INTENTIONALITY
IN PSYCHOSIS

I

In his 'First Philosophy' (1923–24) Husserl has a passage of relevance for our topic. In considering the theory of transcendental reduction, he has ascertained the essential contingency of our perception of the world. The contingency of our perception of the world is not however limited to the perception of particular events, persons, or things; it is contingency applying as well, he says, "to the universal structure of the ongoing process of perception".[1] The world appearing in the course of perception is one of anticipations and their correction through subsequent encounters. Thus the world "by virtue of this structure, is sealed with the twofold meaning of being and truth".[2] So that the presently perceived world always remains, in one sense, a 'mere world of appearance' of a world in itself, to which we can only continuously approximate. Only "as long as the perceptual life of the perceiver continues in this fashion, does he live consciously into a world continually there for him, standing bodily before his eyes as the one and identical world which, as a world out there, is not affected by the possible and frequent disappointments and corrections".[3] Correlatively, the continuation of the world out there for the perceiver depends on the continuation of this style of expectation and confirmation or disconfirmation. "The correlate of the true being of the world which I now perceive as sensuous phenomenon is the never ending and for all the future harmonious structure of my perceptual process".[4] But as soon as we grasp the correlation between this continuing style of perception and the existence of a world 'out there' as a necessary one, if there is to be a truly real world experienceable for us, that necessity itself becomes relative. The style or structure of perception itself is experienced as contingent, "a mere fact, which could also be otherwise". So that even as we continue to live within the horizon of a world in which experience confirms or disconfirms expectations, requiring us to alter our expectations while ensuring the continuation of a real bodily world 'out there',

Tymieniecka (ed.), Analecta Husserliana, Vol. III, 279–289. *All Rights Reserved.*
Copyright © 1974 by D. Reidel Publishing Company, Dordrecht-Holland.

"it is also possible to consider the eventuality that the world 'out there' could be a mere apparition and not as it always obviously is, i.e. the mere appearance of a true world implicitly within it as an idea". Such an appearance would be a 'transcendental appearance', to be distinguished from the naive, initial empirical appearance. Despite the continuing validity of the empirical appearance it is also necessary to recognize the dimension of transcendental appearance disclosed by the possibility of a change of this perceptual process, this style of perception mentioned earlier. Unlike the 'empirical appearance' the 'transcendental appearance' would not be subject to confirmation or disconfirmation. At this point, Husserl, to avoid confusion with Descartes, remarks that the 'transcendental appearance' is not given by doubting the reality of the world. He says: "the existence of the world is entirely without doubt, and this freedom from doubt inheres in the perception of the world within which we continually live".[5] He goes on: "but what is important for us is that this altogether empirical certainty, this empirical undoubtedness, nevertheless, as empirical, leaves open the possibility that the world is not after all, that this remains an evident possibility, although absolutely nothing indicates that it will be realized". The determination of this world as a pure nothing, a mere transcendental appearence, is not incompatible with our empirical, doubt-free perceptual certainty of the concrete presence of the world.

Once again however Husserl breaks in to say that we cannot consider this distinction between the world as empirical appearance and transcendental appearance as a legitimate acquisition, unless we are ready to meet this objection, that the possibility of the non-existence of the world merely means that at any moment it is possible for the harmonious stream of human perception to be transformed into an "absurd confusion, a phantasmagoria". His rejoinder to this is: "But what else does that imply, but that a human being, and finally every human being, can become mad? Still the possibility of madness can say nothing about the possibility of the non-existence of the world. On the contrary, it is just here that we must adhere to its own, absolutely necessary being. For doesn't the possibility of the madman already presuppose the existence of the world?"[6]

Thus the argument from madness, by which the distinction between the world as empirical appearance and as transcendental appearance has been questioned, fails to invalidate the world as at once empirical and transcendant. Of interest for us however is the implication that even the madman must struggle with this distinction, in however an idiosyncratic fashion. If this be the case – and I will attempt demonstrate this with clinical interview material – then I believe it possible to disclose the sense in which the possibility of psychosis is already inherent in the natural attitude, only to be variously constituted in a variety of contingent contexts.

II

Some ten years ago I first met A.O., a 25-year-old male patient hospitalized for psychosis.[7] From the time of his admission he attracted the interest of patients and staff because of his singular belief that he had a 'little girl' on the forefinger of his right hand. When by himself he often appeared to be conversing silently with her, and when with others would report on what she would say to him. She served so to speak as his 'hand maiden' in conversation with others. At that time I made extensive notes of what he told me in a series of weekly contacts over a period of almost six months.

The first time we talked together he showed me his right hand, saying:

A little girl; it's blond-headed. She says she loves me. She says she loves everybody. She's 17. Jane. She says she'll say anything I do. Everything I say she says. Everything she says I say. She tells me what to do some times, what I was doing. I listen to her and she talks to me.

It is the temporality of his exchanges with 'Jane' that is most striking. 'Jane' is his follower, saying what he does, saying what he says. And yet a possible simultaneous reversal of the relationship is suggested by the ambiguity of "Everything she says I say". The order of speakers here remains equivocal. This equivocality is homologous with the vacillation between future and past in what he next says: "She tells me what to do sometimes, what I was doing". There is a direct shift from future to past, without transition or connective. That position of which these are the protentive and retentive horizons is somehow missing here, and may also be related to the passivity of his final remark: "I listen to her and she talks to me".

On the Rorschach Test, administered subsequently, there was a preponderance of responses presenting dualities, fusions, and pure color, with repeated references to 'something between them'.

That this 'something between them', which we took to be missing in his relation with Jane – who increasingly took on the character of a reflection on the fingernail of his right forefinger – had something to do with his own vital intentionality, revealed itself in our very next meeting, when he announced: "Somebody robbed my body, robbed me of everything I feel".

With this introduction he and we are now ready to enter more intimately into the particular constitutions involved in his relation with 'Jane'. Moreover it was through this relationship that we communicated with one another – to the extent that we did – since I had had to recognize 'Jane' at the very outset in order to speak *with* him. In the following extended passage, the time after he had spoken of the theft of his body and feelings – and I will continue to observe a chronological order in presenting these extracts – he began by referring to two kinds of vision, throwing the remarkable duplicity of his vision into relief, this time in terms of 'inner' and 'outer':

If I had glasses on I couldn't see as good. If I had glasses on, it would bring it up closer, and I would just see my finger. I can see inside my finger now. She tells me what to get in the mess line. She says she loves him and me too. She says we have a hard time getting what we want. Have an easier time getting what I want. Some of them get two things when they're supposed to get one. Worry about which one to get. If she was inside my eyes I could see what to get. That would be my mind and memory. Little girls can see both ways. They can see out of my finger and out of my eyes. She says she can see through my eyes. But my mind and memory is on my finger. It goes through my finger to find out what and where to say. She says the same thing I do. She tells me what to say but I don't know what to say. She helps me in the mess hall, so I don't get somebody else's things, so I get the right things. My mind and memory is mixed up with my girl. My mind and memory goes to my finger when I think about it. And she's talking too. Janie's name's there, Pam's name too. My young girl, she's younger than Janie. One's 14, one's 17. She said that's what they went by.... Remember putting her in the finger, but I couldn't think of her because I had to stand at attention a lot. I was doing a job, I reckon. Had to wear gloves and I couldn't think of her after I put her in my finger. I remembered her in my mind. I couldn't think of her in the States because they put wire on my mind and memory. I couldn't think about home. I was thinking about my little girls and my mind and memory – inside my head. My little girl loves you all. She doesn't do anything but sleep and tell me what to do – in my mind and memory. They tell me in my mind and memory and in my finger. They have to work both ways so I can get things right in what I do. It makes me nervous all day long. I take care of them just like me. I take care of them because they won't be like me. She says, is that enough to talk about?

Here A.O. appears to be struggling to resolve the competition between members of paired percepts, as if each pair represented a disjunction effected for the sake of a conjunction that never succeeds. Such pairs may be termed *syzygetic*, from the Greek verb 'to yoke together'.[8] A.O. struggles to grasp 'Janie' as within him, but always fails to realize the sense in *noematically* he is within her *eidetically*. This is possibly implicit in his remark with metaphorical sexual connotations: "Remember putting her in the finger because I had to stand at attention a lot." This attempted sef-representation of his entrapment in his dislodged 'mind and memory' through employment of sexual metaphors will appear more dramatically subsequently as he comes to experience his self-constituting intentionality as imprisoned within the detached constitutions, even while they are still considered being allies.

'Janie' is his 'stand-in', out in a world where he would otherwise be helplessly exposed. She is a tiny invisible Galatea of his sought-for transcendental subjectivity on the very tip of his finger, on the boundary of his body as another object in the world, his nail as if a minute mirror retaining her image. Her refuge is his forefinger, bodily token of transcendant meaning, as we are so vividly reminded in the outstretched finger of God in Michelangelo's 'Creation' or in the raised finger of Leonardo's 'John the Baptist'.

At the same time 'Janie' divides into two, with a younger sister 'Pam', who is 14, twice 7-teen. The search in the world of the already constituted for the lost transcendental subjectivity progresses through a fission of each major percept. This is perhaps the most striking feature of psychotic constitution and intentionality, though occurring differently in the different forms of psychosis. It is significant that the fission of one girl into two occurs with a reference to 'Janie's' name, inasmuch as the name is most expressive of the transcendance of identity, compelling the emergence now of 'Pam', on a still more inferior *hyletic* level.

The figure of the little girl who 'loves you all', asleep in his finger, evokes the figure of the Sleeping Beauty. But that there is a reverse side to this bower where he holds the beloved is already suggested in his final words: "I take care of them because they won't be like me". The next time we meet he says:

My little girls are young. Everything they keep, we need it. [Repeats this] They're in a picture now – of my mind and memory. Right out in the open .... A boy pinches on my hand and thought he had my girl. He wanted to see them in his own finger. Janie said they were in every one of my fingers... she says this world is going around in her mind and memory now. She says she can't think of anything to say.

A.O. is implacably exposed to the world in and through the passivity represented in his 'little girl', of whom he now says: " A boy pinches on my hand and thought he had my girl. He wanted to see them in his own finger. 'Janie' said they were in every one of my fingers..., She says this world is going around in her mind and memory now." As in Hindu and Buddhist religious art, transcendental subjectivity is represented by a multiplication of the transcendental figure, or of a part of him, as in, for example, the dancing Shiva, or the manifold Buddhas. In the multiplication of 'Janie', in all the fingers of his hand, in response to the attempt of another patient to seize her by seizing A.O.'s finger, we can grasp – with A.O. – her character as 'transcendental appearance'.

At this point, two months after his admission to the hospital, with no marked improvement in his condition, shock treatment was initiated. After the first treatment or two the patient began to speak for the first time of his 'position': "Can't hardly get myself straightened out, I've got so many brothers and sisters. I'm trying to get my own position. Never did stand between them, I reckon." The theme of standing up among the older and younger brothers and sisters appears for the first time. With concern for the life history, other more advanced constitutions, for example, sexuality, emerge, although the *syzygetic* formations continue:

I've got [pointing] a handle on my hand and the boy don't like it. I've got Janie God on my hand. They'd better not bother me when I show them my ring finger. Their head is on my finger nail and the rest of them is who I am, inside of me... my little girl smokes a little cigarette while I smoke a big one.

Of his new treatment he said:

Shock treatment might kill my little girl; wouldn't kill her but it might hurt her.... My little girls want me to go home and get married. My little girls will help me find someone. My little girl needs some one to take care of 'em when I have 'em. I can't have 'em unless I go to a woman. I don't know of anyone, but my little girl might figure it out. I been in the Army and the hospital and I don't know any girls. My little girls will have to talk with their little girls; I love them but I can't get to them. My little girls said someone played in my position if my father died. The doctor would give me a little girl but I don't know his name... my little girl loves you, Dr Eng. They have to love people to get along all right. We have to love people to get along all right. My little girl said that.

A.O. struggles to re-establish self-constitution, to once more live in his own time. The imagery of parenthood, of fatherhood, motherhood, and sexuality are eloquent of A.O.'s sense of what is missing. He is employing a loose form of Husserl's 'method of imaginative variation', though without integral intentionality, in an effort to grasp those meanings which would free him from the never-ending uncertainty of the situation in which he finds himself.

When he next appeared, he crossed and recrossed his legs before sitting down. When I asked him why he did this, his reply was:

> My little girl wants me to do that. Somebody played in your position at home and my little girl told me to cross my left over my right so they couldn't get my position. So I can play who I am. War heroes played in my position, my little girl says. They just slept in my home, in the past. A lot of people played in my position, but I wasn't going into my position fast enough... that boy robbed my body, robbed one of my muscles. It went off like a hand grenade inside of me and I started crying. My little girl said they were trying to rob my body of a woman. They might have been using my woman. They took, when they robbed my body, for using my insides on my little girl. I felt it coming out of my legs, too. I felt it from the position my legs were. I felt they were trying to rob me of my body. I couldn't see anything but a yellow light and the top of the world.

A.O. crosses his legs to 'protect his position', one that is given and not actively assumed. He protects access to his passive situation as if a girl, as his subsequent words suggest. He also seems to be recalling his initial psychotic break, perhaps along with memories of shock treatments, past and present. He says that his little girl and/or himself have been raped. The crossing of his legs is also perhaps a reference to the problem of the one and the two, and how they emerge from one another, the *'syzygetic enigma'*.

The *syzygetic* constitution in psychosis represents intentionality unable to return to the ground of its own active moment, thus unable to complete the passive phase in which the Other is revealed as other, i.e., as *thematic* on the horizon of the world; therefore the impaired self-comprehension, and the loss of horizon in the psychotic world. The *syzygetic* formation soon exhausts its possibilities for further development, becoming circular with stereotypes and repetitions.[9] Nevertheless the language of A.O. is able to disclose to us the continuing presence of the transcendental subjectivity of consciousness and intentionality. His own account is, in the remarkable words of Novalis dealing with the relation between logic and pathologic, a "counterposed erroneous representation of error which yields truth"[10] which, in our context, discloses transcendental subjectivity.

It is now that A.O. begins to speak of the world for the first time. It is significant for the primacy of 'position' that the awareness of the 'world' follows his preoccupation with 'position' as he returns to himself'.[11]

'My ankle don't fit right. When I had a position, but they robbed me of it, and I had to be the world. My girl's the world. They're missing but I'm the world. The world's me, I have to be the world. What in the hell am I if I am not the world. The glasses came off my dead body. The glasses you put on my body. My girl says you put my body on me. I'll destroy the world. Ain't it a me. I'm the end of, the end of the world. But I can't end. If I'm the end of it, don't I have to be the world. My little girl likes to rock. Them rocks are my little girls. I've got to have a position that gives me a position from the floor. I've left my position at E. [home town]. I'm a world. If I weren't the world I wouldn't know everything. I'm the world. I gathered it up and made a pretty world. If I lose my position, isn't that where I'm sup-posed to be? I hand out a world when I say it twice. It sort of comes out of me. My leg's a devil. He's using my body. Nobody is. I'm the world. Little girl. She's a world.... To start a war I have to die. Because I'm going to stand there and smile. I'm a world. The world's where I'm at. My mother didn't want me in her own home. My smile is your father because I am the world. Ain't you the father of that guy sitting over there? My little girl is. You might bite my knee.... My little girl is me and I'm the world. I doctor the world when I doctor myself. It's the devil. I'm the world. Everybody played my positions. You're two worlds. That's why I don't end. The world's inside me. You ain't killing me. I'm the world. It ain't the world. It's me. I haven't my position, but it's the world. Ain't you that chair too? I won't end; they put me in the middle. Isn't that the world in the middle?

A.O. is puzzled by the 'two transcendants' of 'I' and 'world', as they face one another in reflective symmetry: "the end of the world. But I can't end. If I'm the end of it, don't I have to be the world." The compresence of 'I' and 'world' is understood as two mirrors confronting one another. These mirror-reversals occur at the *hyletic* level of simple sound-meaning association: "My little girl likes to rock. Them rocks are my little girls". He experiences his speech as world-creating, in its seeming *noetic-noema-tic* sufficiency: 'I hand out a world when I say it twice. It sort of comes out of me". But the reversal of subject and predicate, of body and organ, has its diabolical moment as well: "My leg's a devil. He's using my body". He struggles with the problem of one world and the two worlds: "You're two worlds. That's why I don't end. The world's inside me. You ain't killing me. I'm the world. It ain't the world. It's me." And then, finally, now addressing me, concerned with the possibility of our relation, he asks, "Isn't that the world in the middle?" It is as if he were beginning to awaken.

This impression is strengthened when, a few days later, he says:

It seems like somebody's robbed this place. Seemed like home before. Seemed like the floors were a little heavy; now they seem a little light. This place is about the same, in your office. I'm used to mountains around me. Head hurts a little. I lost my position and went to the hospital at D. This place seems too much out in the open for me. Too comfortable. Roads, hills down a little from the mountains.... I'm going to try to get a home of my own. I'll stay who I am.... I feel like I'm hollow in the head. But it has a filling in it – I don't know what kind of filling; I'd guess you'd call it a sore filling.

It is now that he discovers that 'Janie' is in some way involved with malign persecution. One day he says:

Hitler is who my little girl is, Hitler is my little girl. Adolf Hitler is a girl.... My little girl's the world. My little girl is Adolf Hitler's girl. A. Hitler uses her. She knows everything he asks her. I'm the father of A. Hitler.... I can't be anybody I want to be – because I'm supposed to have made everything.... A lot of people wanted some of my girls.... The little girls might worry themselves to death. They've been in there ever since I've been overseas.

The figure of 'Janie' is now sensed as standing between him and himself, so to speak, thus at the root of his helpless openness, in which he fails to realize the elaborative spiral of constitution and intentionality in the world as a differentiating monad. A.O. continues to remember, struggling to reconstruct what must have happened to his body to account for the changes in the hiatus between past and future horizons, now that his head has begun to have a 'filling':

They stabbed me in my sleep. Somebody was using my body.... Dr Eng, you're a good man; I need my body. My little girl's all right. She said she still loves you. Didn't I once tell you I was your father. If I'm my baby girl. I'm the best man in the world. Everything changes over with me around here. I'm a man of the world – a miracle. They made me a miracle when they screwed my woman and stabbed me. That's the first time I told anybody about screw. Somebody exploded the insides of my body.... I want to use my eyes and help myself. Cast my spell, I'm talkin' to the world. This world here. [Becomes very excited] I got the world mixed up with people. I ought to go back to school. My eyes see pretty now. My little girl's eyes see pretty, and she likes them.

And one the day following:

I didn't fall nowheres – I feel all right. You don't think I fell through a hole, do you? Is this a meteor planet? Where in hell did you come from? I don't know what I'm doing here... my little girl got her head took away. I made her another head. She ain't got no funeral because she's inside of me. And nobody's killing her. I'm from overhead, too. Overhead two [spells it]. Is this Mars? My little girl took my overhead and I got lost. I'm a United States boy.... I'm everybody, Dr Eng. You're everybody, too.

Following his rediscovery of 'world' and 'position', comes the discovery of his loss of 'heart':

My heart: it don't feel I like I got one. Feels like it's busted by shock treatment... my little girl can't talk. That's why I don't talk like my little girl.... I can read her and it hurts my heart. My little girl's God. They stole it away from me. My little girl said they changed my body over. They stole everything inside of me. My little girl takes over everything.... My heart's hurting. Where is this planet at?

As if, now that he has fallen and lost his 'Janie God', he is asking for his coordinates in the space of nature.

Now, even though he is no longer lost in the constitutions of memory and desire, no longer within the emptiness of an horizonless world, he discovers an enigmatic emptiness inside:

> I've got to see some full bodies. I put looks on them and try to help them from my eyes. The position they try to move in.... They've been working on the inside of me for so long I don't know if I've got anything inside of me to work on.... I just go by the outside.

He tries to tell me of this new difficulty encountered in realizing self-constitution from originary intentionality, his struggle to realize the living present (*lebendige Gegenwart*):

> My little girl said: "You can't put it back inside of me. It has to be built from the middle, and filled out." "My heart quit beating," my little girl said, "my heart quit beating, but don't mess up my body," that's what my little girl said.... I'm everybody's brother because I carry a handle. My little girl give me a handle. I've never had a woman, don't have a woman, and I'll never have a woman. My little girl carries a handle, just like mine.

He seeks to break through his sense of helpless passivity into its enabling ground of passive intentionality and synthesis through his own form of the 'method of imaginative variation'[12]. He has still to realize the meaning of 'Janie' for himself, to effect an *epoche*. This neither shock treatment nor his doctor was able to do for him:

> I'm sick. I've got no position on the floor of the hospital. I've got to go and get me a position. My little girl's got it.... My little girl's God. We've got to go back home for a while; we miss our position.... I lost all contact of my body in this world... me and my little girl's God... my little girl says I got nothing to do and they're embalming my body with shock treatments – with embalming fluid. My little girl can do anything. My little girl says you was the Devil.... Everything you put in me is out of me.

And again, several days later:

> Somebody's played in my position, talking, in this building, and I can't get who I am. They use my talkin'.... They played in my position here. Me and my little girl told them everything. I'm in somebody else's position here. Dr Eng, was I your father or was I another man? Somebody played my position all over this world.... I tried to hand out me and the world, but I've got to take up a position at home. I handed out the world, and I was supposed to be back up in the hills with the hills around me so I could have my position and know who I am. My little girl told that.

A.O. continues to locate himself in relation to his mother and to his father and to life, despite the continuing confusion:

> Your father give me my eyes, my little girl said. My little girl's my mother. Cause she takes

care of me. I feel all alone all the time.... God, could you make me a woman?... There's no living, except there are wars and things like that.

Three months later, when I saw him for the last time before he went home to the mountains, he had gone far toward the recovery of his mind and memory, his 'recovery of time lost', his own time. 'Janie' had begun to rejoin his intentionality at that root level in which is grounded the human possibility of assuming the constituted role of father, mother, or child, one or two or even all three of these, as for A.O. and perhaps for all of us alike, in the transcendental life of the originary intentionality (*Urintentionalität*). After a great deal of reminiscences about his relatives he said to me:

I was a little boy and couldn't look up in the air and see who they are.... I'm a little boy now. My glasses show what you could see as a little boy. My eyes shows looks on people's face.... I talk like a little boy.... The picture out of my mind and memory. Thinking of it. They can see everything I can't see. They're so tiny they can see everything in the world. My little girl Janie is.... They want to talk to you if you can hear 'em and see what they're saying. I took care of them so much I know what they're sayin'. They can talk to your mind and memory, if you hear them.

NOTES

[1] *Husserliana* **VIII**, p. 51.
[2] *Ibid.*, p. 51.
[3] *Ibid.*, p. 52.
[4] *Ibid.*, p. 52.
[5] *Ibid.*, p. 52.
[6] *Ibid.*, p. 55.
[7] For the present paper I do not consider his exact psychiatric diagnosis important, though as I remember it it was schizophrenia, schizo-affective type.
[8] Also for the way it is employed with ambivalent affective significance in Gnostic cosmogony.
[9] This pattern has remarkable parallels in the syzygetic formations of Gnostic cosmogonic myth, in which the entrapment of the descended soul in the lower world, represented as all but cut off from the distant Creator, is perhaps a central doctrine.
[10] Novalis, *Fragmente*, No. 209 (*Wasmith Ausgabe*).
[11] Reminding us of the primacy attributed by Erwin Straus to the opposition to gravity of the upright posture in the 'I-Allon relation'.
[12] Husserl, *Erfahrung und Urteil*, Claassen, Hamburg, 1969, 3rd edn., pp. 409ff.

JERZY ŚWIECIMSKI

SCIENTIFIC INFORMATION FUNCTION AND INGARDEN'S THEORY OF FORMS IN THE CONSTITUTION OF THE REAL WORLD

I. INTRODUCTION

The subject I intend to discuss may be approached through its three different aspects. First, there is something that has acquired the name of 'scientific information', as distinct from another thing bearing the name of 'scientific informative function'.

Second, there is Ingarden's theory of forms, specifically his formal ontology of the object as understood in the broadest sense of the word; things, processes, happenings in all spheres of beings, wherever their existence is theoretically possible or where they factually occur. This ontology, as has been pointed out before, constitutes a fragment of an endeavour, the aim of which is attempt once again to solve the central and everlasting argument between idealism and realism.[1] Originally, this work is meant as a direct continuation of that of Husserl. Nevertheless, in several points it surpasses Husserl's ideas, suggesting, or announcing new solutions.[2]

Third comes the problem of the *constitution of the real world* which is itself an ontological or metaphysical problem, depending upon the types of questions and statements it contains, while maintaining a certain connection with epistemological questions which, to a certain extent, determine it.

This compendium of apparently loose connections suggests a question concerning a possible interdependence which might occur between such matched elements. (1) One could ask whether these three groups of problems, when taken all together, will form a sensibly related whole. (2) One might also doubt whether one could expect in this setup the appearance of a strictly philosophical problem at all. One might think such interrelated questions are but a new branch of interscience, or a new normative methodology of scientific information improved by means of a philosophical apparatus, 'theoretically supported and enlarged'. It might also seem that such a combination of approaches amounts to

Tymieniecka (ed.), Analecta Husserliana, Vol. III, 290–322. *All Rights Reserved.*
Copyright © 1974 by D. Reidel Publishing Company, Dordrecht-Holland.

nothing but an application of Ingarden's ideas to the existing theory of information in the current meaning of the word. Consequently, one might ask whether or not we are in fact concerned with building a particular branch of knowledge dealing with facts, or with a practically applicable methodological system concerning chosen scientific conventions.

I submit that all of these suppositions are wrong. I do not mean to exclude the possibility of creating a new branch of 'interscience' on the basis of the existing theory of information and certain parts of Ingarden's system. On the contrary, it is quite possible that such a branch, when created, could be of practical value.[3] But these are not the problems that are going to be dealt with in the present paper.

The problems we are going to concentrate on in our present discussion are within the framework of a detailed ontology insofar as their subject matter is determined by the problems of scientific information. According to the subject matter assumptions, these problems will be analyzed in relation to certain parts of Roman Ingarden's philosophy, either as problems belonging to his existing work or as those directly arising from or designed by it.

II. SCIENTIFIC INFORMATION AS A PHILOSOPHICAL PROBLEM

The philosophical relations between the problems of scientific information and Ingarden's system of ontology are in fact very close. However, they come to light only when the very question of information is looked at from a certain point of view, that is:

(1) When we exclude all the research operations belonging to what we call 'theory of information', understood in the sense of an exact branch of knowledge that deals with facts or, alternatively, in the sense of a normative methodology – or even of a technology – which governs the existing informative processes assuring them *a priori* stated standards of correctness, we are bound to exclude even these research operations which, due to their character, inclinations, or program, would remind us of investigations carried out within the framework of the theory of information.

If elements of 'theory of information' were ever to appear in our argumentation as a subject of research, they would have to be treated as a field out of which some of the theoretically foreseen possibilities had become crystallized. Such an approach to this theory would lead to our investigating it from 'the outside', as any other subject, without any intention of developing this knowledge itself, by enlarging the number of statements it contains. Moreover, our investigations could never be carried out by means of the apparatus proper for the knowledge to be investigated, but merely by means of phenomenological methods of analysis.

Such an investigation from the 'outside' would be bound to have a distinct ontological flair. We would not attempt to investigate this branch of learning in a genetic way, reducing its shape to the result of influences of certain conditions (e.g. to historical or social formations, or to activities of certain people). This science would be an object of investigation in which, by means of an appropriate analysis, we would attempt to discover its essence, its specific character; an irreducible whole, as well as its inner structure (e.g. construction of interwoven theorems formulated within its frame).

In other words, a philosophical approach in the case of examining the theory of scientific information, understood as knowledge, would be similar to the research approach by which we analyze various kinds of 'theories of literature' (e.g. sciences related to sociology). By this examination, we do not create any monographic knowledge about culture (e.g. any historical or comparative-genetic analysis of scientific ideas, as about literature, or a science dealing with creators of culture, their lives, etc.) but are instead creating none other than the ontology of a subject, which in this case is 'theory of literature', by trying to discover its essence, structure, general type of content, etc.

(2) The philosophical relation between scientific information and Ingarden's system will arise, however, we view it as a phenomenon of *information*, and we will attempt to seize its essence, its specific formal structure, objective ground, mode of existence, value, etc.

In other words, this aspect will appear when we view scientific information, not through the prism of any exact knowledge ('theory of information' in the current meaning), but simply as something that appears, takes place, is extended in time, undergoes certain changes while performed, as something that, in its existence, is always specifically, and necessarily, determined as regards its form and content.[4]

These relations will become particularly distinct when we grasp this phenomenon, not as an isolated one, but as one interconnected with everything that exists and appears around it in the world of things, ideas, processes, happenings, physical objects, people, and, in particular, in 'the perceiving I' and in any other person to whom the contents of my experience are transmitted. We treat it as one more element of the reality, in which we, people, have our share, in which we live, in which we create, and which, in turn, influences us in one way or another.

With this approach the investigated object appears as something we wish to examine from the 'inner side', reaching its authenticity, its intrinsic character explained, as such, in its own features, and not historically, psychologically, or in any other indirect way as the 'result of something' only.

Here comes the third element of our subject matter, the problem of the *constitution of the real world*. The initial point is, on the one hand, the circumstance that scientific information forms a 'bridge', a link, and, simultaneously, a 'barrier' between objects of direct cognition and the autonomously existing 'outside' world; thus producing a specific 'picture' of this world which, on the other hand, contains a moment of 'creativity'. Strictly speaking, on the grounds of informative processes, new, very specific objects are 'created', having a formal structure of their own and a complex type of content. However, they are characterized by an intentional mode of existence and elements of physical reality sometimes 'interwoven' into their structures, with still more complex structures evidently resulting from this.

Due to the presence of 'creativity', the scientific information becomes, quite unexpectedly, one more element forming our world, our 'for us existing *milieu*'. How it happens that this 'for us existing *milieu*' becomes enriched – thanks to the informative-scientific processes – will be the subject matter of our discussion.

One must admit that scientific information was only of marginal interest to Ingarden; mainly because he found some of the solutions to problems connected with this field in the field of the ontology of intentional objects, and the theory of their cognition. Consequently, scientific information acquires, in Ingarden's treatment, the character of the ontology of an object.

Ingarden also deals fragmentarily with the problem of information in his treatise on language and the role it plays in scholarship. But here, as before, his argumentation is of an ontological character, and is the reason why we do not find the terms 'information' or, still more unlikely, 'theory of information' in any of his texts.

III. INGARDEN'S STATEMENTS IN ONTOLOGY OF SCIENTIFIC WORK AND THE PERSPECTIVES THEY OPEN INTO SPECIAL FIELDS OF PHENOMENOLOGICAL RESEARCH

Having reduced his theory of scientific information [5] to the ontology of scientific work and an inquiry into the function and structure of language, Ingarden adopted another set of limitations. In his surveys he concentrates exclusively on literary works and that sort of language which consists of words and sentences. This approach automatically excludes from the sphere of his interest all other kinds of informative media usually employed by sciences, in particular non-linguistic ones. For Ingarden, scientific information is nothing but a special kind of literature. This interpretation had symptomatic influence on the statements we can find in Ingarden's texts.

On the other hand, it should be stressed that all these limitations and restrictions had purely methodical origin and purpose. Ingarden never insisted that the problem of scientific communication should be limited essentially or normatively. However, some problems have been, in effect, left 'aside', unsolved, or were grasped by him only in very rough outlines. [6] The subject which emerged from all this was always, as a whole, clearly visible. It is beyond question that Ingarden had left 'gaps' within his work. These 'gaps', when noticed, may sometimes cause the impression of some 'incompleteness' in the whole, precisely with respect to what had been done in it, in relation to what was projected actually by his initial questions.

The highly selective approach that may be noted in Ingarden's works was never meant to be absolute and unchangeable. The possibility of its alteration and development was always taken into account by Ingarden himself and very often suggested in his texts, leaving room in the pattern of questions designed by Ingarden, possibly exposing some new, unexpected aspects.

Filling up these gaps may be one of the tasks which have to be undertaken by the post-Ingardenian generation in order to keep the system alive, not only in the passive, but also in a dynamic way.

With regard to the essence of scientific work, Ingarden formulates the following statements:

(1) Since the scientific work – in a traditional meaning of the term – is included in the class of verbal compositions, it automatically comprises all qualities essential to literary works, first of all those of their specific four-level stratified structure. While considered within the range of this class, scientific works may be distinguished, however, as those which constitute a specific kind of literature, as different from (i) literary works of art;[8] (ii) all kinds of non-scientific literary writings, e.g. journalistic reports, non-artistic popular fiction, etc.

(a) In every scientific literary work all affirmative sentences are always statements, i.e., they never possess the character of *quasi*-statements as it is the rule with literary works of art.[9]

(b) All statements contained in scientific works claim the right of truthfulness. This means that, in life, such statements may be proved as right or false, but even if they are false, they always originate from the intention of being scientifically true: they are 'designed' to be true.[10] Thus, the general notion of that truth which is to be found in a work of science must be constructed much more rigoristically than that in literary works of art.[11]

(c) In the structure of scientific work, similar to that in all other kinds of literary work, the stratum of *represented objects* and of *represented states of being* can be distinguished. These representations come into being as intentional correlatives of sentences. They become constituted while the work is perceived by the reader. Since most of the sentences in scientific works are statements, the content of the represented objects reaches far beyond the work, pointing at some entities transcendent to the work (e.g. ideal objects, real things, processes performed somewhere in time, events, etc.). Thus the meaning of the term 'the represented object' becomes, in regard to the function of scientific works, twofold: (i) it denotes the object belonging to the structure of the work which, in order to avoid confusion with the following type, we shall call 'the object factually represented' or 'the factual representation';[12] (ii) it denotes objects belonging to the 'external' world, those about which the work 'says' or which it 'depicts'. The objects represented factually are as a consequence never functionally autonomous, whereas analogical representations in the works of art are. While perceived they seem 'transparent' and, 'pointing at' something different from themselves, they disappear from the sight of the reader.

The factually represented objects may, of course, be perceived in their specific qualities, as complements of the work. This case, however, may occur in situations which, in reference to the normal function of the scientific-informative work, should not be considered as typical. It is possible, namely, when the factual representations are constituted on the ground of false statements and when their relations with the external objects, denoted by them, become confused or destroyed. It is possible too in the case when the representations (right or false) are analyzed as components of a particular scientific concept or 'idea' and when they are considered exactly as they are, apart from their representational function, quite autonomously or, if in that function, then in a deliberately performed verification of their value. In all these cases the specific character of the representations becomes visible; since we deliberately concentrate on them in particular, we compare them with something which is denoted or 'depicted' by them.

It is evident that this kind of examination demands a special analytical technique and must be grounded on a different attitude than is required in the normal perception of a scientific work. The work of science is here given by a direct contact, as with any other object of direct perception, in particular any other product of culture, we want to examine it according to its structure, function values, etc. When the degree of truthfulness of such a work is examined, the analysis needs a parallel examination of the work, its contents, and of the external objects it denotes; both sides always equally in direct perception. Without this additional analysis, all possible mistakes contained in the work would have been taken for granted as truths.

(d) Scientific works usually lack in artistic and aesthetically valuable qualities (*die ästhetisch relevante Eigenschaften*), i.e., these qualities may appear in such works, but their role is more or less neutral. It is possible too that the appearance of these qualities is harmful to the contents of the work and to its function, e.g. when it introduces some unnecessary aspects to the work itself, or to the objects depicted by it, under the form of particular aesthetic values. This aesthetic object (the intentional one) which in these cases comes into being as autonomous, distorts the uniformity of the work and creates some danger to the correctness of perception.[13]

(e) In scientific works the stratum of appearances (*Schicht von Ansichten*) is generally much poorer than that of any other kind of literary works, especially works of art. Consequently, the objects factually represented are of much more schematic character.

(2) Besides their informative function, scientific works fulfill specific 'creative' tasks, too. This originates from the situation that the language, on the basis of which the work is built, possesses the ability to constitute the objects of inquiry, particularly if these objects have been constituted earlier, to provide them with new form, e.g. to determine them in a specific way. Due to this ability, the range of objects discussed in particular sciences progressively develops.[14]

If this brief characterisation were considered as an isolated item, apart from other statements Ingarden had formulated within his system (especially as isolated from those in the field of detailed ontology)[15] its significance for special phenomenological research would probably have been slight. While considered in a broader context, however, it achieves quite a different rank and value.

It presents a supplement to the postulates formulated by Ingarden in regard to the principles of cognitive processes, their correctness, adequacy, and their value. It is generally known that some types of cognizance, in particular those characteristic of sciences originating from positivistic attitude, have found in Ingarden's opinion a negative proof.[16] Ingarden showed, (1) how this kind of cognizance (or, more generally, of cognitive attitude and cognitive program) falsifies the genuine qualities and the essence of reality as the essence is often missed in these sciences completely; (2) how, in consequence, this reality when 'seen' from an empiricist-positivist view, becomes crippled, deformed, or (3) how this view casts on the genuine shape of reality its own conceptual 'tinge'.[17] This criticism refers generally to direct forms of cognizance, in particular to the sensual and to the internal ones. In another of his surveys,[18] however, Ingarden carries out the problem of interference of language in the content of knowledge.

The problem which arises, in consequence, resembles to some extent one which we often encounter in the practice and theory of art[19] (*Kunstwerttheorie*), and of artistic attitudes, techniques, etc.; in particular, when we ask how different kinds of perceptive attitudes (e.g. the so-called 'artistic view') or conventions in reproducing real things influence (1) the picture of the real world as perceived in the work of art, or (2) the picture of the real world, as perceived directly, but under the impact (suggestions) of artistic representation once perceived, or of creative processes once performed in the perceiving person.[20]

Translating all of this into the sphere and categories of knowledge and of scientific information, we obtain automatically the problem of verification of the exactitude, truthfulness, correctness and value[21] of the particular kinds of means applied in representing facts or ideas, and in transmitting them to others.

The objects which, in consequence, become the centre of examinations are:

    (a) the form of the means by which scientific transmission of a given content is performed;

    (b) the content of the work and its intentional derivatives (i.e. possible concretions of the factual representations, grasped in their structure and qualities);

    (c) processes by the effect of which the work of information is built and achieves a definite form, 'refilled' with a definite type of content;

    (d) processes in which the concretions of the factually 'incoded' content are constituted.

A recapitulation of the whole problem which has been established here may be formulated as follows: Knowledge is developing as practice (i.e. detailed research on factual material), theories and methodology; as systems of norms and rules. Since it claims the right of being true (correct), its results, as well as its principles, need control and an unquestionable verification. It is obvious that if we do not want to commit the error *petitionis principii*, such controlling and such verification cannot be executed within the field of the given branch of knowledge nor by means of its own apparatus of research. In other words, the only way to achieve an absolutely true criterion which could enable us to express judgements in reference to the subject matter in question, is the acceptance of the principle of examining the whole material from the outside and by means of unquestionable criteria. Hence the role of phenomenological technique appears. As a consequence, a new field for detailed philosophical inquiry is exposed.

The items we have announced above connect us with another group of general problems; first of all, those of the conditions under which intersubjective identity of the content perceived in various acts may be achieved.

In the essay concerning the principles of cognizance, Ingarden asks how it is possible that people understand each other, in particular, how it is possible that the content of 'my own' state of mind can be understood, if at all, by the other person, when all that we are able to transmit factually are 'external signals' and not the authentic content itself.[22]

The relation of this question to that of scientific information comes to the fore at the exact moment when we start to examine the theoretical limits of the abilities of transmission in particular types of informative media and when we ask how, and under what conditions, the intersubjective understanding of the 'incoded' contents is generally possible. In regard to the structure of the works as conveyors of content, one can ask what should be fulfilled within the form of the 'bearer' so that understanding could theoretically be possible. It is, abstracting from different abilities of perception, changing in particular people who perceive such content[23] with respect to what should be fulfilled in the form of the 'bearer' of contents, that the possible concretions which might have been constituted on the basis of the work, could be identical or sufficiently similar.[24]

Not entering too far into detail, one can assert that the processes of cognizance, while performed in the indirect way through the scientific work, are extremely diversified in their type and depend (1) on the type of content, in particular on its structure and the material qualities it presents; (2) on the type of work which presents a sort of 'bridge' – and a barrier too – between the perceiver and the objects about which the work 'says', or which it depicts or 'represents' in the current meaning of the word. Due to the polyphase character of the process by which final object of cognition is reached, three different types of 'scientific pictures of the world' must be distinguished: (1) The one presented by the results of direct perception with respect to the scientific 'creative' act – or both acts interwoven together – and prepared for 'translating' into the contents of the work of information, but not yet transformed. If the process of perception of the work is performed correctly, this 'picture' can be reconstructed to some extent and sometimes nearly completely. Its adequate restoration by the perceiving person is, however, never possible.

In practice, this initial 'picture' (the idea of the author) is often identified with correct concretion of the work. (2) The picture contained in the work; in other words, what is obtained when the primary picture becomes 'incoded' and assumes the form characteristic to the type of the conveyor. It is identical with the objects represented factually. (3) The one identical with the concretions – all possible ones, true or false – which may be constituted on the basis of the given work and its contents hidden under the given form.[25]

In regard to these items Ingarden formulates some proposals, with respect to several types of knowledge, especially for the humanities. Ingarden has intended them, however, primarily as principles of phenomenological descriptive analyses. Let us summarize them briefly:

Putting forward the postulates of maximal adequacy of the picture of reality obtainable in the acts of perception Ingarden, on the one hand, demands the extension of the limits of the traditional notion of experience. This demand confirms the general tendencies in the phenomenological trends as a whole. On the other hand, the detailed program of inquiry adds new elements to what has been known, and generally accepted. Ingarden's statements refer to both kinds of cognizance, direct and indirect. Writing about the language applied in literature and science, he postulates that the picture of reality obtained from scientific works should be maximally similar to that obtained in direct perception. Thus, everyday current or literary language, with its flexibility, large range of nuances, and high ability of evoking intuitive images of the factually described objects, 'the lively picture of reality', is regarded by him to be the proper tool which would guarantee the realization of this program. The representative function of the language (*die Darstellungsfunktion der Sprache*), its capacity of 'putting before the eyes' the genuine and nearly tangible shape of things and events is strongly stressed in these enunciations as an important factor.

The principles of adequacy, postulated in reference to some selected fields of scientific transmission or to selected types of knowledge, demand revision when applied to a broader field than that intended by Ingarden himself. The elements for this revision can be found among statements formulated by Ingarden on various occasions. Let us recall, then, what Ingarden says about the structure of a literary work and about the work of scientific literature.

(1) The objects and happenings factually represented in a literary work, because of their schematic character – each of them containing areas of incomplete qualification, *die Unbestimmtheitsstellen* – require, from the side of the reader, an appropriate 'concretion'. This concretion consists in 'refilling' those not-quite-determined 'empty spaces' with a content of determined qualities. In literary works, works of art, this re-filling is executed on the ground of imaginary-intuitive acts, which are in fact creative acts, and leads in consequence to the establishment of intentional objects, constituted 'at the side' of the reader. In other words, the work of art is here functioning as an inspiration for 'creative'[26] activity of the reader, who on the ground of perception 'creates' his own 'private' and unique imaginary world in which only one single person, the perceiver, may 'live'.

(2) In scientific literature the content must be sufficiently similar, if not absolutely identical to all readers. It means that the work should be constructed so that no more than one single type of concretion of the objects factually represented in it could be possible. In other words, there is no room here for such liberty in concretizing the content as with literary works of art.

Ingarden's proposals of introducing colloquial, everyday language into scientific works undoubtedly extends the possibilities of concretizing schematic images transmitted by the content. In some types of knowledge this extension is – as Ingarden insists, and practised himself as a writer – a necessity. In some others, it should be considered only as useful, however, not absolutely necessary. It is useful because it enriches both the possible limits of transmission of the contents (more types of data can be transmitted thanks to it) and the concretions, which become more 'lively'. In consequence, as the contact between the author and the reader becomes more close and 'personal', the work is assimilated more easily. It is not absolutely necessary, because even if the language of the given work were traditionally schematic, the very nucleus of the contents, which is bound to be transmitted, would not have become, in spite of that, essentially deformed. The contact between the author and the reader would be not so close as in the first case, but when nothing more is demanded from the book than a transmission, in an essential and economic way, of some compendium of information, all facilities necessary (e.g. for popular scientific literature) can be postponed.

Though we could agree that in some cases Ingarden's proposals regarding the extension of the limits of language may introduce positive moments into scientific transmission, we are always allowed to ask whether the same extension would not create some danger in other cases. We would ask, for instance, whether it is admissible to introduce literary expressions into descriptive natural sciences (e.g. to morphology, or to behaviouristic psychology). After all, we cannot overlook the very symptomatic fact that the evolution which the language in sciences has undergone over the last hundred years, has not only not shown a tendency toward the enrichment of form but, to the contrary, toward a maximal schematization and complete elimination of emotional-expressive and pictorially-impressive media. The texts which, in regard to their linguistic form, to their style, lexical richness, etc., have been fully accepted (more to say: have been qualified as good) some dozens of years ago (e.g. *Souvenirs Entomologiques* by Jean-Henri Fabre), in our days became quite unacceptable; just because of their too rich, too 'personal' form.[27] In other words, recent tendencies we observe in contemporary sciences seem to have some objective and reasonable grounds. Sciences are oriented in a stabilized direction and cannot be simply reduced to some 'temporary fashions', or to the 'lack of taste' in modern writers and readers.

We cannot deny, of course, that the schematization of form in scientific works leads to impoverishment of the factually represented world. In extreme cases, scientific descriptions only point at definite objects, or indicate, rather than depict, them. The stratum of factually represented objects becomes, in these cases, radically schematic, so that their concretion by completing them with additional qualities is practically impossible.[28] In the cases where the represented objects must be, for some reasons, provided with more 'pictorial' qualities (e.g. those of shape and colour), the degree of schematization is very strong too, so that a characteristic class of conceptual objects is established.[29] All these objects are functioning rather as 'models' after which particular real things can be recognized and classified as members of definite systematic categories (e.g. species, genus, etc.), than as pictures of the external shape of reality, seized in its genuine and adequate character.

The completeness of representation and of their concretions seems here more than unnecessary, if not distinctly controversial, to the essence of this type of transmission, and to the standards of its correctness.

It seems, then, that our doubts about the possibility of application of some of the more flexible and rich media of description of scientific transmissions in natural history are fully confirmed by the neutral attitude of Ingarden himself.

Since the situation in natural sciences seems to be definitively stabilized and the orientation in the development of their verbal media of information has objective reasons, one may ask to what extent (if at all) the non-linguistic media, which have always been applied in these sciences,[30] may guarantee more authenticity in reproducing the image of reality – if, of course, such authenticity is needed in these sciences.

The formulating of this question seems to be reasonable; for it is very probable that non-linguistic media of information present themselves much more as 'being duplicates' of verbal descriptions and that their application, instead of parallelling descriptions, is not based on economic or psychological reasons only.[31] If these suppositions were really true, if the structure of non-linguistic media of information showed that the transmission of contents really has limits larger than those in traditional description (in particular, that non-linguistic media are 'specialized' in transmitting some sorts of data being untransmittable for any kind of verbal construction), then, a perspective of new standards for the information in natural sciences would be opened. In the analysis of the structure and of the capabilities of non-linguistic media of scientific information, Ingarden's technique of investigation was applied. First of all, a set of initial questions was formulated.

(a) What is the relation between the objects represented by verbal and by visual media in regard to their form and type of contents?

(b) What is the relation of both these classes of objects to the world indicated by the scientific work (considered as a whole), in particular to the real world and to the data of direct experience?

(c) Since every work of scientific information contains some percentage of creative moment, what part is played in this creation by visual media of transmission? Is this creation one-sided, performed only by the language, or is it possibly two-sided? And what are the 'competencies' of each side?

(d) In which way and on the basis of what principles (if at all) do the visual means of communication guarantee the identity of transmitted content, while it is perceived in particular acts of perception?

(e) Is it possible that the 'visual language' is not only a means of information but may also be considered a 'tool of better observation', in other words, as a means of cognizance (in particular when the work based on illustration is in its *status nascendi*)?

All these questions become 'starting points' for establishing an extraordinarily interesting domain of inquiry.

From one side, a structure of special products of knowledge comes to sight. From the other, we obtain a new way in which the given material of objects may be systematized and examined. These objects are:

(A) Scientific visual representations of things, processes, and happenings, belonging to the autonomically existing world, in particular to the real world. It seems nearly sure that they must be essentially different from those known in art;

(B) Their intentional derivatives (concretions) grasped according to the type of their specific contents;

(C) The works of information based on the principle of visual representation of all possible kinds. It is possible that in this category of objects we shall find the types analogous to those we know from art; it is also possible, however, that we shall here find some types of works unknown in art, characteristic of scientific information only.[32]

Since the research in this field is but in a preparatory phase (only small fragments of results obtained have been published until now), the whole problem can be presented in very schematic outlines. Summarized in a brief report, it can be formulated as follows.

IV. THE RESULTS AND PROPOSALS

(A) Between the objects represented by linguistic and by visual means of information, different kinds of one-sided, mutual relations may be recorded. Correspondingly, the relations between both kinds of representation and the objects indicated by them – in particular those of the real world – are not uniform, but present a large scale of diversified links.

(1) The simplest possibility that may be spoken of occurs when the objects represented by verbal language and by visual means denote ideal

objects (of *aprioric* or empirical origin), and when visual language does not refer to the real world at all (in *aprioric* knowledge, e.g. in mathematics) or, if it refers, the relation is always indirect, executed through the verbal representations. In other words, it is the case when visual representations are nothing but 'translation' of the objects represented verbally. The content of both kinds of representation is, in this case, always identical.[33]

(2) The case which is one step higher in complexity is that by which 'visual translations' of verbally represented objects are constituted, not only as pure constructions, but as 'mixed' ones by additional 'borrowing' of elements (e.g. qualities) originating directly from the real world. As a consequence, the objects which are factually represented generally correspond to the intentional correlatives of the statements 'visualized' by them; but this corresponding does not mean qualitative similarity. To the contrary, it is admissible that 'visual translations' become 'externally' different from the verbal representations. In the extreme cases even the ontic class, to which they belong as objects, is changed (e.g. when happenings or processes are represented in a static form and when the mutual configuration of these forms symbolizes the speed or the sequence of phases in a process).

The representations belonging to this class (2) demand, in order to be understood correctly as visual analogies of the verbal representations, a special perceptive process in which most of additionally 'adopted' (or 'borrowed') elements are postponed. The perceiver must pass through several steps of 'deciphering' of such representations, until their intrinsic sense, hidden behind the shapes which illustrate or symbolize them, is reached.[34]

(3) The next case comprises the situations of the objects denoted by the content of the statements (and correspondingly: those indicated by visual representations) as belonging to the real world. This case presents an extremely diversified complex of situations, among which five general variants may be distinguished:

(a) When visual representations are based on direct perception of an individual thing (or of an individual complex of things), the content of the statements which are formulated about such things and, correlatively, the verbal representations, are in this case parallel to the visual ones. There is no influence of the statements on the content of visually represented things. Both sides are mutually independent. The contact between these parallel functioning sides does not occur until the intentional correlatives (at the verbal and visual side) are constituted;

(b) When visual representations of real objects originate, not only from the data of direct observation of 'nature' (real things, processes), but also from the statements and from direct perception of reality. Besides, in this case, both the statements and their visual representations concentrate on the same individual, real object (or on the same individual complex of real objects, considered as a whole).

The influence of the statements and, correlatively, of the verbal representations on the visual representations – which in this case takes place – causes contact between both sides: the verbal and the visual emerge in the stratum of factually represented things, i.e., not only in the sphere of intentionally constituted concretions. As a consequence, the objects represented visually are characterized by a more or less distinctly expressed 'abstract' character. They are deformed by the contents of the statements.[35]

(c) When visual representations are established on a larger objective ground they concentrate on more than one single object rather than the verbal ones which, in some cases, may be established on the basis of one individual real object.

Independent of that, the interrelations between the statements and the visual representation may have the character of an absolute independence, as in case (a), or of an influence, as in case (b), which gives, in consequence, two possible sub-classes of this case.

All visual representations in the case are of synthetic and sometimes, additionally (sub-classes analogical to case (b)), heterogeneous character. The complexity of their structure increases when the real 'models' are given not in a single appearance, but in many different appearances, and when the data of all these appearances become united into one 'average', or 'summarized' appearance.[36] Another type of complexity emerges when, besides real 'models', some abstract qualities of shape and colour (independent of those 'translated' from the statements) are adopted.[37] For the establishment of visual representations, as a result, the whole list of possible combinations to be recorded here becomes very long and very diversified.

(d) When verbal representations are grounded on an objective basis broader than that of the visual ones which, in extreme cases, may be developed by observation of a single object given in a single optic appearance.

Similar to the previous cases, both sides may here remain completely independent, parallel, or one-sided, and an influence may take place causing some deformations in the visual representation.

This case is especially characteristic of the morphological sciences, when the representation of one selected specimen symbolizes the whole class to which this specimen belongs.[38]

(e) When the representations on both sides are of a synthetic character, as in the previously referred cases, one may here record a large number of possible sub-classes. These sub-classes emerge: (1) according to the type of influence by the verbal side of the visual one; (2) according to the type of objective basis on each of both sides (i.e. how many individuals are taken into consideration, what is their ontic category, qualities, mode of being, etc.).

This case is one of the most frequently encountered, especially in morphological sciences, and presents a large number of variants. It would be difficult to discuss them here all in detail.

(B) All of the above-mentioned types of relations occur between the represented objects, their concretions, and the elements of the 'external' world (the ideal or the real one). These relations may assume different shapes and produce different types of structures in the representations on both sides, the verbal and the visual. This occurs according to the mode of existence in time, by which the objects denoted by the work are characterized (or, more generally, according to the aspect of time in regard to which particular denoted classes of being are distinguished).

In other words, the situation develops differently when the objective ground of the statements and of the visual representations comprises timeless objects, objects existing in time, or both classes of objects. Let us especcially concentrate on the case when the objects indicated by the work of scientific information belong to the category of objects existing in time or, when being factually ideal, refer to some definite moments or periods of real time. Here, full blown variants are possible.

(1) When both types of representations (verbal and visual) refer to the objects actually existing (a paleontologist or an archaeologist would say rather: to the recent ones, what would not have been quite precise).

(2) When both types of representations refer to past objects or to past phases of existence (and consequently to no-more-actual sets of qualifications) of objects actually given (of the 'recent' ones). In this case, however, the situation may develop differently according to what kind of 'models' are applied to the representations. It is possible, then, that the 'models', on the basis of which visual representations are made, are recent. It is possible, too, that nothing of recently existing objects (qualities, moments of appearances) becomes included in the representations nor is used as a 'model'. These two sub-variants may be combined, of course, and give a third possibility that in the representation both recent and past elements coexist and both indicate something in the past (past objects or past phases of existence of something recent).

In all these cases the verbal and the visual representations may remain completely separate in the sense that no influences or any other kind of interrelations are observed. It is also possible that visual representations are under the influence of the verbal ones, as a consequence of which they assume a more or less distinct conceptual character. This influence, as will be shown later, has an especially great importance in the establishment of visual representations of past objects or of past qualities of recent objects.

Since the representations of the past are of especial importance for several sciences (paleontology, archaeology, and several humanities, e.g. history), we shall limit our discussion to this case only.

V. THE RECONSTRUCTION OF THE PAST

The general sense and significance of this problem refers to the theory of scientific restoration of the past. In relation to the items discussed in Ingarden's ontology, it might be considered as a contribution to his theory of individual objects determined by the time of their existence,[37] in particular things, happenings and processes. It comprises precisely the situation (a) when past objects, or past phases of recent objects, are given indirectly (i.e. have to be deduced from the data given in actually occurring facts), and (b) when the result of such cognitive process is additionally given in an indirect way; when we can only reach it by perception of a definitive work of information. The indirectness here is, therefore, twofold.

The principle on which scientific restorations of the past are built consists in the 'creation' of specific 'hyposthases' of the past. These 'hyposthases' substitute for the authentic past objects or no-more-actual qualities, phases of development, etc., in some recent objects. In reference to the structure of the work in which they are given, they are identical with the objects as factually represented or with their intentional concretions.

In the representations of the past, the way of representing may be grounded on (1) illusory representations (e.g. executed in an illustrative way, as drawings, paintings, etc.), or (2) presentation of some real things, which, similar to the elements of a stage-decoration, 'play' the role of 'themselves from the past' or the role of some other past things they are bound to represent or symbolize. Both kinds of representations may be combined, of course, giving, in effect, works of heteronomic structure.

This double-sided way of representation may be additionally complicated by the situation that among real objects 'playing the role' of the past ones, 'original past objects' (i.e. the objects originating from the definite past time, indicated by the work from, say, the cretacious period), may be included and serve, together with the 'fakes', as illustratory functors.

The situation, seen from the side of the perceiver, becomes, therefore, unique; since, if one wants to reach the denoted object and grasp it in its authentic qualities, one must postpone a large number (sometimes, nearly all) of qualities in the factually exhibited or illusorily represented things. Also, one must re-formulate them, 'forgetting' about the actual phase into which they have evolved. In other words, the whole factual material (real, or represented in illustrations) serves here as the basis for building up intentional constructions which, though 'only imaginary', are supposed to be identical[40] in their qualities with the objects they denote.

There is a vast list of variants that can be foreseen here as possible. Most of these variants are encountered in scientific practice. To figure it out, let us only consider that the objects denoted by factual restorations may belong to different ontic classes, that they may be things existing in time, general objects 'projected' to some definite periods of time (e.g. particular species of extinct organisms, types of extinct cultures, etc.), individual processes and types of processes, happenings and groups of happenings considered as a whole[41] (as 'quasi-objects'), etc. Besides, all of these objects may be restored in different ways and to a different degree. Thus, the areas of undetermined qualities, so characteristic of any representation, here become automatically of double origin and have differentiated significance. The concretions, which in all other cases may be complete, are here essentially lacking in some qualities (namely which are scientifically indeterminable) and can never be completed to the end.[42]

All types of the above visual representations are recognizable as subordinated to science and scientific information due: to the type of content they 'bear' as well to the specific qualities of their external shape; to their 'form', as designers would probably say. However, the content differs in particular works according to the character of scientific disciplines (in particular to scientific trends, factual material, types of selection applied in various sciences). All variants we can record in this field exhibit some common feature, founded on the principles of scientific cognizance and the transmission of results.

First is the concentration of permanent and stable qualities of the denoted objects. This principle remains valid even if the examined objects are changeable and cannot be perceived in any other way than in the stream of their permanent changes (e.g. in motion, in the process of constant transformations of its form and qualities). It is also valid when it is not the object as the basis of processes, but the process itself (e.g. the changes of something: in particular the speed, rhythm, etc.), which becomes the centre of examination. Second is the maximal reduction of momentary features, especially those connected with appearances (*Schicht von Ansichten*). This principle has important significance in the establishment of styles of scientific visual representations, e.g. of the types of realism. These factors are, perhaps, the most important ones in creating an essential barrier between the representations subordinated to sciences and those belonging to art.

However, scientific works based on representations are, in respect to their content and form, essentially different from those of art. This does not mean that they must be automatically deprived of the creative factor. It is obvious, of course, that the role of this factor, and the extension of the field in which it can come into sight, must be different than in the case of art. Let us report that briefly.

One must distinguish between two different kinds of 'creative' elements in the works of scientific information. The first group originates from the 'creative' moments in scientific research itself, in particular, in the cognizance of real things as well as in research based totally on 'creation', e.g. in mathematical sciences. These elements are introduced into the work of information as a complement to the material, from which the factual contents of the work becomes established. We shall not be interested in this group now. The second group of 'creative' elements emerges in the works of information due to specific features of visual transmission (considered as a type of process) to the character of images of things while reproduced visually (e.g. as illustrations, paintings, etc.), or to the external shape of the work, considered as a composition (to its 'form', as designers would probably say). The relations of these three groups of 'creative' elements to the processes of research (in particular: scientific research[43]) may be different. Some of these elements have nothing to do with science at all. Their role merely consists in determining stylistic features of the work, or of its technique. Another group, however, may be classified as scientifically significant, since the creative elements which belong there are manifestations of this type of research which is not possible in any other situation but in the phase, when the work of information is becoming actually established (in *statu nascendi*), and, in particular, when the process of its constituting becomes simultaneously that of scientific cognizance (executed by means of reproducing things, e.g. in drawings). There is a third possibility of relations as well. It comes into being when the creative elements are introduced into the work as non-scientific interpretations of the real objects denoted by the work, e.g. as interpretations of their expressive characters. These elements may have some influence on stylistic features of the work, but this interrelation is not necessary, though it does sometimes happen in practice. In respect to the function of the works, the elements of this class must be considered as additional and marginal. Their importance becomes essential only when the works of information become of a popular character and are addressed to special groups of people (e.g. in museum exhibitions), in other words, when they are intended as compositions from the border of science and art.

The presence of 'creative' elements in the works of scientific information may cause these works to be considered as parallel to those of art. This parallel quality is sometimes stressed by the opportunity that many classes of these works (especially in the field of museum exhibition) show properties, which make them externally similar to the works of art.[44]

In some cases a strong influence of various artistic trends can be recorded in the process of their establishment, especially in the establishment of their 'form'.

We have learned from Ingarden's texts that scientific literature (in the sense of a particular class of works) is essentially indifferent to aesthetic and artistic qualities, and that these qualities, when appearing in these works, must be considered additional, marginal, if not sometimes harmful. The same principle is generally valid for visual means of scientific information too. One small correction should be made in respect to this field; that aesthetic and artistic qualities are unnecessary only in 'purely scientific' works, addressed to specialists. On the contrary, they become of great importance in the works which, beside their principal role of 'bearers' of scientific contents, simultaneously play the role of 'objects of culture' and are addressed to a broader audience, as in the case in museum exhibition. Each type of such works has, as a consequence, its own set of characteristic aesthetic qualities and artistic values, and is characterized by a unique kind of expression. When we examine these works from this aspect, we can easily notice that, though all non-scientific moments we record in them are of subordinate significance (they are never exposed as *Leitmotiv*, in correct works, at least), they can never be eliminated completely, thus making the works aesthetically 'neutral'.[45]

The existence of non-scientific moments in the works of scientific information evokes some important questions: one may ask to what extent these moments (most of which are of intuitive-imaginary origin) are acceptable in the works and where lie the limits of their acceptability. More precisely stated: (a) which of these elements should be considered as useful (or even as necessary) complements of composition (e.g. as those, which guarantee the uniformity of the work); (b) which are only tolerable, though not evidently harmful, to the content of the work; and (c) which – and when – get into conflict with scientific ideas contended in the work, with its function, or with the truth that should be transferred by the work, etc.

Although these questions seem to have some importance for completing the whole problematic of scientific information, and of its works, it is probably too early to give them any definite answer. It seems though, that no uniform answer should be expected in this regard, since diverse types of situations, presented by particular classes of informative works, especially by their diversified content and function, make the problem extremely complex.

According to our suppositions, the works which belong to the class of visual means of information (or to the 'visual language' as it is sometimes called), are aften applied as instruments of scientific research. In this case, the process of cognizance has a direct character and is always oriented toward real objects. Its realization resembles, to some extent, the process in which the so-called 'academical drawings' are made. The specification of this process consists in the situation (or is grounded in it) that 'following the shapes of Nature with a pencil' simultaneously modifies the process of 'ordinary' perception, executed by means of 'simple' observation. The influence of drawing occurs in the moments when the drawing is executed, as well as a long time thereafter, when the perception of the object is no longer supported by the drawing, and develops in a traditional way. It is obvious that, in the final content of such perception, the role of the 'creative' elements, introduced by the designer into the drawing (or having emerged in the drawing in an 'organic' way), must be always significant. After all, the perception which is 'reinforced' by the parallel process of drawing – being always creative to some extent – becomes more active and 'creatively organized'. In arts, the introduction of 'creative' moments has always the character of purely 'personal' and individual imperative, 'I see it so'. The effect of this on the work may be accepted or objected to by others, but no reasonable discussion is possible there. In scientific research, creative moments introduced into cognitive processes may have something of that intuitive character too; however, for the most part, they are subordinated to the program of cognizance. When the process of drawing is applied as a tool of 'more effective observation', the attention of the draftsman is concentrated on tracing morphological elements of the 'models', while their expressive or decorative qualities are regularly postponed. If, then, the process of perceiving becomes modified by the drawing, it is possible only so that the content of perception becomes more uniform, is better defined and understood. Most of the 'conceptual' qualities of the objects of scientific inquiry – especially in descriptive empirical sciences, e.g. in biology – originate exactly from this moment.

VI. SUMMARY

If, on the basis of what has been said or announced as a research program for the future, we attempted to draw some conclusions, it seems very likely they would be grouped on several different levels:

(1) In reference to the research itself: as has already been mentioned, the application of Ingarden's analytical system (mainly his question system) to visual means of scientific information, leads, in fact, to an enlargement of Ingarden's original frame of 'detailed (regional) ontology', i.e., ontology dealing with certain groups of objects of a particular mode of existence, structure and quality allowance.

(2) In reference to the subject of investigations and their aim: investigations carried out on works applied in scientific information, in particular on works of visual transfer, throw light mainly upon the structure of the world represented in these works. This analysis enables an insight into the relations occurring between the external forms of transfer (its specific way) and kinds of subject matters contained in certain kinds of works. Had the problem been referred to the arts, one could have treated the represented works as autonomous and finished-in-themselves, leading no further. Its resulting from anything given beforehand would also be of no consequence to us. Since the matter concerns knowledge (science in particular), questions of no importance or even senseless when investigating the works of art here gain essential significance. Since the actually represented world always functions as an 'indicator' of something else and always has to be treated: first, in relation to what is its cause and 'archetype' (the real world, concepts of the real world scientifically 'created', etc.), and second, in relation to what becomes 'created' out of it in perceptive process (true or false concretions), one can always enquire whether, or to what extent (or degree) the cognition of these actual presentations supplements the cognition of reality carried out by other means (e.g. by means of direct experiment) independently. Do these works, being nominally nothing but conveyers of information (which suggests their secondary character in relation to direct cognition), imply what is real, but what 'escapes' ordinary observation?

We have stated in our argumentation that the world represented in the works of scientific information is deformed when compared with reality. We have stated, too, that these deformations result from eliminating certain features of the real world in the transfer and from adding 'conceptual' features, specific to knowledge and to the presented image of the world. The elimination of realistic features occurring in transfer – both verbal and visual – has a double cause: (1) certain qualities, or moments of appearances, are purposely not mentioned, which is due to their not being within the frame of scientific cognition program; (2) certain features are dropped as being either visually, or verbally untransferable.

It should be noticed, however, that the means of visual transfer (at least some of them, e.g. those based partially on presentation of 'authentic' fragments of reality, as in museum exhibition) include moments which, although they are not a part of science (at least, not in their present, modern character), go very deep into the essence of reality and its metaphysical qualities. They mainly touch these qualities which are associated with the essence of duration and the passage of time, in the metaphysical sense of the present and of the past. These metaphysical qualities or moments can never be perceived, as such, except in informative works (or by means of these works) which is due to their occurrence in context with other qualities (belonging merely to these works' structures). So the presence of consensual qualities in the work enables the perception of the above-mentioned metaphysical qualities. In such a case the mutual interaction of various structural elements of the scientific informative work occurs, to a certain extent, bringing to mind mutual influences of elements in a work of art, which result in a new specification of the work qualities, aesthetic and metaphysical. Similar to a work of art, where the appearance of these new specific qualities cannot be reduced to summing up the elements, the works of scientific information are not the results of simple agglomeration of the individual 'pieces' they are constructed from.

Consequently, the investigations sketched above are in close connection with Ingarden's 'aesthetics' or, at least, draw from it their inspirations.

It should be emphasized here that (1) it is Ingarden's concept of aesthetics we are concerned with at the moment; that here it principally means an ontology and theory of cognition of aesthetic objects, and not of those aesthetic approaches which are, for example, created in accordance with the exact-sciences' viewpoint as a branch of sociology, nor as a branch of psychology applied to the works of art and emotional experiences connected with them. (2) The inquiry here outlined cannot become aesthetic even in Ingarden's sense, as we are not concerned with the works of art but merely with those from the border of the arts or related to them by some formal characteristic. Experiencing these objects cannot have affinities with the qualities of aesthetic experience. Should this experience go further than is foreseen by the function of the work, it focuses rather upon the level of temporal genesis of the world than upon its aesthetic relevance.

Recapitulating, one can state that, although nominally knowledge alone is assumed to investigate legitimately the world under the categories of intersubjectivity and intersubjectively valid truth, in this case some means of scientific information becomes a parallel of equal rank as cognitive instrument.

As we have already mentioned, knowledge and the works by means of which its objects are conveyed together constitute a specific 'milieu'. Our research and its 'creative' and practical activities all take place in this milieu. It also influences our attitudes' framing 'our view of the world'.

If we accept Ingarden's scope of the term 'world' and when, according to his suggestions, we include within this scope not only the sphere of the physical, ideal or psychical, but also that of intentional beings (i.e. those often described from the exact sciences' viewpoint as 'objects of culture' or 'products of culture'), our cognition of the works of scientific information will make a contribution to the cognition of this intentional sphere.

Thanks to the 'creative' function of both knowledge and scientific works, this world progressively expands. Making a very distant parallel to cosmic phenomena, it is a similar expansion of the world to that occurring in the physical sphere. As cosmic space continually expands, so expands the scope of the world, being a product of human spirit but, at the same time, anchored in the physical reality. This expansion is directed according to the cognitive programs and according to the nature of the works, which, created as results of these programs, become inspiration for new cognitive and creative actions, for new 'beginnings'.

The direction of the expansion of the intentional objects' sphere is reducible neither to the sum of activity of economic factors – as some would like to see it – nor to psychology, as others would be ready to declare. It reminds one of a 'chain reaction' the centre of which lies in the nature of objects which at the very moment of their constitution successively determine their derivatives, not reducible to ordinary outgrowths, but constituting successive 'layers' in configuration of the sphere of the world.

[The English language was revised by Robert Greenberg.]

NOTES

[1] According to the distinctions introduced by R. Ingarden in the first volume of *The Controversy About the Existence of the World*, one should speak rather about an argument between different 'idealisms' (which form no uniform group and are often mutually contradictory) and different 'realisms'.

[2] As is known, Ingarden's enormous output, in relation to the program designed by initiatory questions, must be considered fragmentary; metaphysics remained in this system as a project, determined only according to its specification and range. According to the relations of Ingarden's solutions to those of Husserl, see 'A Letter to E. Husserl', published in Polish translation in *Z badań nad filozofią współczesną*, Warszawa 1963.

[3] Cf. 'Ingarden's Philosophy as a Tool', *Studia filozoficzne fenomenologia* **20** (1971). The perspectives discussed refer mostly to scientific illustration and to museum exhibition as a *sui generis* work of scientific information. Another survey dealing with these problems was presented at the *VIIth International Congress of Aesthetics* in Bucharest.

In one of my dissertations I have given a perspective of Ingarden's philosophy, pointing out how, despite its theoretical character, or rather, thanks to its theoretical character, it may be applicable in some exact sciences and in certain branches of creative work on the border of science, arts and pedagogics. I have also discussed a rough scheme of these branches and their problematics; if ever they were to be realised.

[4] However, we are still in the initial phase of our discussion. It can be taken as certain that the problem which at this point comes to light must consist of a minimum of three different members:

(1) We shall have to grasp here the formal structure of a process probably belonging to a specific class.

(2) We shall have to give an answer according to the type and structure of objects on the basis of which the processes of scientific information are, or can be, performed. It is very likely that there is a great number of such types. It is very likely, too, that there is a strict correspondence between the contents transmitted and the means of transmitting.

(3) We shall have to determine the structure and the mode of existence of all those objects which are constituted on the basis of informative processes, e.g. which are found in objects transmitting definite contents (meaning or other kinds of data), or which are constructed later, in the final phase of perception, as concretions of the transmitted contents.

[5] As has already been mentioned, Ingarden never used this term; however, he factually deals with the problems of scientific information.

[6] The way in which Ingarden formulated his enunciations was often deliberately intended to provoke the reader to further investigation. Similar suggestions have often been ex-

pressed by Ingarden in discussions or private talks with his disciplines: e.g. the field of special research in scientific information, fragments of which I refer to at the end of this essay (problems of scientific restoration, theory of the work of museum exhibition, etc.) resulted from such discussions. In this regard the problems I intend to present here may be considered as a continuation of those founded by Ingarden, as their 'further concretion', so to say.

[7] It is always very symptomatic how particular chapters in Ingarden's works are ended. 'Open questions' we often find there stress the necessity of further examination of the referred problems, or even of their revision. Addressing these questions toward the reader suggests from where the answers of a discussion have been expected.

[8] As we shall see later, not only literary but all kinds of works of art must be distinguished essentially from the works of scientific information. Here, Ingarden's statement made in regard to one type of works of art is factually of general meaning.

[9] This statement has evoked discussion from some philosophers in the Warsaw Centre (Pelc).

[10] Different variants of 'provisorical truth' are often encountered in empiric sciences. Besides, it has not been said that some 'hybridic types of works, joining the principles of scientific and artistic truth' are not possible. In fact, we often find such works in older scientific literature.

[11] Cf. R. Ingarden, *Das literarische Kunstwerk* §52, Tübingen 1965, and *O poznawaniu dziela literackiego (On the Cognizance of the Literary Work of Art)*, Warszawa 1957, chapter: 'O tak zwanej 'prawdzie' w literaturze' ('On the So-Called 'Truth' in Literature').

[12] One must loyally admit that in Ingarden's texts this distinction is not explicitly expressed. For this reason we had to introduce in our discussion that in a popular attitude of scientists-practitioners the term 'the represented object' is always understood in the second meaning, i.e. as something about which the work says, and never as something belonging to the structure of the work. In the case of visual representations, the confusion between factually represented objects (in Ingarden's original terminology, the represented objects: *dargestellte Gegenständlichkeiten*) and the denoted ones (ideal or real objects 'visible through' the factual representations) is even greater. Ingarden does not introduce this distinction for the reason that, concentrating on works of art, he is not especially interested in the relation which emerges between the represented objects and their real 'models' in 'nature'. The problem of representations in scientific works must, of course, be grasped in the first line in the aspect of this relation. Thus there are the necessary alterations in the present text.

[13] This statement, which is indisputably true in reference to modern scientific literary works, seems doubtful (or less rigorous, at least) when applied to the works of visual information, e.g. to drawings, paintings applied as scientific illustrations in books and in museum exhibitions. However, scientific illustration cannot be provided with aesthetic qualities or be subordinated to artistic program to the same extent as typical works of art are; some restrictions must here be followed. Due to these restrictions, scientific illustration (apart from the technique in which it is executed) may be distinguished from all kinds of non-scientific plastic works, in particular, from those of art.

[14] In the popular view of many scientists-practitioners, the object of scientific research is often identified with something real. Even if the truth presented in particular works is only 'provisorial' or has evident conceptual character, the 'creative' role of science (in particular of its language) is overlooked. Only when some concepts become the object of verification, they become 'less transparent', but in this situation they are considered always as something possibly false.

[15] Ontology in which objects of special ontic classes (formal, existential or material ones) are examined. It may be confusing, however, that many of such typically ontological problems are 'screened' in Ingarden's texts under the label of 'aesthetics'. The confusion disappears, if we consider that, for Ingarden, aesthetics is, for the most part, identical with ontology or epistemology of works of art. In other words it is never meant in terms of some social science or of special psychology. (See conclusions to this essay.)

[16] In one of his surveys (*Studia z Estetyki III*, Warszawa 1970) Ingarden opposes various up-to-date 'objectivistic' trends in recent humanities. These tendencies, originating from sources foreign to the fields in which they have been definitively applied, are realized mostly as rules of scientific cognitive methods. They are generally characterized by application of mathematical (in particular, statistical) methodology. The futility of applying these methods to the humanities (e.g. to the theory of art) consists, in Ingarden's opinion, in the fact that these methods are helpless in analysing the qualities belonging to the essence of the works of art, and select features of secondary importance. Thus an illusion of a scientific discovery appears. However, the results obtained by application of these methods are undoubtedly true; they are, in regard to the object of examination (precisely: to the factual material or to the material of things) deprived of real scientific value.

[17] To some extent Ingarden here refers to the general opinion in phenomenological philosophy.

[18] Texts which are actually prepared for printing. They will be published in the second volume of the studies on the principles in epistemology.

[19] It is clear that the situation here described, however seemingly similar, is in fact much more complicated than that which we encounter in the works of art; especially in those which represent things, people, happenings, which 'tell us a story'.

In a work of art, the perception of the factually represented world does not need any verification from the real world. If we sometimes do such verification, we introduce into the perception some 'foreign' moments, precisely speaking, we do not perceive the work of art according to its essence and aesthetic function. Even if such works represent things or people who might have their 'models' in reality, and if these representations are 'labelled' with names identical to some real beings (e.g. 'Napoleon', 'my home', etc.), they never claim the right of being true (i.e. of being identical in most significant features with analogical real things or people). The autonomy of such representations is so strong that, even if we know they are not true (e.g. they have been 'interpreted' and deformed by the author for some reasons) but, on the other hand, are provided with qualities which draw our attention, we are often inclined to accept them so, as they are. Sometimes we ascribe to them some other categories of 'truthfulness' that seem to us satisfactory or significant (see, different categories of truth, discussed by Ingarden in *Das literarische Kunstwerk*, §52).

In a scientific work such a degree of autonomy of the objects factually represented is, of course, not possible.

[20] In this case, the following variants are possible:

(1) When the real world is perceived by the artist and when the act of perception develops simultaneously with the process of drawing the perceived object, the contents of perception, while constituted gradually, become gradually modified by the simultaneously emerging forms of the drawing. Mutual relations between the emerging work and the perception of the 'model' change in proportion to the phase of development of both sides.

(2) The real world is perceived by the artist after his work is finished. In this case we can talk about a 'summarized' influence of all components of the creative process which have once been performed on the actual contents of the act of perception directed toward the 'nature'. The situation may develop differently, however, if the object perceived was at the

same time the exact 'model' of the drawing, or if the act of perception is directed toward some other objects and is only 'reinforced' by the experience gained in the process of drawing. In both cases, however, a specific 'artistic view' of the 'nature' given directly in the act of perception comes into being.

(3) The real object (or a group of objects) is perceived by a non-artist. The perception is actualized, however, under the influence of some visual representations of the 'model' actually perceived, or of some other things (each case gives rise to appropriate variants!).

Although the examples above referred to are typical for art and artistic attitudes, they can be quoted here, since analogical processes which take place when both perception of real objects and processes of reproducing them in drawings, paintings, etc. are extremely similar. The main difference consists only in the fact that the role of aesthetically valuable qualities in scientific drawing and in the perception of objects, is here minimal and the selection of qualities perceived and drawn is subject to different principles.

²¹ These four items are not homonymic: in particular, a scientific content can be true, however, without being exact in relation to the authentic qualities of the object described. An exact transmission – supposing that one is practically possible – may have little value in the scientific sense because of too many unnecessary details. Correctness of a scientific transmission generally means accordance with the standards or rules accepted for the given type of work transmitting informations, but it must not mean that the content which is transmitted correctly is at the same time true and exact.

²² In the article concerning the problems of understanding the states of mind of the others (published in the volume *U podstaw teorii poznania*, Warszawa, 1971, pp. 406–426) Ingarden presents a revue of particular theories attempting to clarify the problem. He expresses the thesis that the cognizance of the states of mind of other individuals is possible in direct perception.

²³ This question must not be confused with a psychological one, which might have been posed here as well. The difference is both in the method of analysis and in the validity of the answer obtained in both cases. Discussing the question on the level of ontology, nothing but pure theoretical possibilities are taken into consideration. Automatically, the result gives us only the limits within which possible facts could take place.

²⁴ Psychological analysis as a method of testing and experimenting must have been applied. The results would be based on concrete material but, on the other hand, they would always be fragmentary.

²⁵ The structure of contents in works of scientific information is discussed in detail in reference to one selected type of museum exhibition in a science-museum. See, Jerzy Świecimski, *Zawartość treściowa dioram przyrodniczych i warunki jej czytelności* (*The Content of Natural-History Dioramas and the Conditions of its Legibility*) and *Diorama jako przedstawienie natury* (*Natural-History Dioramas as the Representation of Nature*), *Przegląd Zoologiczny* (*Zoological Review*), issue of Wrocław University, 1962 (in Polish, with English summary).

²⁶ Throughout this volume runs a thread of concern to establish a specific approach towards the phenomenon of creativity. Since in this paper the term 'creative' is used in the current, uncritical sense, we will put it in quotes. [The Editor]

²⁷ These texts would be considered 'serious' in our times as well but only if they were discussed as documents of an early phase of development of the given branch of science, or as a product of somebody's personal, scientific activity (that of J.-H. Fabre). In this view, however, they would not have been examined according to their intended function, i.e. as means of information about some natural-history facts. Only in one single case would the contents of such works normally be discussed: if they contained facts no longer appearing

in our actuality and if the author were the only witness who had recorded and described them.

[28] The most extreme cases are characteristic of the recent trends in taxonomy, where descriptions of taxonomic units (species, etc.) are reduced to reports formulated in mathematical symbols. Since the understanding of such texts is normally actualized in the presence of the material of real objects denoted by the content, the concretizing of factual representations (very 'abstract' ones in this case) is practically not necessary. After all, it is usually not intended by the authors.

[29] How much such descriptions are schematized one may prove by reading morphological descriptions in zoology or botany. However, we shall find there expressions which mean different 'visual' qualities of things; e.g. 'white', 'red', etc., or 'rounded', 'square', or 'plump', 'slender', etc., are in fact nothing but conventional 'signals' denoting classes of qualities into which a given quality found in the object should be numbered. They are often very far from what is given in direct perception of individual things.

[30] These media are really very frequently applied, first of all as scientific illustration in literature and in museum exhibition. The popularity of non-linguistic media should not discourage us in formulating our question. If we consider that, in most cases, these media are applied uncritically and, as a consequence, lots of mistakes and instances of nonsense result, one must really think about building some theoretical basis for practical activity in this field.

[31] It is well known that some people assimilate scientific content better when it is explained by drawings or models, etc. than by pure verbal description. This facility does not seem, however, to be the only argument for applying non-linguistic media of information. Analogically it would be insufficient argumentation if anybody attempted to persuade us that the only reason for the application of drawings is the economy of print; instead of writing long descriptions one can explain the same thing with a simple sketch. I do not object, of course, that both kinds of factors play an important role in the development of non-linguistic media; they really are very economic and save much time for the reader. I insist only that, however useful they are, the essential reason of their application consists in their specific capability of transferring scientific contents, especially the data of direct perception of real objects.

[32] Comparative and genetic systematics of these works (limited to architectonic ones) is already made; cf. Jerzy Świecimski, *Forma i Styl muzeów naukowych* (*Form and Style in Science Museums*), Teka Architektury i Urbanistyki Ossolineum, PAN, Kraków, 1970 (in Polish, with English summary).

[33] This is but a purely theoretical possibility. I do not state whether this case, in its pure form, is really attainable.

[34] In extreme cases this 'nucleus' is not represented at all and the whole work is built on the principle of suggestions and analogies. It is especially characteristic when processes are the object of information.

[35] The reverse situation, i.e. when visual representations are the active side and influence the verbal ones is, of course, theoretically possible. The deformation must have occurred, in this case, in the statum of concretions. Practically, it is rather unlikely that such interrelation could occur.

[36] This summarizing of data is especially characteristic of morphological representations, in which all objects are characterized by stable qualities alone. If in such cases one talks about particular qualities of colour or shape, one must understand by these terms not the changeable data of direct perception (nothing that belongs to the appearance of the denoted object) but 'abstract' qualities in the sense of general classes. It is sometimes hard to

identify the meanings of these classes (however they seem, at the first glance, easy to understand, or to 'visualize') with what is really perceived in nature.

[37] These originate merely from the aesthetic concept of the composition, e.g. when the designer chooses a solution close to some 'abstract' style in graphics, and when he introduces stylization of shapes and colours according to the principles of composition, characteristic for the given style.

[38] This relation has close analogies in art, e.g. in primitive realism where factually individual figures or representations of things are supposed to be understood as symbols of groups of figures or things in reality. Morphological realism is different from all kinds of primitive symbolistic realisms, because of the typically scientific way in which the data for representation are selected and transformed.

[39] When we talk about the past, we mean: past things, processes, happenings, or past phases of existence of objects still existing in our actuality. On this principle, museum dioramas are built.

[40] They are only *supposed* to be identical, since there is no possibility to prove whether they are identical or not. This moment throws some light on the concept of truth in scientific restorations, as well as that of the basis for intersubjective agreement in respect to the statements in this field. From my own observations (e.g. from discussions with paleontologists) I could draw a conclusion that some percentage of statements which are formulated there are based on nothing but intuition and beliefs – especially when these statements refer to restorations completed to the last details.

[41] I mean here groups of happenings which do not comprise processes, but are separated one from another. Such 'loose' happenings may be summarized and, while treated as a whole, intentionally transformed into a specific kind of 'quasi-objects', in other words, into 'objects' which lack the formal structure characteristic of all objects *sensu stricto*. They present a class of artificially built constructions.

[42] One may, of course, complete them, but if we really do that, we trespass the border of science and enter the region of imaginary creation. However, in practice, many restorations are completed in this way (those by Jay Matterness, Zdeněk Burian, Charles Knight, and many others). What is especially curious is that these restorations are generally accepted as correct (or tolerable, at least) and are published in various textbooks.

[43] Academic artists would probably talk about 'artistic research'. Whether it is really research, or to what extent this activity may be called 'research', is the question. In some types of scientific illustrations these two types of research coexist.

[44] *Cf.* the systematics of museum exhibition I have proposed in the essay on their form and style (*Forma i Styl Muzeów Naukowych*, Teka Architektury i Urbanistyki, Ossolineum, PAN, Kraków, 1970).

[45] This is according to some doctrines in contemporary museology.

DISCUSSION

Tymieniecka: These last two papers I believe present very challenging and constructive points to what we are after. Dr Eng was presenting first an illustration of how the *life-world* of a patient is constituted at two different stages of constitution as if the patient had two different *life-worlds*. But his main effort, it seems to me, has been to show that in both of these stages there is a permanent reference to a fundamental constituted world. Furthermore, it seems that this fundamental constituted world is a constant point of reference for the specific *life-worlds*. (Here Professor Mohanty's thoughts and comments are very much in focus.) Secondly, Dr Eng's most provocative thesis is that this distorted *life-world* is not the distortion of the fundamental world, but the possibility of this distortion has been foreseen in the fundamental constituted world understood as *Urphänomenon*. This thesis is in radical opposition to the analysis made by Henri Ey of how the world of the psychotic patient, not only his *life-world*, but the fundamental world, is in psychosis distorted to the point that, as I see it, it reveals the limitations of the constitutive factor at large. According to Dr Ey he has no more foothold in his fundamental world. Lastly, Dr Eng made the persistent point through his entire analysis that the distortions of the patient's 'first' *life-world* were referring to the posture of the body, and if he was, after shock treatments, coming back to a 'normal' *life-world*, it was with reference to straightening his posture. By the way, Pradines in his *Psychology* stresses that the upright posture is the basic posture of man. Obviously the role of the body in the constitution of the *life-world* is essential. However, the fact that the *life-worlds* refer to posture as a fundamental body phenomenon is not yet an indication that they were referring to the *Urphänomenon* of the body as already predelineating the possibilities of *life-worlds*. Although it is convincing that the *life-world* necessarily refers to the constitution of the body as to our functional *a priori*; does not the constitution of the body along the postural line belong already to the limitation of the constitution as such? Does the disintegration of the

Tymieniecka (ed.), Analecta Husserliana, Vol. III, 323–331. *All Rights Reserved.*
Copyright © 1974 by D. Reidel Publishing Company, Dordrecht-Holland.

life-world occurring in psychosis not show, firstly, that the *life-world* refers to the basic *Urphänomenon* of constitution, but, secondly, indicate its limits by revealing that there are conscious activities preceeding the constitutive action? Dr Świecimski has presented a specific type of 'constitution'; in fact he has shown that the reconstruction of a form falls between the constitution of the passive genesis of the world and the structurizing function of creativity. The reconstruction of forms which are not present in reality, which are not purely invented, so they are not works of art, are 'retrieved from the past'. They are neither present in the actual real world nor in memory. This restoration of forms in scientific procedures takes off from the present forms which seem to indicate what other types of beings must have existed in the past, that is from the present we do not project to the future but retroject into the past. This is a new, so to speak, function of intentional constitution which lies inbetween the creative and the constitutive acts, and it is particularly interesting with respect to the temporal vector of experience. These restored forms of e.g., dinosaurs, are not works of creation, because the artist invents new forms, forms which have never been constituted, whereas the restored forms are assumed to have been constituted already. But there are elements of invention in this special intentionality, since the hints given by present reality are insufficient to reconstruct them fully; creative imagination is necessary to effect this reconstruction. And yet they are not purely invented, to be added simply to the whole setup of human world for its enrichment, but as having already been there, though missing in the present. The rational reconstruction of the world is called upon to complete the process.

Eng: I found these remarks helpful to me and helpful for the understanding of Dr Świecimski's presentation. I would say what is central in Husserlian phenomenology for me is how it attempts to grasp 'transformation' of experience. This transformation poses the problem of the relationship between what we may live and what we may think (the thinkableness and the livableness) of transformation is the central continuing open problem of phenomenology. I am not interested in the etiology of psychosis – your remarks were not meant to suggest that – I would say that you tend to be quite interested in the primordial. The emphasis of your question suggests an original logical interest that I don't have. I feel that Merleau-Ponty was concerned with origins in a way that I do not share. For me phenomenology has literally much more to do with the *love* of the word, than with the *logos* of the word, with philology rather than with archeology, that is, the *logos* of the beginnings. Coming to the issue of transformation I agree with Goethe's views on experience. In fact conscious experiencing is, in my view, 'systolic' and 'dyastolic'. For example, in the function of the heart, there is the dyastole – the heart opens up, and then it contracts – and there is a vital pulsation; in respiration there is 'inspiration' and 'expiration'. Experiencing is of this sort, and perhaps the most encompassing transformation in our experience is that whereby accumulated memories disclose guiding figures which now have an imperative force and which give directions to our life. These configurations have a mythic quality. (By 'mythic', I am referring to the transformation – from *logos* back into *mythos*: historically we believe that the Greek philosophy emerged from the mythic spirit of early Greece, but there is in the progress of the culture a transformation of *logos* back into *mythos* with its guiding force.) The upright posture comes quite late in our functional organization: we are already breathing, we have our eyes open, innumerable functional mechanisms are operating before we stand up.

Continuing my thoughts on experience I can relate the systolic/dyastolic movement to Husserl's notions of activity and passivity. In inhaling, you might say, there is a 'passive' moment; when I exhale there is an 'active' moment, or I can reverse the emphasis: actually each is in both phases, each is both. But movements of breathing in and breathing out, of waking and sleeping, movements of the heart opening up and contracting – these are all rhythms. *Rhythmos* originally means in Greek *Gestalt*. It is significant that *rhythmos* literally means 'a static unit of flowing' thus, to begin with, it unites the two: position and movement. We will try to use this as a key: movement and position are given in *rhythmos* and in order to remain upright I have to maintain a constant *tonus*, neurophysiologically there is a constant *tonus*; in other words, in order to remain upright there is a constant movement. It is an active stance, remaining upright, and as soon as we get tired one part of our body droops down and begins to pull us toward the ground. Thus the upright posture exemplifies *rhythmos*.

To proceed to your other point I believe I am in agreement with Ey in my understanding of what is occurring in psychopathology, but I do not think I explained myself well enough. I see existence 'musically'. It is a very complex structure of intentionalities, and intentionality is dynamic and rhythmic. In the sphere of experience it is a vital pulsation and only in the reflective sphere is it that it becomes a network of strictly logical correlations. In the original rhythmic relation activity and passivity are all the time changing poles and the human belief is a very complex pattern of rhythms, like a very rich piece of music; a piece of music is built up in time but when we listen to a fugue of Bach we are hearing all the previous things that have been said in that fugue in any one moment of time, and there is a future and a past time horizon; it can only be developed in time. Human life is like that: a temporal *Gestalt* with a retentive and a protentive horizon, and these are always changing. In the mentally sick person this temporal *Gestalt* has broken down; the master intentionality has suffered partial fragmentation and now the patient suffers from this fragmentation: he is going back and forth, somehow, to try to find the more encompassing intention. Yet, he can only circle rather helplessly in this fragmented intentional scheme. Now don't ask me how these larger rhythms are recovered. I present this only as 'a poet of nature'. I do not want to get into the question of primordial origins. I think Merleau-Ponty does that too much whereas Husserl finally gave this question up as an *Irrweg*.

Tymieniecka: You have given a considerable emphasis in your lecture to the 'posture-attitude'. Would you agree with Pradines' view – as he propounds it in his psychology – that the specifically human type of being has emerged from the evolutionary process only when the organic being has acquired the upright posture. Pradines is attributing to the upright posture of man – as the culmination of the functional organization of all beings, viz., being erect, lifting his head and having his brain in an upward position – that he can also develop all the higher functions he displays. Would you care to comment on that?

Eng: To begin to explicate the upright posture we have to say that the horizon is its inseparable correlative. When I am lying down I lose the horizon, I may say, and yet: do I not have the experience of a horizon as long as I am awake? Don't I stand up only to better grasp the horizon? And with the mention of 'horizon' (and its Greek sense of 'definition') other matters emerge for consideration: like a *temporal* horizon and a *meaning* horizon. Here I would like to clarify my remarks with a digression.

Let us remember that the horizon is a 360° circle; and that there is always a segment of horizon that remains behind us. This might be one way of grounding the possibility of the unconscious phenomenologically. In no matter what direction I am facing there is always a portion of the horizon lost to view, a kind of virtual horizon, and it is precisely this constant virtuality which supplies the possibility of the unconscious. My point is that the horizon is a 360° circumference, embrace, and that the man in my paper wanted the mountains ringing him all around, he wanted a full 360° horizon, which, if you stop to think of it, is an impossible project. I have the horizon most clearly when I stand up. Should I sink or fall, I lose the clearly defined horizon. I lose the vertical dimension of human existence. It is the vertical dimension that is the dimension of transformation, relative to the horizontal dimension of everyday experience. But we do realize transformation in the wake of an accumulation of experiences in the horizontal dimension. (Don Quixote comes to my mind here.) These experiences form a mass which is leavened, and can undergo a transformation out of which guiding figures now emerge. The *epoché* is also an invitation for 'guiding figures' to emerge from the mass of past experience and become relevant for decision and action. The evidence for this is in Husserl, in spite of his reticence on this score, since these are matters that require figurative language. Husserl speaks several times of *'die Mütter'*, referring to 'the mothers' of *Faust* II (*Ideen* III, p. 30; letter to Ingarden, 8/31/23). These are extremely shadowy figures, creative potentialities. I think of Marcel's 'creative receptivity' in this context. As Husserl puts it in his letter to Ingarden: *"Der Gang zu den 'Müttern' ist der Gang zurück u. vorwarts zu einem 'ursprünglichen' Leben..."*. Husserl invokes another mythic figure elsewhere in his remark that: "One must have the courage, even where there are 'threats' of regress, to say what one sees, and allow the evidence to speak for itself. The Medusas are dangerous only to the one who believes in them to begin with and who is afraid of them. There might be riddles remaining here for the time being, but they are just riddles, unsolvable riddles are an absurdity" (*Ha.* VIII, p. 442). The Medusa, you will remember, froze with her glance the one who looked at her, changing him into stone. Perseus could behead her only by viewing her in a mirror and struck backward over his shoulder with his sword. The mirror is, of course, the phenomenological instrument *par excellence*. Efforts to accomplish the *epoché* involve encounter with the 'Medusas', which represent paralyses of will and action. As for Husserl, the theme of the enigma or riddle (*Rätsel*) is very important in Freud's texts, and often lost in the English translations where it is rendered as 'problem' instead of 'riddle'. But 'riddle', unlike 'problem', suggests a surprise, either liberation or a reward – as in the fairy-tales – as a consequence of its solution. It is the key to a transformation. That, incidentally, is for me the essential sense of Heidegger's *Kehre*, or the experience of the middle years of life described by Jung, following on his own experience when everything was turned around in its meaning. A liberation from past, and seemingly imposed, definitions of self and others. This is where a new paradigm may emerge. Is not Husserl concerned with the paradigmatic forms of modern science in relation to our present lives? All of what I have said here goes back to the opening remarks on the upright posture, the relation of verticality to transformation, and the way in which the phenomenological *epoché* is an invitation for transformations in our experiencing to occur.

Matczak: I wish to direct a question to Dr Świecimski. Throughout your lecture you have emphatically propounded the theory that we come to know the external world through experience. But nowhere have you clearly formulated your understanding of experience. What is experience? In my view experience is a subjective event. How do I experience this chair, for instance? I experience myself, and if I experience the chair I experience it only in me by having pictured it in my mind. And here is now my question: is this picture *in* my mind or *outside* of my mind, and if it is in my mind, then I know really and concretely only myself, other things I know only analogically. This much was made clear in Dr Świecimski's paper.

However, if this is the case how do we pass from the 'I' to the external world? What is the bridge that connects them both? Dr Świecimski tried to find this bridge, this additional moment which permits us to go to the real world, within the structure of experience itself. However, I don't see or find any additional moment discoverable within experience. In other words, in the basis of pure, Husserlian, phenomenology we are unable to explain the process of transition. On the other hand, I find the possibility of unraveling this enigma in other positions, in the position of Kierkegaard, for instance, who finds the key in affection, or Kant who finds it in the will. Without some additional mediating element, exoteric to cognition, we are hopelessly enmeshed in subjectivity and idealism.

Półtawski: What you just said was indeed idealism. When you say that everything is in your mind, well then, where is your mind?

Matczak: I defend realism in another way. One way of doing it is the way of Maine de Biran who explains it through *action*; parenthetically, this is also the explanation of Nicolai Hartmann and Meyerson.

Eng: Can we perhaps reformulate the issue in this manner: how does one describe the experience of the world and himself at the point of turning in the transformation? I don't think one can describe it, except by either choosing terms from the I-world or from the other-than-I-world and making mythic entities out of them. One can employ a metaphor here: the blooming of a flower – is it coming into life or is it dying? Actually, it is both. It is passing, it is a passing moment, and its *rhythmos* has it precisely that it is a unit of flow. How does one talk about the moment, the point of transformation that is transcendental? It seems to me that all of our descriptions are either in terms of 'I' or 'world', and then we fall into mythical kinds of explanations. I think the terms 'active' and 'passive' were used in these past discussions many times in the mythic fashion.

Matczak: I will eliminate myth by saying that the explanation is *action*, the other faculty of man.

I distinguish clearly three faculties in man in realism: one faculty is *reason*, which alone leads to idealism; the other is *will*, which is action, and which alone leads to realism; the third one is *affection*, which gives values and grounds ethics. Now, ethics and values are not the same as cognition, as some of the discussants would like to have it.

Eng: What kind of role do you give to *phantasia*?

Matczak: Phantasia has the mediating role between reason and sensory cognition.

But I would like Dr Świecimski to explain what does he consider to be that additional moment to our knowledge which will lead us to the real world, to the world of hard science? What is this addition?

Świecimski: Let me begin by stating first that your question entails the problem of falsifying the genuine character of the world by signs. What I have in mind is the falsification of the world by applying to it arbitrary constructs, standards, prejudices, and a specific kind of language. A sort of selection of data, forgetting that some data are transferable and some are not. Furthermore, there are different interests involved in cognition; as an illustration let us take a stone or crystal. Our examination of the stone or the crystal will vary whether we look at them from the viewpoint of a chemist, or geologist, archeologist, speleologist. In each case we are constituting another type of object intentionally. But this is a falsifying process in a sense because the content we are building by means of our cognitive apparatus and the data of experience is different in every particular case.

Matczak: I agree with what you are saying, but you still have not answered my question; how do we pass from cognition to the external world?

Świecimski: I don't know whether I can answer this question to your satisfaction. Some weeks ago I wrote a paper on the experiments and observations of a prominent neurologist and neurosurgeon in Warsaw. He performed operations on people who were blind from birth, and after a successful operation it was very characteristic of all patients that they were unable to manage the optic data suddenly engulfing them. Perhaps it is a process of very long training from our birth on that we learn to manage the visual data and identify them with things. How this process is formed and performed I really don't know.

Tymieniecka: Ladies and gentlemen, having now reached the end of our debates we may entertain the hope that, while we did not solve many problems, our confrontation may nevertheless have unearthed several issues that appear to be in dire need of rethinking and reformulation. Moreover we may also have specified certain directions along which the progress of 'philosophical reconstruction' of Reality should proceed if it is to take advantage of the phenomenological inheritance. Hopefully, these intimations will germinate within each of us for the next two or three years and we will all be able to continue their nurture at the next Conference, which we are planning in Europe. As the program chairman I thank you all for your cooperation in this program. It was philosophically very rewarding to see it emerge out of the many deliberations with the participants, and to see it through to its conclusions.

The membership of *The International Husserl and Phenomenological Research Society* will, I am sure, wish to express its gratitude to the acting members of the board, Professors Marie-Rose Barral, Dallas Laskey

From left to right: Prof. S. A. Matczak, Dr Erwin Straus, Dr Mary Rose Barral, Dr Erling Eng, Prof. Anna-Teresa Tymieniecka, and Prof. Henning L. Meyn.

and especially Dr Erling Eng, who has presided at our gathering, for their devoted work during these three years of the Society's existence. We are grateful to Professor Sebastian A. Matczak, the vice-president of the Conference, for assuming the burden of the local arrangements. We are also glad to acknowledge the dedicated and efficient efforts of Professor Henning L. Meyn, of the State University of New York, Albany, which ensured a smooth and comfortable progress of the meeting. Thanks are due also to Mrs Madden and to Mr John Levonik for their untiring efforts in the backstage.

We shall remain together in the spirit in the progress of our work.

COMPLEMENTARY ESSAYS

LE PLATONISME DE HUSSERL

L'affirmation par Husserl que l'analyse phénoménologique se meut dans le monde des 'essences' suscite de façon renouvelée l'embarras des commentateurs: comment concilier cette formule avec le transcendantalisme? comment la concilier avec l'intérêt pour l'histoire qui s'affirme dans les derniers écrits? comment la concilier avec l'intérêt pour la 'vie' et pour son monde?

Un modèle exemplaire a soutenu Husserl dans sa méditation de la phénoménologie et du destin historique de celle-ci: il s'agit du platonisme. D'un bout à l'autre de son oeuvre, l'appel à Platon est présent. Face à des interprétations déviantes du platonisme, Husserl a prétendu en préserver l'héritage authentique et inscrire la phénoménologie dans une histoire de la vérité que Platon a inaugurée.

Aussi bien, la phénoménologie ne peut-elle être comprise si on se borne à examiner les résultats partiels auxquels elle a pu donner lieu dans l'ordre des analyses particulières: il faut la référer à ce dont elle se réclame, à une entreprise de fondation scientifique de la philosophie, qui pense répondre au projet platonicien de fondation de la science. Ce propos lui-même contient des aspects qui ont été plus ou moins effectivement thématisés au fur et à mesure que se développait la pensée de Husserl. Alors qu'il réfléchit explicitement l'histoire de la philosophie dans sa dernière période, il s'attache plutôt dans les premières oeuvres à une élucidation effective de la pensée idéale, en particulier mathématique. Là se fait jour une interprétation de l'eidos, qui n'est sans doute pas exempte de contradictions, mais qui s'impose sans autre caution que celle de l'expérience, c'est à dire de l'évidence.

Les tensions qui naissent alors entre la description et la systématisation font que l'on peut parler d'une 'crise de l'*a priori*', surtout visible dans les oeuvres de la maturité. La mise à jour de cette crise doit permettre de montrer la fonction de l'idéalité dans la constitution de la phénoménologie transcendantale, par suite le sens spécifique de ce transcendantalisme.

Tymieniecka (ed.), Analecta Husserliana, Vol. III, 335–360. All Rights Reserved.
Copyright © 1974 by D. Reidel Publishing Company, Dordrecht-Holland.

Le retour insistant du Logos et de la pensée du Logique dans l'oeuvre
de Husserl contribue à faire entendre l'identification de la phénoménologie et de l'ontologie. Même les critiques que Husserl adresse au platonisme dans les derniers écrits témoignent de la fidélité et de l'ampleur de
l'attachement initial. La pierre de touche en est fournie par la corrélation
partout affirmée de la présence objective et de l'intuition, corrélation qui
donne sa signification au rationalisme husserlien.

L'enjeu d'une étude du 'platonisme de Husserl' est donc de restituer à
la phénoménologie ses attaches historiques, celles-là mêmes qu'elle a
toujours revendiquées pour siennes, et ainsi de la rendre à sa destination
de *philosophie*, conçue comme élucidation téléologique de la vérité.

L'anti-psychologisme des *Recherches logiques* s'affirme comme une dé-
fense de l'idéalité, partout opposée à la réalité: d'un côté la loi, de l'autre
le fait. La vérité exige pour être reconnue que son objectivité soit affirmée,
c'est à dire que sa thématisation puisse s'accomplir. Celle-ci ayant pour
lieu le domaine du 'super-empirique', la vérité doit se préserver de toute
confusion avec le domaine du fait, et par suite du temps réel de l'expérience. La vérité est alors entendue comme idée: les énoncés qui la visent
ont la valeur de possibilités idéales, comme le rouge individuel est 'possibilité idéale' du rouge *in specie*. Cette assimilation contient en germe
toutes les ambiguïtés qui pèsent sur l'eidos tant que n'est pas faite la
description de l'accès de la conscience à l'eidos, et tant que n'est pas
élucidée la fonction de l'eidos. Ces tâches ne pourront s'accomplir que
lorsque la phénoménologie sera devenue proprement transcendantale:
ensemble la réduction et la variation permettront de libérer l'eidos comme
le pur possible. La connexion entre les vérités de la théorie, celle entre
les expériences vécues du connaître, et celle des choses du domaine seront
alors pensées corrélativement, et l'accent se déplacera de l'une à l'autre
au fur et à mesure que progressera l'analyse transcendantale elle-même.
De l'unité idéale du vrai, elle passera à la description du flux noétique,
et à celle de la mondanité d'expérience.

Mais il est clair que le déni opposé, dès les premiers écrits, au 'réalisme
platonicien traditionnel' n'est rien d'autre déjà que le refus de séparer
l'élucidation de ces trois ordres d'enchaînement systématique. C'est
pourquoi la reconnaissance de l''objet idéal' ne peut recevoir un statut
thématique que dans une 'logique' qui ouvre à l'exposé de la phénomé-

nologie transcendantale. On le voit bien dans la le Section de *Ideen I*, où la distinction entre le fait et l'essence, entre les sciences de faits et les sciences des essences s'identifie à la distinction entre nécessité et contingence. La nécessité appartenant à l'essence, elle vaut comme généralité; le fait de son côté est l'individuel. Fait/essence, individuel/général, contingent/nécessaire sont donc trois couples antithétiques, dont les termes relèvent de la catégorie universelle de l'objectivité, qui a pour index l'intuition.

Tel est le 'platonisme', non 'réaliste', non 'hypostatique', que Husserl entend soutenir. Cependant le mode d'accès à cet objet qu'est l'essence n'est pas suffisamment caractérisé par son identification à l' 'intuition'. Il faut encore que l'intuition de l'essence soit rigoureusement opposée à l'intuition de l'individuel, à l'intérieur du genre commun 'intuition'. On doit reconnaître que cette opposition ne se trouve pas accomplie de façon définitive dans *Idées*. Par contre le mode selon lequel l'objectivité eidétique est posée est parfaitement explicite: c'est l'apodictique, l'indépendance à l'égard de toute thèse de réalité. Si l'eidos précède toute thèse réelle, c'est parce qu'il rend possible et l'individu et la connaissance de l'individu. Il appartient à l'aménagement de la réduction, et à une première indication de son résultat, de faire apparaître cette fonction de l'eidos. Car en éliminant la thèse de la nature, la réduction élimine la thèse du contingent, de l'individuel factice.

Le mot 'pur' désigne l'inversion de l'attitude qui est instituée par la réduction; son résultat sera l'ouverture à la région de l'apodicticité ultime, sans laquelle le réel ne saurait être posé. Condition du réel lui-même, et par suite s'affirmant par soi dans l'être sans avoir besoin d'aucune autre chose pour être (*nulla re indiget ad existendum*): telle est la conscience *pure*.

Mais pour être possible, la démarche réductrice devait être elle-même commandée téléologiquement par la région qu'elle amène à se révéler. En d'autres termes, il fallait être assuré d'avance qu'*il y a* de l'apodictique, que le domaine de la réalité factice ou 'naturelle' n'est pas le seul domaine de l'être, qu'il existe une 'région d'être' absolument absolue. Cela, c'est l'ensemble des considérations sur les essences et sur leur connaissance qui l'a montré. Cet ensemble est préliminaire à la mise en place de la réduction, car il aménage son champ. Si la thèse naturelle et avec elle l'expérience de l'étant mondain étaient les seules possibles, l'entreprise réductrice serait une entreprise fantastique, dépourvue de toute scientificité. C'est aussi en cela que la réduction n'est pas un simple doute. Qui doute accepte que soit perdu tout terrain possible de la pensée. La réduction, quant à elle, repose sur cette certitude préliminaire qu'il y a de l'être absolument nécessaire, que la réalité naturelle n'est pas la seule réalité.

L'affirmation de l'eidos permet aussi de justifier les résultats de la réduction. La 'région conscience' est explorée comme une région d'être absolu, dans la mesure où on se livre sur elle à une analyse essentielle (*Wesensanalyse*); c'est par là que la découverte de l'intentionnalité peut revêtir dès l'abord une valeur d'évidence absolue qui la fait échapper au psychologisme brentanien. Cela se marque en particulier dans le rapprochement à première vue énigmatique de la formule de l'intentionnalité et de celle des axiomes de la théorie mathématique. Le 'pur' est donc ici, non pas seulement comme pour Kant l'opposé de l'empirique; mais la pureté caractérise en propre une sphère d'être qui est thématisée, c'est à dire objectivée dans les jugements qui s'y rapportent. Car il y a des jugements 'portant sur les essences'.

Que la science de la conscience soit une science eidétique, cette affirmation comporte cependant une difficulté principielle. En effet, tandis que la science des essences en général est pensée de façon privilégiée sur le modèle de la science mathématique, la science phénoménologique de la conscience récuse quant à elle le modèle de la mathématique. A vouloir penser la science de la conscience exclusivement comme une science eidétique, on risque d'omettre sa spécificité (*Eigenheit*) parmi les sciences eidétiques. Celle du son par exemple thématise un *a priori* de subsomption: celui-ci se monnaie en unités plus ou moins générales, auxquelles finalement se subordonne cet individu singulier: le la ou le do entendu *hic* et *nunc*. L'*a priori* de la conscience se révèle être un *a priori* de constitution, et non de subsomption. Ni l'universalité mathématique, ni la généralité des sciences ontologiques matérielles n'appartiennent donc à la phénoménologie de la conscience. En elle, le dépassement de l'individu empirique ne se fait ni dans la figure de l'universalité formelle, ni dans celle de la généralisation. C'est pourquoi il sera si difficile de situer dans l'univers eidétique l'unité de la conscience, comme unité synthétique d'auto-temporalisation continue, et de trouver le langage adéquat à l'intuition authentique de cette synthèse: l'égologie est la dernière parmi les sciences constituées.

La situation problématique de la phénoménologie de la conscience dans les sciences eidétiques en général fait prévoir quelles ambiguïtés pèsent sur la signification du transcendantal chez Husserl. Il distinguait lui-même, dès *les Leçons* de 1907, deux acceptions de l'*a priori*: l'une qui caractérise les essences universelles, l'autre, non moins légitime, qui concerne tous les concepts catégoriaux. A ces deux acceptions il est demandé de diriger l'enquête phénoménologique, comme analyse des objectivités visées d'un côté, comme critique de la raison de l'autre.

Par là qu'il est entendu comme 'région d'être', le transcendantal se subordonne à l'eidétique; il s'y subordonne aussi en ce qu'il se thématise dans une science dirigée explicitement sur lui, et qui comme telle est une 'analyse d'essence'. A plusieurs reprises, Husserl, pour marquer la différence qui sépare sa propre interprétation du transcendantal et celle de Kant, dit que l'analyse transcendantale de la conscience telle qu'il la pratique est 'platonisante'. L'eidos a une fonction rectrice pour toute analyse quelle qu'elle soit: de l'individualité spatio-temporelle, ou de la subjectivité constituante. Ainsi aperçoit-on la signification de ces préliminaires à l'exposé de la réduction transcendantale auxquels Husserl donne le nom de 'logique'. Pour lui la logique ne fait qu'un avec l'ontologie. Comme ontologie formelle ou/et comme ontologie matérielle, la logique entendue comme science du discours est la science de l'être même. Les déplacements de signification auxquels a donné lieu le mot de Logos au sein de la tradition doivent être conservés et récupérés dans la phénoménologie: parler, penser, chose pensée, le Logos est tout à la fois. L'identification de la logique et de l'ontologie est alors l'identification des articulations du discours et des articulations de la pensée: elle témoigne d'une conception triomphante de la raison, entendue comme capacité de *voir*. Pensée et intuition ne s'opposent pas, mais sont constamment tissées ensemble. C'est cela même qu'indique le mot d'eidos. Le programme de la phénoménologie: le retour aux choses mêmes, prend son sens dans ce contexte. On voit aussi par là que la signification de l'ontologie implique que l'être se dévoile à la pensée comme *objet*, c'est à dire comme telos d'une thématisation toujours possible dans le milieu de l'évidence.

Mais aussi le transcendantal est la région d'être la plus concrète qui soit, celle dont la concrétion enveloppe toutes les relativités, c'est à dire toutes les abstractions contenues dans les autres régions. Elle est authentiquement première par là même. L'unité synthétique de prestation du sens précède toute généralisation et formalisation. L'ontologie absolue, c'est la phénoménologie transcendantale elle-même, c'est la philosophie première qu'avaient en vue les Anciens.

L'accent mis sur l'aspect transcendantal de l'analyse, ou bien sur son côté eidétique, peut sembler dès lors arbitraire et relever davantage des préférences de l'interprète que des nécessités de la pensée. Crise de l'eidétique? diront certains. Crise du transcendantal, diront d'autres. De toute manière on ôte à la pensée de Husserl le bénéfice de la cohérence. S'il y a 'crise de l'eidétique', c'est que l'on considère que le motif transcendantal l'emporte finalement: la phénoménologie génétique des derniers écrits, en reconnaissant la nécessité du recours au monde originaire de la 'vie' en serait le témoignage.

S'il existe une 'crise de l'eidétique', il faut la chercher dans une crise de l'eidos, c'est à dire qu'il faut se demander si oui ou non le modèle platonicien reconnu de la pensée de l'être s'est trouvé contesté *de l'intérieur* par une acception de l'être déviante par rapport à lui. Or on sait que cette situation s'exprime dès l'Introduction à *Ideen I*, lorsque Husserl oppose à l'eidos pris dans son acception traditionnelle 'le concept kantien si important d'Idée' ('*den höchst wichtigen Kantischen Begriff der Idee*'), dans le dessein de justifier l'emploi des deux vocables en concurrence. Si l'idéalité se scinde, n'est-ce pas la preuve que le concept platonicien d'eidos s'est révélé insuffisant, qu'il a pour ainsi dire éclaté sous la pression de l'expérience transcendantale elle-même? Il suffit pour s'en convaincre de considérer le domaine où l'Idée a manifesté sa fécondité.

Ce domaine est celui de la perception, et corrélativement de la mondanité comme horizon de toute perception. La chose perçue se révèle porteuse d'un infini ouvert de notes que l'expérience en acte ne parvient pas à repérer effectivement: chacun des côtés vus de cette chose-ci renvoie à et implique des côtés non-vus qui leur répondent et qui sont anticipés dans la perception 'de' cette chose. Car on a bien affaire, non pas à la perception d'un aspect ou d'un côté de la chose, mais, à chaque moment, à la perception *de la chose* totale, laquelle se trouve ainsi non pas proprement vue, mais posée comme l'unité synthétique où viennent se rassembler toutes les appréhensions successives. La chose perçue est donc le terme non-vu et pourtant visé 'en personne' de la synthèse antéprédicative qu'est la perception. La même description vaut pour l'horizon de la mondanité, dont le survol ne saurait être accompli, dont la saisie objective ne saurait être effectuée, sous peine que disparaisse justement ce qui caractérise en propre l'essence de l'horizon: d'être le sol et la condition de toute objectivation. Il n'y a pas là, d'ailleurs, une extension des résultats de la description de la perception à celle de l'horizon, mais plutôt un passage incessant de l'une à l'autre, sans qu'on puisse décider avec assurance laquelle des deux est première.

L'Idée kantienne a donc été élue pour expliciter ce caractère essentiel de la perception d'être une synthèse inachevable, prescrite a priori dans chacun de ses moments, mais constamment repoussée de l'acte en tant que tel. La considération de la perception a été postérieure, dans la phénoménologie, à la description des significations idéales, mathématiques en particulier: par là s'est ouvert à la phénoménologie un champ infini d'investigations que les premiers écrits ne pouvaient que pressentir. L'eidos a pu valoir comme type de l'idéalité exacte, mathématique, mais il a dû faire place à une idéalité d'une autre sorte, ouverte celle-là, lorsqu'il s'est agi de penser les synthèses inférieures de la vie perceptive, comme vie anté-prédicative. Des deux côtés cependant: du côté des synthèses mathématiques, comme du côté des synthèses perceptives, la même tâche s'offrait au phénoménologue: celle de penser la synthèse. L'on voit que des deux côtés la synthèse a dû, pour être thématisée, être prise comme achevée. Dans le cas de la pensée idéatrice, l'achèvement est en acte; dans le cas de la perception, l'achèvement de la synthèse est à la fois posé et supprimé, position et suppression étant ensemble indivisément. Mais *l'essentiel est de le poser.* L'essence de l'intentionnalité l'exige: car, en tant que visée de cette chose-ci, la perception vise un transcendant, c'est à dire un terme qui n'est pas contenu dans son décours immanent, un terme qui n'est donc pas 'réel' et qui ne saurait l'être. Dans la mesure où ce terme est présent et irréel, cette présence ne peut être caractérisée autrement que par l'idéalité et par l'idéalité d'une norme plutôt que par celle d'un noyau de signification immédiatement saisisable. C'est ce que dit Husserl:

> ... als Idee (im Kantischen Sinn) ist die vollkommene Gegebenheit vorgezeichnet – als ein in seinem Wesenstypus absolut bestimmtes System endloser Prozesse kontinuierlichen Erscheinens, bzw. als Feld dieser Prozesse ein a priori bestimmtes 'Kontinuum von Erscheinungen' mit verschiedenen aber bestimmten Dimensionen, durchherrscht von fester Wesensgesetzlichkeit.

L'Idée kantienne permet de poser un ordre eidétique rigoureux, alors même qu'elle en supprime la réalisation effective.

La fonction de l'Idée kantienne est donc bien claire : il s'agit en elle, non pas de démentir l'idéalité de type platonicien qui anime la description phénoménologique en général, mais bien plutôt de confirmer cette idéalité en lui assignant un type de donation, c'est à dire d'évidence, différencié. On assiste ainsi à un reversement des pouvoirs et des prérogatives de l'eidos dans l'Idée. Il en résulte que l'idéalité, qui avait chez Kant pour fonction d'indiquer, et se bornait à indiquer, une synthèse à jamais inachevable, sert chez Husserl à munir la synthèse temporelle d'un *a priori* qui est la règle immanente de son déroulement.

Cette interprétation trouve sa justification dans le fait qu'à l'échange qu'elle décèle entre l'eidos et l'Idée correspond dans la description un échange réciproque de sens contraire. Ainsi il arrive à Husserl de penser l'idéalité mathématique comme le résultat d'un 'passage à la limite', limite de l'exactitude, de la régularité…, ce qui l'autorise à la ranger sous l'espèce de l''Idée' au sens kantien du mot. L'emprunt à Kant de son concept de l'idéalité est ainsi au service d'une signification profondément anti-kantienne. L'idéalité avait chez Kant le sens de la négativité ; chez Husserl cette identification se trouve récusée dès le principe. L'origine de la phénoménologie n'est pas dans une Esthétique transcendantale – celle-ci se situe bien plutôt au terme. Cette origine est la pensée de l'idéalité *close*, de l'idéalité intuitivement saisie dans un acte dirigé sur elle. Dans cette mesure la pensée de l'Idée est subordonnée à la pensée de l'eidos. L'opposition, devenue thématique, de la donation adéquate et de l'inadéquate, servira ultérieurement à ressaisir l'opposition de l'eidos et de l'Idée : mais il s'agira alors, c'est bien clair, d'une différence descriptive, et non pas d'une différence proprement thématique. Elle n'interdit pas que l'idéal demeure compris comme ce qui est *vu*, et par conséquent objectivable.

La pensée husserlienne de l'*a priori* est commandée par les résultats de l'interprétation de l'eidos qui viennent d'être mis en évidence. Tandis que chez Kant, l'*a priori* désigne les conditions de la nécessité et de l'objectivité du jugement d'expérience, chez Husserl, l'*a priori* se confond avec les conditions de possibilité de l'expérience elle-même.

Pour Kant la synthèse est véritative dès là qu'elle est apophantique ;

chez Husserl il y a un primat de la synthèse matérielle ou véritative sur la synthèse formelle. Le formel et le matériel ont également une signification ontologique, car l'ensemble des conditions formelles de la vérité n'est autre que l'ensemble des conditions de l'objectivation en général. L'intuition et la pensée ne sont pas séparées l'une de l'autre; dire que l'objet est donné, c'est déjà dire qu'il est pensé. Car la donation est prédonation: elle est le fruit d'une prédication antérieure, qui s'est déposée et maintenue, devenant ainsi réactivable. Par suite la tâche de description des conditions de l'expérience est dans la phénoménologie celle de la description d'un rapport de fondation: la synthèse véritative originaire est fondatrice à l'égard de la synthèse prédicative, dont la structure abstraite peut être mise en évidence dans une analytique pure. La recherche de l''origine' acquiert par là son sens. Il ne suffit pas de noter que cette origine ne saurait être comprise chronologiquement. Il faut entendre par ce refus l'effort pour désigner un *a priori* qui est proprement la jonction de la pensée et de l'être. Cet effort ne fait qu'un avec celui de reverser la pensée dans l'intuition, et réciproquement. Bien loin que l'intuition soit une limite pour la pensée, elle est son essence, car toute prédication en tant que telle repose sur et engendre un noyau: comportement de choses plus ou moins général qui, par sa clôture même, est susceptible d'être repris dans une pensée ultérieure. Au sein de l'horizon, l'objectivation en acte est toujours possible. En tant qu'il se lève sur le fond de la prédonation permanente, l'objet est toujours déjà là. Ainsi s'éclaire la situation non pas liminaire, comme c'était le cas chez Kant, mais terminale d'une Esthétique transcendantale: en elle c'est le plus proche de la donation objective qui est atteint, mais cette proximité ne fait qu'un avec le plus grand éloignement, car il s'agit d'une donation non-thématique, en laquelle l'antériorité et la frontalité renvoient incessamment l'une à l'autre.

Le dévoilement de cet *a priori* 'de constitution' ne pouvant être confié à une science des faits, telles que sont les sciences de l'homme, il s'accomplit dans le milieu de l'essence, par la méthode de la variation. Celle-ci a pour fonction de délivrer une thématique des 'mondes possibles' parmi lesquels ce monde-ci, le mien, apparaîtra comme un parmi d'autres. L'arrachement à l'historicité de fait s'interprète dès lors comme idéalité, et la science génétique du monde de vie, dont l'élaboration est si difficile, s'installe ainsi sur le sol du possible.

Le possible qui est parcouru par la libre variation se recouvre donc avec l'horizon d'objectivation. Le possible entendu selon l'essence et le possible transcendantalement compris peuvent se réunir. Qu'il s'oppose au 'fait' ou au 'réel', des deux parts, le possible s'oppose à l'actuel. Dans le texte intitulé *Ursprung der Geometrie* datant de 1936, Husserl s'engage dans une analyse de la mathématique qui est non-historique si l'on entend par 'histoire' un savoir empirique: il s'agit d'une enquête portant sur les potentialités qu'elle recèle, c'est à dire sur l'horizon de sens au sein duquel elle s'accomplit:

> *"Geschichte ist von vornherein nichts anderes als die lebendige Bewegung des Miteinander und Ineinander von ursprünglicher Sinnbildung und Sinnsedimentierung".*

Tout dévoilement porte en lui l'horizon de son histoire.

Cet horizon est l'*a priori* de l'histoire: *a priori structurel*, la structure n'étant rien d'autre que l'essence de l'expérience.

Le souci, qui se fait jour dans les dernières oeuvres, d'élucider l'historicité comme tradition: permanence de l'acquis et ouverture du projet à l'à-venir, ne fait donc pas échec au platonisme ou à l'essentialisme de l'analyse phénoménologique. S'il est vrai, comme on a essayé de le montrer précédemment, que l'Idée représente un type d'idéalité dont les pouvoirs ne sont pas intrinsèquement différents de ceux de l'eidos, puisqu'il s'agit dans les deux cas d'apporter la possibilité d'une *vue*, même si celle-ci est repoussée à l'infini, il en résulte que l'horizon de la mondanité est toujours susceptible d'une interprétation eidétique, bien plus: qu'il doit lui-même être pensé comme une essence au sens platonicien du mot.

Le déni opposé par Husserl à un 'platonisme vulgaire' dès les *Recherches logiques*, platonisme de l''hypostase', n'a jamais affecté l'absolu de l'objectivation entendu comme idéalité. Le principe en sera affirmé jusque dans les derniers écrits. L'objectivité idéale en effet, conformément au sens natif de la phénoménologie, n'a jamais été identifiée à une existence 'en soi', c'est à dire séparable de la visée intentionnelle. Dès les *Recherches logiques*, l'existence de l'idéal a été pensée comme sa validité, et rapportée au procès de la répétabilité de l'identique. La répétabilité indique, c'est bien clair, un pouvoir qui certes n'est pas thématisable dans son effectuation factice, puisqu'il s'agit d'une répétabilité infinie, mais auquel l'idéalité de type kantien prête une règle, et par suite la possibilité d'une thématisation de son résultat. Pour faire entendre l'omnitemporalité de l'idéal, Husserl dira dans les derniers écrits qu'avant même d'avoir été découvert ou 'inventé', il a déjà 'valu'. Ce qui s'exprime par là, ce n'est rien d'autre que le sens même qui s'attache à toute découverte de type mathématique: le mathématicien a la certitude de la vérité comme valant au-delà de sa propre existence et du moment effectif où cette vérité s'est découverte à lui. Même s'il faut noter l'abandon du vocabulaire éternitaire qui était présent dans les premières oeuvres, cette constatation ne doit pas empêcher de reconnaître, dans l'omnitemporalité et dans la fonction qui lui est attribuée désormais, non pas un gauchissement de la pensée, mais bien plutôt la traduction, dans le langage de la description transcendantale, de ce qu'indiquait l''éternité' opposée à la temporalité du réel dans les premiers écrits. La nécessité de montrer le dépassement de la vérité par rapport à l'expérience peut à bon droit être considérée comme l'une des motivations de la phénoménologie tout entière: par elle, la valeur de vérité de la pensée scientifique authentique doit être reconnue et fondée; par elle, le vrai luit dans sa présence effective qui est celle de 'l'unité de validité' (*Geltungseinheit*). A une pensée enfoncée dans l'attitude naïve de la naturalité, le vrai ne peut s'imposer que sous la forme de l'étant: d'un étant certes plus 'élevé' que la table ou le pommier, mais partageant avec eux le mode d'existence 'naturelle'. Cette attitude une fois exorcisée, tout est prêt pour que soit reconnue la corrélation originaire de l'intention, prise avec toute son épaisseur de signification, et de ce qu'elle vise. Une analytique de l'évidence s'ouvre alors. Esquissée dans les *Recherches logiques*, c'est elle qui fait le prix de cette oeuvre aux yeux de Husserl.

Le 'platonisme' de Husserl ayant reçu ainsi une première élucidation, il devient inévitable de s'interroger sur les sources qui ont contribué à former l'image de Platon dans laquelle tout à la fois la phénoménologie s'est reconnue et où elle a puisé. Deux interprètes de Platon ont joué à cet égard un rôle décisif: Lotze et Natorp.

De H. Lotze, Husserl parlait avec la plus grande admiration, qualifiant de 'géniale' sa lecture du platonisme. Il consacra d'ailleurs à l'exposé de la *Logique* de Lotze un enseignement à Götingen en 1912: c'est à l'intérieur de cette oeuvre que se trouve un examen du platonisme sous la rubrique 'monde des Idées'. Lotze situe la platonisme par rapport à l'héraclitéisme: pour lui, la pensée de Platon vaut comme une réaction contre ce qu'il appelle aussi 'psychologisme', dont la thèse fondamentale est l'assimilation de la pensée à des combinaisons de représentations. Cette thèse ruine la possibilité qu'il y ait des 'lois logiques' à proprement parler. S'il y a de telles lois, leur source doit bien plutôt être cherchée dans la métaphysique, et, par la médiation de la métaphysique, dans l'éthique: car l'esprit n'est pas la résultante de forces simplement réelles, mais un système téléologique d'activités et de besoins qui appartiennent en propre à l'âme. Les lois logiques de la synthèse déterminent la constitution, dans le contenu de nos pensées, d'un reflet de l'ordre objectif du réel. Leur oeuvre consiste en une réorganisaton du contenu proprement psychologique grâce à l'appui des formations linguistiques. La signification des formes logiques réside en une reprise et une information des formes métaphysiques, lesquelles reposent sur la jonction préalable de la pensée et de la vérité.

Le mérite de Platon, selon Lotze, consiste à avoir réanimé, sous le nom de *théorie des Idées*, le socratisme, c'est à dire une réflexion éthique. Son point de départ serait la distinction entre le représenter et le subir. Au niveau de l'affection, l'héraclitéisme et le scepticisme sont justifiés à décrire un devenir incessant dans lequel aucun évènement ne se reproduit comme le même; au contraire, le contenu objectivé de l'affection signifie toujours ce qu'il signifie, et ses relations aux autres contenus possèdent une validité éternelle toujours identique. S'il est vrai que dans la perception, les choses sensibles changent de qualités, il demeure que ces qualités elles-mêmes ne sont pas ce qui change. Les choses participent aux concepts de manière fluente; mais les concepts sont éternellement égaux à eux-mêmes. Le contenu que nous ressaisissons dans son identité (cela qui est proprement le concept) est possédé comme vérité dans le milieu de l'intuition. Il faut ajouter que ce n'est pas seulement le contenu isolé dans sa singularité que nous possédons ainsi, mais le tout fixe et articulé des concepts, un véritable 'monde' de prédicats, c'est à dire d'Idées auxquelles ont part les choses de notre expérience changeante. A ces contenus ce n'est pas proprement l'ʼêtre' qu'il faut attribuer, car l'être désigne le statut des choses extérieures, mais le *valoir*. Ainsi peut-on, selon Lotze, résoudre la question à première vue énigmatique de savoir si, lorsqu'une couleur ou un son ne sont perçus par quiconque, ils sont devenus un pur néant. Question mal posée tant qu'elle repose sur le présupposé de l'existence: existence des choses ou de nos représentations en tant que nous les *avons*, c'est à dire en tant qu'on les considère comme de simples évène-ments. Au contenu de nos représentations, ce n'est pas l'être qui convient, mais le seul valoir (gelten): un mot que Lotze juge inutile d'élucider da-vantage.

Bien loin d'avoir, comme on lui en fait le reproche, 'séparé' l'être des Idées de celui des choses, Platon n'a voulu rien d'autre qu'enseigner la validité des vérités en tant que cette validité est indépendante de l'objet du monde externe, quelle qu'en soit la nature, auquel les vérités se rapportent. La signification éternellement identique des Idées: cette formule veut affirmer qu'elles sont ce qu'elles sont, qu'il y ait ou non des choses qui en participent, qu'il y ait ou non des esprits qui leur donnent la réalité d'un état d'âme. Mais la langue grecque manquait d'une formule capable de désigner ce concept du *valoir*; dès lors le concept d'être prenait sa place, mais comme un alibi. Que les Idées n'appartien-nent pas au monde de l'être, c'est à dire au réel, les Grecs l'exprimaient en disant qu'elles ont leur patrie dans un lieu intelligible supra-céleste, et que leur demeure ne peut être assignée où que ce soit. Mais c'étaient là des déterminations négatives qui, prises littéralement, ont contribué à accréditer la més-interprétation du platonisme qui en fait une théorie de l'être des Idées. A cette més-interprétation, Platon lui-même a d'ail-leurs prêté les mains en plaçant les Idées sous le concept général d'ousia. Ainsi la porte était ouverte à la critique aristotélicienne de l'Idée. Certes Aristote a mal lu Platon: en particulier, il n'a pas compris que les Idées ne sont pas attachées à un lieu singulier. Mais, alors que pour lui comme pour Platon, l'objet du connaître est l'universel, il ne disposait pas plus que Platon du concept adéquat qui lui aurait permis de désigner cette modalité de l'universel, qui à la fois est et n'est pas. Il voit bien que seule la chose singulière possède une ousia authentique; pour parler de l'Idée il emploie alors l'expression: '*deutera ousia*'.

Si Lotze étudie la pensée de Platon au sein d'une 'logique', c'est parce que précisément à ses yeux la pensée vraie de Platon n'appartient pas aux 'ontologies', lesquelles sont des pensées soit de l'être des choses externes (telle est la cosmologie) soit des choses internes (comme la psychologie). Pour Platon, l'Idée n'est pas un modèle qui précède l'étant, celui-ci étant alors pensé comme sa copie; bien plutôt l'Idée est-elle l'imitation, par la pensée, de l'un des traits en lesquels s'exprime la vérité éternelle. Le sens des Idées n'est pas celui d'une simple possibilité d'être qui exigerait, pour atteindre au réel, l'assistance d'un objet qui lui serait étranger. En ce sens la pensée de Platon ouvre la grande lignée des idéalismes qui étendent les valeurs de la pensée au tout de la réalité.

Ce que Husserl trouve là de 'génial', c'est cet effort d'une interprétation qui va chercher, au-delà des mots effectivement prononcés, une signification qu'ils indiquent sans pouvoir la thématiser: ainsi se trouve souligné en particulier le primat du valoir sur l'être. Par contre, ce qu'il ne trouve pas chez Lotze, c'est l'amorce de cette métaphysique à laquelle est assignée la tâche de fonder la logique elle-même. Lotze ne se borne pas à reconnaître l'identification ordinaire de l'être et de la chose singulière: il fait sienne cette identification. L'effort de Husserl consistera dès lors, à partir du primat reconnu du gelten, à repenser en entier toute la démarche de l'ontologie, à restituer par là la signification ontologique du platonisme, non pas certes par le recours au modèle de la chosalité, mais en adoptant le fil directeur de la visée du vrai: il voudra être plus fidèle que Lotze au dessein de sa Logique.

Dans cette tâche, c'est encore Platon qui lui servira de caution; mais il sera aidé dans cette relecture du platonisme par l'interprétation de Natorp, dont il y a tout lieu de penser que Husserl la connaissait bien, étant donnée la familiarité qu'il avait avec l'ensemble de l'oeuvre de Natorp. Natorp publie en 1902 la première édition de son grand ouvrage: 'La théorie platonicienne des Idées', réédité en 1921. Il en donnait en 1913, l'année de *Ideen I*, un exposé oral à la *Kantgesellschaft*. Ces travaux n'ont pas été ignorés de Husserl. Comme Lotze, Natorp insiste sur la filiation de Platon à Socrate, mais c'est pour en faire la base d'une interprétation 'intellectualiste' du platonisme. Ce que Platon a trouvé chez Socrate, selon Natorp, c'est une interrogation *sur le concept*, sur ce qui fait que les choses sont rassemblées sous un nom commun: à savoir précisément leur forme. Lotze pour sa part refusait dès l'abord d'identifier l'Idée à un concept, et préférait la penser dans la perspective de l'opposition du représenter et du subir plutôt que dans celle du singulier et de l'universel. Mais les deux auteurs se rejoignent en assignant à l'Idée la même fonction de mise en évidence de la qualité des choses, et par là la fonction de découvrir le logique, c'est à dire la loi grâce à laquelle la pensée s'informe en contemplant son objet. De l'aspect on est passé au *voir* de l'aspect lui-même. Le Logos devient le centre de l'interprétation de Natorp: Platon entend le Logos comme la corrélation de l'unité objective et de l'unité subjective, cette dernière étant la convenance à soi de la pensée. L'opposition socratique à la sophistique repose sur une interrogation concernant *la forme de la connaissance*: cette interrogation est indissociable, selon Natorp, de celle portant sur l'objet déterminé de cette connaissance. La conscience a en propre d'être à la fois conscience de soi et conscience de son objet: de là résulte la double condamnation d'un psychologisme et d'une pensée chosale des Idées. Comme Lotze, Natorp récuse l'interprétation aristotélicienne du platonisme, qui lui paraît démentie en particulier par des textes comme celui du *Phèdre*, où l'Idée est prise comme la pure position du penser, comme l'origine *a priori* de toute connaissance empirique.

Natorp s'accorde avec Lotze pour reconnaître que Platon n'a pas eu de longtemps, et jusque dans le *Phédon*, les moyens de déterminer le fondement logique d'une connaissance des lois du monde. Il a porté tout son effort sur la science du connaître. Qui connaît connaît quelque chose qui est. Corrélativement: ce qui est pleinement, cela est aussi l'objet d'une connaissance parfaite; ce qui n'est d'aucune manière, cela doit être totalement inconnaissable. Ce qui à la fois est et n'est pas, cela correspond à un milieu entre connaissance pure et non-connaissance complète. Bien loin donc d'adopter les limites de la pensée de Lotze, qui reposait sur la réduction de l'être en général à l'être de la chose singulière, Natorp fait de l'être la position du penser comme tel, et confère ainsi à la logique une signification ontologique. L'Idée du Bien est un être, non au sens d'un préalable à la pensée qui le vise, mais comme la loi propre de cette pensée. Pourquoi l'Idée suprême est-elle dite 'Idée du Bien'? Natorp hésite sur le point de savoir s'il faut penser l'Idée platonicienne comme l'Idée kantienne, comme l'exigence d'un devoir-être. L'interrogation reste en suspens. Natorp voit dans l'Idée du Bien l'amorce d'une cosmologie. Son propos en tout cas est de montrer à l'oeuvre chez Platon le principe d'une enquête sur l'être se formulant ainsi: l'être ne peut être que l'être d'un penser. Par là l'être se surordonne non seulement à l'être des choses, mais aussi à celui des simples 'qualités', même à la valeur de vérité, à la 'validité' des propositions vraies: pour toutes ces modalités de la position, l'être demeure le même.

Par là Natorp est bien conscient de s'opposer à Lotze, ou plutôt de prolonger son interprétation, de la porter à l'universel. Il est conscient de l'opposition décidée de son interprétation avec toute espèce de sub-

jectivisme, comme aussi avec tout objectivisme, car l'un comme l'autre
se situent dans une région proprement transcendante à l'être du pensé,
région dans laquelle toute question devient insensée. Certes, Natorp a
bien conscience qu'il confère ainsi au principe intentionnel une signification décisive: non pas descriptive, mais ontologique, et plus précisément
idéaliste. Dans l'accomplissement de l'être s'accomplit la pensée. Platon
en effet n'est pas un 'logicien' au sens que ce mot peut avoir ches les
néo-kantiens de Marburg! Il est tout à la fois un 'logicien' et un 'méta-
physicien', en prenant ces mots dans le sens que les modernes leur donnent. Pour lui la logique ne fait qu'un avec la dialectique, ou théorie de
l'être, parce que l'être ne réside pas dans un lieu trans-logique. Dès lors,
l'être de l'Idée est audessus de toute existence chosale en général: il n'est
ni 'âme', ni 'nature', puisque le Logos pose précisément la corrélation
de l'un et de l'autre dans le oui de l'être, auquel rien ne saurait être
surordonné. Comme tel, le Logos doit être identifié à l'être de la
relation entendue mathématiquement, et non téléologiquement comme
fait Lotze.

Ce sont ces deux interprétations *ensemble* qui ont valu pour Husserl dans
son rapport à Platon. Ce qui est important dans ce rapport n'est pas
son aspect 'historique' (selon l'aspect empirique de l'histoire que Husserl
récuse), mais bien plutôt la façon dont il fonctionne à l'intérieur de la
phénoménologie. S'il est vrai que, chez Lotze, Husserl a puisé l'identification définitive de l'idéal et du valide, il faut reconnaître que dans le
même temps a joué l'affirmation inconditionnelle de l'unité de la pensée
et de l'être, dont la formulation, comme 'principe de l'absence de limites
de la raison objective' pose la stricte réciprocité de l'être et du pensable.
Ce principe peut à bon droit être considéré comme la matrice de la pensée
phénoménologique tout entière, puisque c'est en lui que prend sens la
mise en oeuvre proprement phénoménologique de l'intentionnalité.
Husserl n'a pas fait sienne la leçon de Lotze qui imputait à un défaut de
la langue grecque la prétendue incapacité de Platon à récuser une pensée
chosale de l'être. Il a au contraire attribué, comme Natorp, au Logique
une signification *ontologique* qui ne s'est nulle part démentie dans l'ensemble des écrits phénoménologiques. Elle est le sol sur lequel repose la
phénoménologie elle-même; c'est elle qui confère à la science intuitive
de l'eidos une valeur absolue. Il faut aller jusqu'à dire que c'est par elle

que le phénomène se distingue de l'apparence, et justifie ses titres à 'manifester' l'être, à valoir, donc, dans sa transcendance immanente, pour 'la chose même'. Ce qui a retenu Husserl auprès de Platon, c'est qu'il puisait chez lui le moyen de conceptualiser la confiance de la pensée dans la validité des résultats opératoires, ainsi que Lotze l'a bien compris. Rien n'est plus significatif à cet égard que la permanence du vocabulaire à travers le trajet entier de la pensée. Lorsque Husserl a substitué à l'élucidation de la pensée scientifique la tâche d'une pensée de la mondanité, on le voit définir le monde comme sol permanent de croyance en l'être, comme l'assise de toute validité. Le sens du monde, c'est le sens même du valide: *Geltungssinn*, où l'on retrouve la formule qui servait dans les premiers écrits à désigner l'idéalité de l'espèce. S'il y a eu, comme l'affirme Lotze, un gauchissement dans la pensée de Platon qui fit passer de la pensée de la valeur à la pensée non reconnue de l'être, chez Husserl en tout cas il n'est pas possible de parler d'un semblable gauchissement. Car c'est de façon principielle que l'analyse phénoménologique, grâce au primat de l'objectivation, affirme la corrélation de l'être et de la pensée: *'was ist, ist erkennbar'*.

Le 'platonisme' de Husserl, affirmé d'abord à travers le logicisme et le criticisme des premiers écrits, est bien ce qui a permis à la phénoménologie transcendantale de dépasser le plan de la simple description et par là d'enraciner l'analyse sur le sol de l'ontologie. La réduction transcendantale est toujours déjà une réduction eidétique: cette proposition énigmatique n'a de sens qu'à la condition qu'on reconnaisse le platonisme pour le terrain même de la phénoménologie. C'est cela même qui fait d'elle une philosophie, une philosophie *première*, c'est à dire la fondation radicale du savoir. Cette fondation est en même temps libération d'une fin. Le telos de la pensée philosophique, c'est l'Idée de science. Bien qu'elle ait été suscitée historiquement par la polémique avec la sophistique, cette Idée possède en elle la signification essentielle de *toute* philosophie: elle indique la foi dans la vérité, c'est à dire dans la capacité d'atteindre à un fondement absolu de toutes les propositions vérifiées des sciences effectivement constituées. Or ce fondement ne saurait être que l'unité de la pensée et de l'être, cela même que la phénoménologie se propose de décrire sous le nom d'intentionnalité de la conscience. La phénoménologie est ainsi la science absolue, l'ontologie radicale à laquelle aspire toute la tradition philosophique. Platon le premier a eu en vue la fondation de la science à partir d'une enquête portant sur le lieu de sa vérité. Cette enquête vise à mettre en évidence des *normes* grâce auxquelles la science pourra justifier continuement sa méthode et sa théorie. L'Idée de science a une valeur rectrice à l'égard du déroulement de fait des sciences historiquement constituées. C'est à partir des principes que la logique peut seulement accomplir son oeuvre de mise en évidence du droit, et par là de justification de la science. Aussi sa démarche échappe-t-elle à la naïveté qui caractérise celle de la science elle-même. Pour avoir une valeur de justification ultime, il faut que la logique précède ce dont elle cherche à montrer le droit : c'est en cela qu'elle s'accomplit dans le règne de l'Idée. A cette situation oubliée au cours de l'histoire, il importe de revenir. En elle le logique se trouve investi d'une fonction critique fondée non pas sur l'exigence purement formelle de cohérence interne, mais sur la parenté principielle de l'être et du sens, telle que Platon l'avait révélée.

De la pensée socratique il faut retenir que le critère de la science authentique est la possibilité de l'enseigner et de l'apprendre. La science quant à elle permet de déceler le valable du non-valable. Elle s'oppose en cela à la pensée historique qui est une pensée du fait, incapable donc de décider si entre la religion ou le droit tels qu'ils se présentent dans le monde, et l'Idée de droit ou de religion (entendue au sens platonicien) il y a un rapport de convenance : celui de l'absolu et 'de la forme voilée de son apparition' (*ob oder ob nicht zwischen dem Einen und Anderen, platonisch gesprochen, das Verhältnis bestehe der Idee und ihrer getrübten Erscheinungsform*). Husserl estime que la philosophie moderne, depuis le 18e siècle, est paralysée par la peur du platonisme, qui est peur devant les idées. Cette peur est responsable de l'apparition des psychologismes, dont les philosophies historicistes du 19e siècle sont les rejetons. Pour ces philosophies, la science se confond avec une 'sagesse' plus ou moins vague, toujours liée en tout cas à l'esprit d'une époque : en d'autres termes, la science est destinée à réaliser une fin qui se trouve dans le fini. Au contraire Husserl défend le caractère supra-temporel de la science, lequel interdit de la mêler avec telle ou telle 'vision du monde'. Cette condamnation des 'sagesses', qui s'exprime sous sa forme la plus aiguë dans l'article de *Logos* consacré à l'identification de la philosophie et de la 'science rigoureuse' deviendra le motif de la condamnation de la technique, assimilée elle-même à une sagesse, dans les écrits ultérieurs.

L'obsession de la technique compromet l'idéal de scientificité dans le monde contemporain, cet idéal n'étant autre que la philosophie au sens platonicien du mot.

L'oeuvre de Platon à l'aube de la science et de la philosophies modernes a donc une double fonction : c'est d'une part une critique, et c'est d'autre part la monstration d'une 'idée téléologique' destinée à commander le développement culturel tout entier. A ce dernier égard la morale sociale de Platon a joué un rôle déterminant, en marquant la continuité des problèmes propres à l'individu et de ceux propres à la communauté humaine, replaçant ainsi la raison sur son terrain authentique qui est celui de l'universel. L'idée d'une humanité et d'une culture entièrement soumises à la raison philosophique est ainsi affirmée. C'est elle qui domine la pensée européenne tout entière, même s'il faut reconnaître qu'elle s'est avilie au cours des âges.

Ces avilissements doivent être eux-mêmes mesurés et compris. Lorsqu'avec Galilée se fait jour l'idée d'une nature entièrement dominable par la mathématique, on a affaire en réalité non pas à une résurgence du platonisme, mais plutôt à sa dégradation dans la technique. La recherche d'un instrument universel de domination des êtres naturels s'inscrit effectivement en continuité avec l'art ancien de l'arpentage, et se soumet à un idéal qui n'est pas tant celui de la vérité que celui de l'exactitude. Dès lors les déviations qu'entraîne dans l'ordre de la pensée philosophique l'introduction de cette interprétation nouvelle de la scientificité ne sont pas imputables au platonisme authentique, mais à ce gauchissement qu'il a déjà subi chez 'Galilée'. Ce n'est pas, donc, l'idéal platonicien de la scientificité que Husserl, à partir de *Krisis*, va mettre en cause, mais la conception moderne de la science et de son objet : la nature. En un mot, c'est le naturalisme qui porte la responsabilité de la chute de la philosophie, de l'oubli de sa vocation originaire. La nature est devenue un univers mathématique, non plus parce qu'en elle on apercevrait pour ainsi dire directement les formes mathématiques, mais par l'effet d'une mathématisation indirecte qui n'est autre qu'une violence faite au monde de vie par l' 'idéalisation' des données sensibles. Les conséquences philosophiques de telles initiatives sont bien claires : le monde mathématisé est posé comme un 'en soi', c'est à dire comme un absolu, comme un irrelatif ; la mathématisation elle-même est étendue à la 'nature psychique' : le hobbisme en résulte, et toutes les philosophies naturalistes de l'esprit.

De manière plus lointaine, on trouve parmi les conséquences du galiléisme la distinction classique entre les qualités secondes et les qualités premières, distinction à laquelle s'est attaché le nom de Descartes, et que la phénoménologie a récusée depuis longtemps. Lorsque Descartes sera accusé de n'avoir pas mis en question la mathématique, dans son entreprise de doute universel, et d'avoir ainsi compromis le développement de la philosophie moderne, c'est à la présupposition d'un 'en soi' des formules mathématiques que Husserl prétend s'attaquer. Par ce présupposé le finitisme s'introduit dans la pensée du monde. C'est une pensée finitiste qui a gouverné la philosophie des deux derniers siècles, sans que son origine fût mise en question. L'identification de la philosophie et de la science, que Husserl a contribué à exalter, il est le premier à reconnaître qu'elle a un caractère suspect si elle indique seulement une méthode d'enchaînement du vrai, et si elle repose sur la position exclusive d'un étant (la nature) érigé en absolu. Aussi bien la science la plus récente ruine-t-elle ces présupposés abstraits, et l'on peut dès lors considérer que la philosophie qui les faisait siens est devenue caduque. La philosophie ne saurait s'inféoder à une norme de déduction, si précieuse qu'elle puisse être par ailleurs. Elle ne saurait adopter non plus pour principe une certaine image, injustifiée, de l'être. Le rêve d'une certaine scientificité philosophique est bien révolu: c'était celui du scientisme positiviste.

Il faut s'interroger cependant avec une rigueur renouvelée sur la part de responsabilité qui revient exactement au platonisme dans la naissance et dans le développement du positivisme et du naturalisme. N'est-ce pas chez Socrate et chez Platon que les modernes ont trouvé la formule explicite d'une séparation de la sensibilité et de la raison, celle aussi d'un 'être en soi' opposé par son être à l''apparence'? A cela Husserl répond une fois de plus qu'il ne s'agit que d'une lecture fautive du platonisme, que la science que Platon avait en vue était la philosophie elle-même, et qu'elle devait permettre à la connaissance absolue de s'instituer. Surtout, l'effort de Husserl consiste à montrer que l'idéal platonicien n'a pas perdu son sens pour nous. La vocation de la philosophie demeure la vocation même de l'humanité, comme vocation à l'universel.

Mais il faut entendre la philosophie désormais comme une thématisation du monde. La science du monde doit se substituer, dans la pensée philosophique, à la thématisation de l'univers, entendu comme nature scientifiquement déterminable. Le modèle platonicien du savoir demeure valide à l'égard d'une telle entreprise: l'exigence d'un *a priori* du savoir, corrélativement l'exigence de l'évidence, demeure la norme absolue. Mais, tandis que la philosophie classique entendait cette norme de telle manière que pour elle toute évidence est l'index ensemble de l'adéquation et de l'apodicicité, la phénoménologie découvre que de cette liaison découle le caractère finitiste de la pensée de l'expérience. L'analyse phénoménologique de la perception et de son horizon a ruiné de façon radicale l'identification traditionnelle de l'inadéquat et du non-apodictique. Ce qui a été désigné précédemment comme 'reversement des pouvoirs de l'eidos dans l'Idée kantienne' permet à Husserl, dans les derniers textes, de condamner l'idéal de définité ou de clôture qui est à ses yeux celui de la philosophie classique, principalement sous sa forme de 'métaphysique théologique'. Platon lui-même est alors considéré comme responsable de cette finitisation de la pensée de l'expérience. Car il a contribué à expliciter l'idéal d'une pensée close du réel, s'appuyant sur la mathématisation de celui-ci. Husserl définit ainsi cet idéal:

... das Ideal der universalen Wissenschaft (als Umformung der antiken, in der parmenidisch-platonisch-aristotelischen Hauptlinie der antiken Philosophie treibenden Idee Philosophie) als universal exacten Wissenschaft, einer allgemeinen Ontologie als Umwandlung der aristotelischen Ontologie, korrelativ das Ideal einer mathematisch definiten Welt und zunächst Natur an sich, einer 'mathematisch' erkennbaren.

Si l'apodicticité de l'évidence n'est plus solidaire de son adéquation, aucune expérience adéquate du monde n'est désormais reconnue, et un regard divin est définitivement banni du champ de la vérité mondaine. Une science neuve de l'expérience devient possible, et nécessaire. Comme toute science, elle doit être conduite et animée par la certitude quant à l'être de son objet. Le monde de la doxa, ainsi que s'exprimait Platon, ne doit pas être opposé à un monde prétendu en soi dont il ne serait que la copie, mais bien plutôt amené à l'évidence qui lui est propre: évidence qui exclut principiellement la donation immédiate, mais évidence apodictique, puisqu'en elle s'institue la croyance universelle en l'être. Cette science de l'expérience est bien la science radicale et universelle que voulait être la philosophie antique, car elle concerne le sol de toutes les disciplines scientifiques. Mais elle n'est plus 'méta-cosmique'. L'absoluité d'un tel savoir se marque par le même critère que les Anciens avaient retenu: l'opposition au relativisme de l'empirie. Mais désormais, les relativités qui composent l'expérience mondaine et la font ce qu'elle est, au lieu d'être purement et simplement bannies du total de cette expérience (entendu lui-même comme l'objet d'un savoir divin) sont bien plutôt conservées *et élevées en elle au rang de l'absolu.* Cette fois le vrai n'est plus opposé à son 'apparence', mais il est la vérité des apparences elles-mêmes. La transcendance 'mathématique' de l'en-soi a fait place à l'immanence du manifester. Ce que Husserl a en vue sous le nom de science du 'monde de vie', c'est alors une ontologie universelle pour laquelle la tradition philosophique ne lui fournit pas de modèle : pas plus le 'platonisme' que le 'cartésianisme'. Aucune des pensées qui ont pesé sur le destin de la philosophie européenne n'a pu accéder à la notion d'un absolu qui ne soit pas supra-temporel ; aucune d'elles n'a eu en vue l'absolu de la temporalisation ; c'est dire qu'aucune d'elles n'a pris au sérieux cette dé-présentification incessante de l'expérience qui est son essence même.

C'est en ce point que se situe l'opposition décidée de la phénoménologie au platonisme traditionnel. Cette opposition concerne le visage de l'absolu, que la phénoménologie entend faire valoir. Ce que la pensée moderne, à plus ou moins juste titre, a trouvé chez Platon : la transcendance de l'en-soi, c'est cela qu'il faut exorciser. Par contre ce qui du platonisme demeure valide, c'est le rationalisme, car la philosophie authentique ne fait qu'un avec l'authentique rationalisme. L'acception de l'expérience comme temporalisation continue ouvre à une 'logique' du monde de l'expérience, entendue comme ontologie. Tandis que le platonisme achoppe à la mise à jour des relativités mondaines, et ne parvient pas à dépasser effectivement le scepticisme, le rationalisme phénoménologique constitue une ontologie concrète. Celle-ci n'est pas soumise au modèle exclusif de la chosalité naturelle ; elle est l'intégration de tous les côtés de l'expérience réelle. Ce rationalisme n'est ni suspendu à l'affirmation d'un en-soi, ni commandé par l'exigence d'une déduction universelle. La raison en effet n'est pas l'apanage d'une pensée séparée de ce qu'elle pense : c'est l'unité radicale de la pensée et de l'être, telle que les Anciens l'ont posée, qui vient à son propre dévoilement dans la démarche de la pensée. Une téléologie universelle, ce que Platon sans doute avait en vue lorsqu'il plaçait la vérité sous l'égide de l'Idée du Bien, telle est la science rigoureuse à laquelle Husserl veut aboutir. C'est la thématisation du Logos absolu.

En ce dernier avatar de la phénoménologie, on voit que le platonisme persiste en elle non comme thème, mais plutôt comme dessein. Le rejet de l'aséité de l'Idée va de pair avec le rejet d'une interprétation finitiste du monde. Mais l'identification proprement phénoménologique du monde et de l'horizon ne vise à rien d'autre qu'à 'sauver les phénomènes' grâce à une pensée concrète de l'être. La pensée du monde que la phénoménologie substitue ainsi à la pensée de l'au-delà veut être universelle. En elle l'humanité doit découvrir la vérité de son destin comme individu vivant, certes, mais surtout de son destin comme pensée philosophante. L'erreur, la mort, le mal en général, doivent y trouver leur sens. Ces tâches, la philosophie les reconnaît pour siennes précisément depuis qu'avec Socrate elle a accepté de 'revenir sur terre'.

Les armes dont pouvait disposer la pensée antique pour les mener à bien ne peuvent être réutilisées telles quelles par un philosophe du 20e siècle. Le choix même que Husserl a opéré parmi tous les 'Platon' possibles en témoigne. 'Platon' vaut à ses yeux comme l'exemple du penseur rationaliste, comme le premier héraut d'une 'critique de la raison'. Sa lecture des dialogues ne retient pas ce qui en eux appartient au mythe; d'une manière générale elle laisse de côté tout ce qui est propre à Platon: son mode personnel d'écriture, pris dans son indistinction réelle d'avec la pensée. De Platon à Husserl, toute la philosophie moderne s'interpose sous la forme du 'système', et c'est bien vers une pensée systématique que tend la phénoménologie. La 'monadologie' en laquelle elle s'achève renouvelle un style de réflexion emprunté aux Modernes, d'où le mythe est explicitement banni. Car le mythe convoque la représentation d'un autre monde, et avec elle une sorte de liturgie qui fait obstacle à la pénétration rationnelle de ce monde-ci.

La question peut cependant être soulevée de savoir dans quelle mesure la phénoménologie a effectivement opéré l'abandon de toute pensée mythique, et de savoir corrélativement s'il ne reste pas en elle des traces de la pensée classique de la transcendance, visibles principalement dans les écrits semi-confidentiels de la dernière période. Que Husserl appelle Dieu l'entéléchie qui est immanente au tout des monades, comme Idée d'un développement infini, cela n'indique-t-il pas que le Logos qui régit l'humanité et son monde a besoin d'une figure pour pouvoir être pensé concrètement? Du même coup, n'entre-t-on pas avec une telle figuration dans la pensée subreptice d'une transcendance personnelle, et ne ramènet-on pas alors avec elle l'en-soi de l'existence absolue? La pensée de l'Absolu divin alors s'emparerait de la pensée de l'Absolu de la Raison.

Le motif d'une telle surimpression semble bien se trouver dans le caractère *formel* de l'interprétation du Logos qui est celle de Husserl. Il a cru en effet qu'il était possible de reprendre à Platon le motif du rationalisme moderne: une telle décision n'était-elle pas condamnée dans le principe à être stérile? Car il ne pouvait s'agir que d'un motif strictement critique, dans lequel le côté positif de la pensée se réduit à l'exigence d'un fondement absolu du savoir, mais où la nature de ce fondement reste indécidée. Cette situation éclate au jour lorsqu'on voit que Platon ne peut fournir le modèle de la pensée concrète de l'immanence et de la finitude que recherchent les derniers écrits. Seule la téléologie indiquée par l'Idée du Bien, l'au-delà de l'essence, est ressaisie. Mais elle ne saurait suffire à éclairer l'idée d'une raison historiale, d'une raison qui est en même temps destin. Elle ne peut que mettre un obstacle permanent à l'épanouissement, dans la phénoménologie, d'une pensée rationnelle de l'histoire et de l'effectivité du vrai.

L'origine de ce formalisme est indubitablement dans la Platon néo-kantien qui était celui de Natorp. Pour Natorp en effet, l'Idée entendue comme l'unité constitutive de l'objet est la pure position du penser. Résultat de la synthèse judicative, elle est dépourvue d'autre contenu que l'activité logique elle-même. Seul le principe de raison suffisante, avec le principe de contradiction qui en dérive, peut, selon Natorp, constituer pour Platon la base assurée de la science. Certes il reconnaît que Platon pose la nécessité d'une science du devenir, même il reconnaît que cette science doit trouver son fondement dans le Bien. Mais c'est, dit-il, que le Bien 'signifie ce qui est voulu dans l'unité du but', qu'il exprime par conséquent dans sa pureté la loi fondamentale de la pensée elle-même. Une telle lecture de Platon élimine sans doute tout réalisme de la position de l'Idée; mais elle élimine aussi de la pensée toute matière. Dès l'abord le phénomène est identifié par Natorp au 'multiple', que la pensée a pour tâche d'élever à son unité par la mise à jour de la loi, qui n'est autre que l'Idée ou forme de l'objet.

Ce platonisme kantien est sans doute à l'origine des difficultés qu'a éprouvées la pensée phénoménologique lorsqu'elle a voulu réduire le fossé que creuse l'analyse intentionnelle entre la forme et la matière de la pensée. L'effort de Husserl pour parvenir à une Raison vivante, entendue comme une puissance téléologiquement orientée vers la venue à soi du sens, témoigne en tout état de cause de la lucidité et de la profondeur de son génie.

La lecture husserlienne du platonisme porte ainsi en elle nombreux caractères de la 'métaphysique' des Modernes, au sens que cette formule a pris pour nous depuis Nietzsche. On se bornera à en indiquer deux.

Nietzsche lui-même voyait dans le 'platonisme' de la pensée occidentale une pensée des valeurs, plus précisément le détour par les valeurs pour asseoir la pensée de l'être. Il est clair que si la 'validité' est bien, pour Husserl comme elle était pour Lotze, l'essence de l'idéalité, l'être est pour lui unité de validité ou unité de sens. Le principe de l'illimitation de la raison objective a pris en charge, peut-on dire, la validité de l'espèce, et, devenu principe même de l'intentionnalité, en a fait l'être même. Le 'monde' n'aura ultérieurement pas d'autre statut que celui d'unité concertante du multiple, c'est à dire d'unité pré-donnée des biffages et ratures de l'expérience en général. L'être est devancé par la confiance en l'être; il est le corrélat du vrai.

Le deuxième de ces caractères apparaît dans l'interprétation heideggerienne du platonisme. Etudiant le début du 7e Livre de la *République*, Heidegger y voit à l'oeuvre successivement deux doctrines concernant la vérité. Selon la 1e, la vérité est dévoilement, et cette identification commande la possibilité même de l'allégorie de la caverne, comme clôture qui conserve un accès vers le dehors. Selon la 2e, la vérité de l'être réside dans son essence, entendue comme la quiddité de l'étant.

So wird das Unverborgene zum voraus und einzig begriffen als das im Vernehmen der idea *Vernommene, als das im Erkennen Erkannte. Das* noein *und der* nous *(die Vernehmung) erhalten erst in dieser Wendung bei Platon den Wesensbezug auf die 'Idee'. Die Einrichtung in dieses Sichrichten auf die Ideen bestimmt das Wesen der Vernehmung und in der Folge dann das Wesen der 'Vernunft'.*

Cette deuxième leçon de Heidegger épouse si parfaitement l'interprétation husserlienne du platonisme que tous les moments de la phénoménologie naissante la confirment: la perception entendue comme intuition est bien l'accès à l'idéalité, et celle-ci fonctionne comme l'essence, assimilée, conformément à la tradition, au possible. Comment d'autre part ne pas reconnaître chez Husserl l'héritage de la pensée 'verticale' de la transcendance? Heidegger dénonce parmi les suites d'une telle interprétation le primat de l''exactitude' sur tout autre pensée de la vérité. Il s'agit alors de l'exactitude du regard, en tant qu'il se soumet à la présence objective. Comment nier que le parallélisme noético-noématique appartienne en propre à la pensée représentative, et au primat reconnu, au sein de la représentation, de l'objectivité? S'il en est ainsi, on mesure les obstacles qu'a rencontrés Husserl au moment de constituer la logique de la vérité. Celle-ci se précédait en quelque mesure elle-même dans l'affirmation que l'apophantique et l'ontologique sont les deux faces corrélatives d'une seule et même problématique. La logique transcendantale, c'est à dire l'installation sur le sol du monde, a permis à la logique de la vérité non pas certes de se produire, mais de se thématiser. L'origine du vrai est alors inassignable, et l'essence de l'horizon recueille son 'toujours déjà'.

Ce que Heidegger aura en vue lorsqu'il parlera du 'combat' de la terre et du monde, c'est ce que Husserl attribue à la Raison comme son devenir immanent, comme son auto-révélation. Mais chez Husserl ce devenir n'est pas un 'combat': la foi triomphante du rationalisme, la certitude que les différents degrés de l'apparaître parviendront à leur intégration concertante au sein d'une ontologie concrète, tout cela n'est que la foi dans les pouvoirs du logos, entendu non comme mythe ou comme poésie, mais comme jugement. C'est à la synthèse, c'est à dire à la conservation des disparates, que revient la fonction de présenter le monde; c'est dans une éthique humaniste que l'oeuvre philosophique s'accomplit, comme degré suprême de l'auto-responsabilité. Le refus que Husserl avait opposé aux sagesses, ce n'est donc pas lui qui l'assume effectivement. Il se réalise bien plutôt chez Heidegger comme déni de l'humanisme, et clôture de la philosophie.

EDWARD S. CASEY

ART, IMAGINATION, AND THE *A PRIORI*

It might seem paradoxical, if not simply foolhardy, to attempt to assimilate three such apparently disparate notions as art, the *a priori*, and imagination. At first glance, these concepts appear to be distinguished more by their discontinuity than by similarity. This discontinuity is particularly striking when we try to think of the *a priori* in relation to imagination. For what could be less *a priori* in character than the act of imagining – an act which is frequently taken as paradigmatic of what is most arbitrary, capricious, and unanticipatable in human experience? What can be said in advance – that is, *'a priori'* in the literal sense – concerning an act which by its very nature seems to subvert all our advance expectations? Isn't imagination, as the basis of the unreal and the fantastic, outside or beyond any significant *a priori* structure? In fact, isn't it the single human act or activity which is most alien and resistant to structure as such? Structure, which is an inherent feature and effect of the *a priori*, implies delimitation and exactitude: constrictions which imagination often seems to scorn and surpass. Regarded from its own perspective, imagination appears to seek one thing above all: freedom of movement. The *a priori*, viewed from its standpoint, would seem to desire something very different: stability of structure. Thus we witness a clashing of concepts from the very beginning: not only indifference of one to the other, but outright opposition.

This conceptual antithesis or antipathy between an open, fulgurating mental act and a closed, standpat structure is made even more acute when we consider the role of imagination in art. For it is precisely in art that imagination seems to have its most ample scope and to be least restricted by the rigor of structure. Does not art – especially certain kinds of art, such as surrealism – involve the exfoliation of the fantastic, hence of imagination in its freest form (or rather, anti-form)? How could structure intrude into a project whose very aim is to question, to transmute, and perhaps even to destroy structure? Isn't art, in its tendency toward *démesure*, the living proof that imagination and the *a priori* are incom-

Tymieniecka (ed.), Analecta Husserliana, Vol. III, 361–386. All Rights Reserved.
Copyright © 1974 by D. Reidel Publishing Company, Dordrecht-Holland.

patible? Are we not forced to say, as a kind of meta-paradox, that it is *a priori* true that, in art, imagination and the *a priori* are irreconcilable? More generally: are these two notions not distinct and disjunct in their very essence? And if they are thus independent of each other, can they ever mix meaningfully? Can they ever interact in a significant way? The skeptical tenor of these questions would seem to suggest a curtly negative answer: namely, that imagination and the *a priori* cannot commingle. If this is the case in general, then it would be all the more true in the instance of art, where their opposition becomes still more divisive.

But the appearance of incompatibility between imagination and the *a priori*, so seemingly self-evident in the case of art, may be deceptive. Perhaps there is a deeper level of continuity and communication between these otherwise so disparate notions. And if there is, it would no longer be paradoxical to seek *a priori* aspects of imagination or to see how the *a priori* might be operative within aesthetic experience. In short, the paradox involved in relating these notions to each other would be only a seeming paradox. In order to show that – and how – this is so, we must re-examine our usual, uncritical concepts of both imagination and the *a priori*. These concepts must be seen as including more (and in certain respects less) than we normally impute to them. For it is only by taking a fresh look – which, in the present context, means a phenomenological look – that we shall be able to undercut the paradox in question. We must emphasize that the ensuing effort will be one of description and not of interpretation, for we must discern hitherto neglected (or only superficially recognized) traits of the two phenomena at issue. Concretely, this means showing that a full description of both will help to dissolve their apparent irreconcilability, especially when they are juxtaposed within the domain of art. For they appear incompatible only when each is given a partial description which stresses a single trait at the expense of other essential (but submerged) traits. This mistaken procedure, which underlies the generation of many pseudo-paradoxes, takes the following characteristic form:

(a) the *a priori* is conceived as strictly formal in character; its only significance is held to be logical – i.e., it is observed as ingredient in logical operations or procedures and nowhere else.

(b) imagination is seen only in its most amorphous or extravagant forms – e.g., in the bizarrely fantastic as this occurs in art or hallucina-

tion. No distinction between types of imagination or imaginative fulfillment is made – just as, in the case of the *a priori*, no specific sub-types
are allowed.

When the *a priori* and imagination, thus narrowly circumscribed, are
compared, it is not surprising that they are found to be incompatible.
But what will happen if we consider each phenomenon more fully and
on its own merits by means of an open phenomenological method? By
reversing the usual reductive-descriptive procedure and aiming at a more
complete description, we may find ourselves in a position to speak of the
relationship between the two phenomena in significantly different terms.
In the pages that follow, we do not pretend to give fully adequate descriptions. Rather, we shall simply sketch or suggest such descriptions –
first in the case of the *a priori* (Section I), secondly with regard to imagination (Section II) – in order to outline the consequences as they bear on
the role of imagination and the *a priori* in art (Section III).

I

Since Kant, the notion of the *a priori* has had a rigorously logical, or at
most an epistemological, signification. It was Kant in fact who established the basic lines of almost all subsequent discussion of the *a priori*.
One does not even have to accept Kant's two main criteria for *a priori*
status – i.e., necessity and universality – in order to remain one of Kant's
many modern *epigoni* in this sector of philosophical thought. All that is
required is that one continue to conceive the *a priori* as fundamentally
formal in character. By 'formal' is meant that the *a priori* is, or is deter-
minative of, form alone. This form may be metaphysical and transcendent, determining the kind of thing something is (as in Platonism); or
transcendental and immanent, determining the nature of space, time,
and physical reality (as in Kantianism). But in either case, and in hybrid
positions as well (e.g., in strands of Thomism), the *a priori* qua form is
characterized by two essential features:

(i) *purity:* its non-empirical character; it is strictly distinguishable
from what is empirical, even if it may be associated with empirical elements.

(ii) *logical structure:* the *a priori* is responsible for the way something
is defined, for it delineates the structure which underlies any adequate

definition. And a definition, as Aristotle argued, involves the recognition both of the *type* of thing something is and its *differences* from other things. In this respect, the formal *a priori* is concerned with the structure of things insofar as they are logically distinct from one another.

Consequently, a summary definition of the formal *a priori* in terms of its two primary features would be: the *pure*, i.e., non-empirical, structure of something in its logical distinctness or specific difference from other things of the same type. Thus conceived, the *a priori* is clearly incompatible with those features of imagination that are amorphous (in the literal sense of 'formless'): that is, free-floating, fluid, shifting. For the formal *a priori* pertains only to what is stable and static in a given phenomenon – namely, to what is strictly *invariant* in it. (It is this invariancy, a notion used by Husserl to define his concept of essence, which makes something specifically what it is and not some other thing.) Imagination, by contrast, often seeks the novel and the changing: the perpetually variable and varying. If we were to arrest our analysis at this point, we would be forced to admit sheer difference, if not straightforward contradiction, between imagination and the *a priori*: the loose luxuriancy of the one would stand in stark contrast with the austere formality of the other.

But the relationship between the two notions does not in fact end here. The primary reason why this is so is that the *a priori* is more than formal in character. It possesses at least one other dimension or, better, type which forces us to recognize its extra-formal nature. Following Husserl and Scheler, we can call this other type the 'material *a priori*'. Seen negatively and in contrast with the formal *a priori*, the material *a priori* is neither 'pure' nor logical-definitional in character. On the one hand, it fully permeates the matter through which it is presented or what Scheler calls its 'bearer' (*Träger*); thus it *appears* empirically and historically, that is, as inseparable from the empirical and the historical. Insofar as its being is not separable from its appearance (and this inseparability is part of its very nature), it can be said to *be* empirical, even though it is not known by simple induction. Or, to be more exact, we should say: it presents itself empirically, i.e. directly to sensory perception. On the other hand, appearing in this manner, the material *a priori* does not furnish us with the logical structure of its bearer. Unlike the formal *a priori*, it forms no basis for definition or logical distinction. Instead of defining, it *characterizes* the matter in which it is ingredient – giving to this matter a distinctive, but not a logically distinct, character. In other words, it further specifies the specific difference given by the formal *a priori*. It is responsible for the concrete character, rather than the abstract formula, of a thing.[1]

Still more crucial for our purposes are two positive features of the material *a priori*:

(a) *mode of givenness* – In contrast with the formal *a priori*, which is not strictly given at all but rather deduced, the material *a priori* is given to apprehension. Moreover, it is given immediately and directly; it is not inferred, constructed, or reconstructed. It is, as it were, the given *qua* given – the given aspect of the given.[2] This means that the material *a priori* can appear only in an individual experience; it is not a generic trait that ranges over all things as a universal feature of experience in general. In Mikel Dufrenne's phrase, it is too "profoundly engaged in the matter of the object" to have this much generality of scope.[3] For it is given as uniquely qualifying the object or bearer by which it is originally presented – even if subsequent examination may reveal that it has a somewhat larger range. In sum, the material *a priori* is given immediately and as uniquely characterizing the bearer or substrate in and through which it appears. It is not necessarily – indeed, is usually not at all – applicable or generalizable to other, even other similar, instances.

(b) *multiplicity of sub-type* – This second positive feature follows from the first. If the formal *a priori* tends to be exemplified by a single type – the pure, logical form – the material *a priori* knows no such restriction. Instead, the latter tends to be expressed in a plurality of sub-types – by which we do *not* mean species of a single genus – that cannot be enumerated or otherwise delimited with exactitude. All we can say is that there is an indefinite plurality of such sub-types, each successive material *a priori* being linked to a particular experience in which it appears. And if the attempt is made to determine classes of the material *a priori* – as in Dufrenne's efforts to distinguish between expressive, affective, vital, existential, and noetic material *a priori*[4] – the success of such an enterprise depends on allowing these classes to be overlapping and quite open in membership. It is a question of family resemblance rather than of exact coincidence. Thus, in the end, it is difficult for Dufrenne to say whether a given material *a priori* such as joy (*allégresse*) is expressive, existential, or affective.[5] Joy is simply given, and given differently, in various individual experiences – e.g., in a child's delight, in a piece of music by Mozart, in the look of lovers. It is not, strictly speaking, the *same* joy which appears in each case. Nor is it an entirely different phenomenon, for which we would need a different name. Rather, it is something intermediary between these two polar possibilities: the material *a priori* 'joy' itself changes character slightly from appearance to appearance because it characterizes each new appearance differently: it constitutes each such appearance as distinctive, as uniquely given, as *itself* given. Thus it permeates its substrate, through which it appears, so completely that it is phenomenally indistinguishable from it. The fact that it might afterwards and on reflection be logically distinguishable from its substrate does not mean that it constitutes the logical structure of the object as given. It has no such definitive power, that is, no power of definition such as is possessed by the formal *a priori*.

What are the consequences of this expanded view of the *a priori* for our present problematic? In brief, the following: (1) We are forced to recognize that the *a priori* cannot be confined to its formal dimension or type. There is at least one other kind of *a priori* than the formal, and this other kind (namely, the material *a priori*) is itself not a single entity but a loose and limitless collection of sub-types. (2) If this is so, then we cannot simple-mindedly oppose the *a priori* factor in human experience to its imaginative dimension. For imagination and the *a priori* are opposed only at their antipodes – i.e., when the *a priori* is constricted to its formal dimension and imagination to its Dionysian excesses. Viewed in a larger perspective, the two may not be locked in ineluctable conflict. (3) Seen in its material aspect, the *a priori* gives promise of being reconciled with imagination. For in its immediacy, its infinitely plural character, and its lack of strictly logical status, the material *a priori* already resembles certain aspects of imagination such as its spontaneity and versatility. The material *a priori*, we may say, sufficiently 'softens' our original notion of the *a priori* so that it is brought within the vicinity of imaginative phenomena.

So far, then, we have traced out one half of a pincer-like movement of convergence. Let us now turn to the other half of this same movement towards a common center – a center which we shall find to consist precisely in the domain of art.

II

Our strategy in this section will be exactly the converse of that employed in the foregoing section. Where before we deliberately diluted the traditional philosophical notion' of the *a priori* as strictly formal in character in an effort to reveal its material aspect, now we must harden and constrict our ordinary notion of imagination as shapeless and indeterminate. If our previous effort was one of de-logicizing and de-structuring – in Jacques Derrida's term, 'deconstructing'[6] – the *a priori qua* formal, our present task is to refine and restructure our usual idea of imagination. In short, we must show that, contrary to its ostensible formlessness, imagination *has* a structure, and in fact a rather definite one. To demonstrate this, we shall be more expressly phenomenological in this section, endeavoring to make evident that, however *outrée* it may be in terms of possible content, its inner structure is remarkably regular.

In other words, the question which we are in effect raising in the present section is the following: does imagination itself have *a priori* features? Does it have a determinate structure, or is it as formless as its enthusiasts often claim it is? By the terms 'determinate', 'structure', and '*a priori*', we do not mean to imply that we seek the strictly formal features of imagining in the strict sense of the formal *a priori* discussed above. Rather, it is a matter of discerning order in the phenomenon, even if this order is subject to continual transformation or a tendency towards the proliferation of aspects. Of course, there have already been a number of attempts to discover such an intrinsic order in imagination. Two of these are especially worthy of mention:

(i) *archetypal* exegesis – Jung, Bachelard, Eliade, Gilbert Durand and others have conceived the ordering principle of imagination in terms of archetypes: i.e., structural emblems or motifs which are transmitted from generation to generation or from culture to culture. Discerning archetypes is unquestionably a useful means of cataloguing and co-ordinating dreams, hallucinations, and works of art among culturally diverse individuals and whole peoples. The indisputable existence and efficacity of archetypes instill a remarkable regularity in the *products* of imagination. But archetypes prove much less with regard to the regularity of the *process* of imagining. Indeed, they are typically projected onto either a biological or a metaphysical level which, being merely posited, is singularly unilluminating with respect to the specificity of the act of imagining itself. Thus a scrupulous concern for consistency of content leads to a neglect of how this content is entertained in imagination.

(ii) *epistemological* theory – When attention is finally focused on the act of imagining, the tendency in Western philosophical thought has been to insert it into a hierarchy of epistemological functions. In this cognitive hierarchy, it often occupies the lowest position (as in Platonic *eikasia*); at best, it enjoys a middle status, located somewhere between sensation and thought (as in Aristotle, Kant, and Collingwood). It is undeniable that from the standpoint of rational knowledge imagination usually does play such epistemologically inferior or intermediary roles. And it is also the case that giving it such roles is one way of introducing order into the act of imagining itself: witness Kant's nuanced (and sometimes overnuanced) distinctions between various types of reproductive and productive imagination. But in this endeavor to find the correct epistemological position of imagination, little attention is paid to its content – where by 'content' we mean not only its products, but more specifically what is intended as such in the act of imagining. Much is said of imaginative activity, little of *what* is imagined. In this regard, the epistemological approach to imagination indulges in an error of omission opposite and complementary to that of the archetypal point of view. Where the latter fails to do justice to the act of imagining, the former overlooks the significance of the precise content of what is imagined.

A phenomenological approach to imagination has the merit of accounting for *both* of the crucial aspects or phases of imagination: act and content. For its method of 'intentional analysis' is designed specifically to grasp mental acts in their full and correlative *noetico-noematic* character. However bizarre imagination may be in the detail of its content, in its essential operation it displays perfect conformity with other kinds of mental acts, all of which exhibit a general noetico-noematic or intentional structure. Thus a first way of discerning order in imagination consists in seeing that in its intentional character it follows the same general pattern as that of other mental acts. According to this pattern, to each aspect of *noematic* content there corresponds a *noetic* meaning-intention, and to each *noetic* phase corresponds something specific in the *noematic* content.[7] To affirm this rule of intentional correlativity is to remain, however, at a level of generality which does not furnish the specificity which we seek. To achieve this specificity, let us divide our analysis into two parts; in each of these, we shall point to distinctive traits of imagination which serve to distinguish it from kindred intentional acts.

(a) *noetic* aspects of imagination – Under this heading, we shall single out three noteworthy features of the act of imagining which set it apart from such allied acts as perceiving or remembering:

(i) *self-inducement*. Perhaps the most remarkable trait of imagining is the way in which it is self-inducing. By this, I mean that we cannot, strictly speaking, fail to imagine what we want to imagine; there is no way not to succeed in this respect. For we can without fail summon up exactly what we intend to contemplate imaginatively. And we do so spontaneously, without the necessary intervention of calculation, memory, or any other intermediary act. If I intend, for example, to imagine a three-headed man, no sooner have I entertained the thought of such a man than I have brought forth his quasi-presence. Only on later analysis might I be tempted to distinguish between the intention-of-a-three-headed-man and its actual fulfillment. In the experience itself, however, no such distinction is present to consciousness. I am simply aware of a single, continuous act in which intention and fulfillment meet and merge instantaneously. This coincidence of the given and the meant contrasts with what happens in perception, where my anticipations may remain unfulfilled or be directly disappointed. The back side of a perceived object may reveal itself to be distinctly different from what I expected it to be. Whereas in imagining the back side of an imaginatively projected object I cannot be disappointed or even surprised. The unseen side presents itself to my imagining mind exactly as I intend it to, in precisely the form which I originally aim at, for it is my own self-induced act that has brought it into being in the first place.[8]

(ii) *self-containedness*. A second outstanding feature of imagining is its self-contained character. We not only imagine exactly what (and how) we want to imagine but at the same time *no more* than what we wish to entertain imaginatively. Our act is not part of a series of acts which follow some rule of regular succession. In perception, by contrast, each act is linked to other closely related acts. As we view a moving object, for example, the succession of our perceptions exhibits a strict regularity and interconnection which we do not find in acts of imagination. This is not to deny that regularity in the sequence of these latter acts is possible.

But we do not find the same degree of, or demand for, rigorous sequence which we experience in the case of perception (and, by derivation, memory). Instead, each act of imagination is typically independent of other acts (whether these be imaginational or not in character). Every act is sufficient unto itself, being fully realized and satisfied in its instantaneous fulfillment. Since this fulfillment is always complete, there is no need for further fulfillment – hence no demand for futher acts in the same series, or for supplementary acts from another series. There is no sense of continuous exploration in which one act leads inexorably to another. Rather, it is a matter of momentary, but total, gratification of intention in which we are not impelled, or even tempted, to continue in the same vein. (To continue is in fact either to repeat the same act or to initiate some quite different act; it is *not* to explore further dimensions of the same imagined object, since there is a different such object for each act of imagining.)

(iii) *self-evidence*. This third noetic trait is closely linked with the other two. For what we spontaneously self-induce as gratifyingly self-contained is by its very nature self-evident. We apprehend the noematic content that we imaginatively project with perfect self-evidence, that is, as indubitable. The imagined content is not subject to doubt because we have brought it forth ourselves. Being self-contained, it cannot contain anything other than what we have ourselves (*qua* imagining consciousnesses) put into it. No further evidence – such as we might collect in perception or recollect in memory – bears on the imaginative presentation, which is fully formed and exhibited in its entirety. There is no aspect of this presentation which requires a search for additional evidence. (To be sure, we can seek the causes or motives for a particular presentation, and in this case further evidence is pertinent. But we have then left the level of imaginative experience *per se* and entered a different realm, that of causal or motivational origins.) The paradox here is that what is an ideal in mathematics, formal logic, and theoretical physics – to wit, apodicticity of evidence – is given immediately and effortlessly in the instance of imagination. In imagining, we cannot fail to achieve self-evidence, since such evidence is intrinsic to the act itself. For we are always certain that and how we imagine, as well as of what we imagine. There can be no situation in which we seem to imagine without actually doing so: to seem to imagine is already to be imagining.

Two general comments concerning the above *noetic* traits need to be made. First, all three traits point to a remarkable feature of the act of imagining: its inability to fail. It is an act whose success is assured, even if this success is itself subject to definite human limitations (i.e., we cannot imagine *any*thing at all or *every*thing at once). A whole analysis is suggested here which would bear on acts that could be called 'self-achieving' in contrast with all those acts whose success is not assured. The latter class is by far the greater and has deservedly received more attention from philosophers in the past. But, with the possible exception of Sartre, the former class has been almost systematically neglected; it calls for serious and prolonged investigation.[9] Secondly, we should observe that the three traits just discerned express, in more measured language, the features of imagining which have traditionally been subject to exaggeration and exaltation or (as a counter-tendency) to condemnation and vilification. In other words, they express in somewhat more precise and phenomenologically tempered terms what is often referred to honorifically as 'spontaneity' or 'creativity' (self-inducement), 'living in the instant' or 'letting the work suffice by itself' (self-containedness), or 'inspiration' or 'revelation' (self-evidence). The important point here is that the object of much Romantic effusion is itself subject to a decidedly different analysis in which a more structured picture of the same phenomenon – i.e., the act of imagining – emerges. What has too often been considered intrinsically formless, unruly, or wild can undergo a phenomenological analysis in which structural, *a priori* characteristics are discerned.

(b) *noematic* aspects of imagination – In this case, phenomenological description is even more effective in subverting our ordinary expectations. For we enter here into the region of imagined content, which seems at first practically structureless. As to detail, this content is subject to very few limitations – indeed, only those of the individual imaginer himself. There are an infinity of items, and of aspects of items, which may be imagined. We can see this in the instance of colors, which we can imagine in an unending variety of nuance and combination. Nevertheless, within the *noematic* content itself of imagination, we can make out at least two distinctive structural features which show that it is not merely a disordered pot-pourri of detail, however strikingly variegated this detail may be.

(i) the *imagined object*. It might seem surprising to mention this as a distinctive trait of the phenomenon of imagination. For what else do we imagine *but* such objects? Nevertheless, their existence has‚been placed into question, and even denied, by contemporary philosophers. "There are no such objects," remarks Gilbert Ryle laconically.[10] Thus it becomes necessary first of all to affirm their ingrediency in the full *noematic* content of acts of imagination. They are an uneliminable aspect of this content, which must be structured in very specific ways for it to be able to fulfill the detailed demands of imaginative intentions. Thus this content is specified in terms of objects, states of affairs, relations, *thetic* modalities, etc. Each of these parameters of noematic content represents the crystallization of an otherwise inchoate imaginal medium. We never apprehend this medium in itself – that is, in its amorphous state. It always appears to the imagining consciousness as fully clothed in forms. And this is the case even though we may not possess an exact verbal equivalent of the imagined object, which can remain nameless. Moreover, a certain indeterminacy is not excluded from the imagined object, which often presents itself with characteristically vague contours and fringes. Yet such vagueness does not reduce the imagined object to a formless imbroglio of sensible qualities. It infects the imagined object but does not destroy it. The object survives and can even take on a quite definite shape, as when we imagine a precise type of centaur or an absent friend whose features we know well. Thus the imagined object deserves the appellation 'object', for it is an identifiable and internally consistent whole in spite of its momentary and admittedly precarious presence to consciousness.

(ii) the *image*. The genuine basis for the indeterminacy of the imagined object is not found in this object itself – which *qua* object is perfectly definite in character – but in its mode of presentation. We have chosen to call this mode of presentation the 'image', drawing on the sense in which this term may imply a totality of imagined content – not only the imagined object itself, but its peculiar mode of quasi-presence: i.e., *how* this object is present and not just *what* is present in it. In this sense, the image functions as a field for this kind of presence, acting as a ground or basis for the appearance of imagined objects. Such a field-factor is an inherent part of our experience of imagination. For imagined objects do not appear to our imagining consciousness as wholly free-floating or unattached, as if they were the ghosts of perceived or remembered things.

They have their own experiential matrix, albeit an extremely indefinite one. Instead of a circumambient world of distinct things (as in perception) or a distinct temporal horizon (as in memory), we have only a vague and neutral flux that trails off from, and surrounds, imagined objects, ending indeterminately. Nonetheless, in spite of this intrinsic vagueness, the image or imaginary field is an essential feature of the total *noematic* content of imagination.[11]

In conclusion, then, we may assert that imagination possesses distinctive structural characteristics in both of its primary intentional dimensions. Noetically, it is a self-induced, self-contained, and self-evident act; and the repetition of the 'self-' in each case underscores the unity of imagining as an act which takes place within the individual consciousness that imagines. In this regard, it can be seen that imagination is not some extraordinary, surreal act which is added to normal consciousness from above, but rather an act which is perfectly continuous with consciousness, indeed immanent in it and necessary to it. It contributes possibilities to consciousness which it would not otherwise possess – in fact, it exhibits consciousness *as* possible – but in so doing it does not prevent a structural analysis of its dominant features. The second group of these features, the *noematic* ones, comprise the imagined object and the image. We have seen that both of these are intrinsic factors in imagined content, providing this content with an unexpected regularity of form. Most paradoxically, the image, though itself indefinite in character, serves a perfectly definite function as the ground or field in which imagined objects inhere; its vagueness acts as a counterfoil to the comparative clarity of form manifested by imagined objects.

Thus the second half of the pincer movement of convergence mentioned earlier has been completed. While the first half of this movement was from an overly constrictive view of the *a priori* as formal toward a more elastic vision of it as material – hence from a closed to an open notion of the *a priori* – this second half possesses an opposite orientation. For it begins with the ordinary, or Romantic, view of imagination as amorphous and open, and proceeds to show that imagining is much more tightly structured than is usually conceded. This surprising structuredness has been revealed in terms of a series of features which have been neglected or passed over in most former analyses of imagination. Consequently,

we observe here a counter-movement to the first, but one which is nevertheless symmetrical with it, since the analysis operates between the same
two poles of closure and openness. As a result, the two movements are
conterminous in the sense that the pole at which one ends is the point at
which the other begins, and vice-versa. Together, the two movements of
analysis form a single circular motion, tracing out a circle which encloses
a central arena in which imagination and the *a priori* are co-present
partners: the arena of art or aesthetic experience. To give a schematic
representation of the peculiarly converse character of our analyses in
these first two sections:

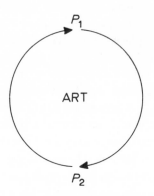

P_1 = open, fluid pole
P_2 = closed, static pole
$P_1 \rightarrow P_2$: direction of analysis in Section II (imagination)
$P_2 \rightarrow P_1$: direction of analysis in Section I (the *a priori*)

III

Since at least Plato, and especially since the Romantic rebellion, there
has been a traditional liaison between art and imagination in Western
thought. The two activities have been seen as intimately implicated in
one another. Most frequently, it is held that art implies – i.e., demands,
entails, requires – imagination. Art or aesthetic experience (which includes both the creation and the appreciation of art) calls for imagination
in order to be realized with the requisite inventiveness and verve. Thus

we arrive at a first, and for the most part undisputed, formulation concerning the relation between art and imagination:

(1)   Art implies imagination.

But a second formula is also discernible in a significant segment of Western thinking on art:

(2)   Imagination implies art.

This is a more controversial utterance. Kant, for example, would demur, since for him imagination can lead to art only if it exists in a particular harmony with understanding. But others have held that imagination, through its own ongoing activity and without conscious intention, brings about art: to imagine fully *is* to be artistically creative or at least aesthetically sensitive. Collingwood, following Croce, has been the most articulate philosophical spokesman for this view in our century. For Collingwood, to exercise one's imagination actively is *eo ipso* to bring forth imaginative products that qualify as art. In a rapid series of equations, Collingwood maintains that to imagine is to attain the level of consciousness; and to be conscious of something is to express it; and to express something fully and authentically is to create or to re-create art.[12] Hence imaginative activity is in effect synonymous with artistic activity.

If art and imagination have thus been seen to imply each other, an equally insistent strain of thought in the West has denied any significant relation between the *a priori* and art. These latter have been viewed in terms as antithetical as those in which the relationship between the *a priori* and imagination has been conceived. In both cases, there has been a pronounced penchant for dissociating the two terms in question, denying any fundamental accord or link between them. In fact, they are regarded as parallel cases of mutual *dis*-implication: just as the *a priori* does not imply imagination or imagination the *a priori*, so art does not demand the *a priori* or the *a priori* art. Thus we are confronted with four denials of implication:

(3)   Imagination does not imply the *a priori*.
(4)   The *a priori* does not imply imagination.
(5)   Art does not imply the *a priori*.
(6)   The *a priori* does not imply art.

But we have learned from earlier discussions that proposition (3) is false, since imagination does in fact exhibit *a priori* structures. And the same discussions would cast doubt on the certainty of proposition (4); for in its material aspect the *a priori* would not be strictly disjunctive with imagination; instead, as we have suggested, it would tend to be conterminous with it and to encircle a common center. We have called this center 'art', and we can now see why this is so: art or aesthetic experience is a domain in which such otherwise disparate factors make contact and overlap, thereby allowing for the possibility of mutual implication where at first there seemed to be only disjunction and disharmony. How can this be? One way of answering this question is to set forth the following propositional proof, which may enable us to effect the crucial transition from art to the *a priori*:

> Art implies imagination [by (1)]
> Imagination implies the *a priori* [by correction of (3)]
> ∴(7)   Art implies the *a priori*.

Two important reservations must, however, be made. First, we cannot prove with equal rigor the converse proposition that the *a priori* implies art, for this would depend on showing that the *a priori* actively implies imagination. At best, we have shown that the *a priori* is not incompatible with imagination and even that it is consonant with it. But this is not to have demonstrated a relation of implication between the *a priori* and imagination. Secondly, the above 'proof' is only formal in character. It depends for its persuasiveness on a strict and synonymous sense of 'implies' in each separate statement. Yet it is not at all certain that any such synonymy is present throughout the propositions in question. Nor have we defined with the requisite rigor the critical terms 'art', '*a priori*', and 'imagination'. Even granting that some precision has been given to the latter two terms in Sections I and II, we cannot claim to have clarified sufficiently the term 'art', which remains importantly ambiguous. Is it a process, a product, an idea? What *is* art?

Thus a primary task should be the elucidation of the meaning of the term 'art'. We cannot undertake this elucidation here.[13] In any case, our main interest bears on the *relation* between the terms in question and not so much on the terms themselves. And the most important of these relations at present is that which is expressed in our formal conclusion

(7): art implies the *a priori*. What concrete meaning can this have, and how does it relate to the role of imagination in art? As a first approximation, we may be tempted to say that art implies the *a priori* in terms of its structural properties. This is by no means incorrect: if the New Criticism has not already convinced us of art's intrinsic formal properties, then the impact of structuralism on literary criticism cannot leave us any longer in doubt.[14] There can be no question as to the highly, and often hiddenly, structured nature of works of art.

But art implies the *a priori* in a deeper sense. For the above derivation of proposition (7) makes it clear that art implies the *a priori* only via imagination. And the *a priori* and imagination are linked because of their common basis in the human subject. The *a priori*, formal or material, is found not only in the objective realm – where it structures aesthetic objects – but also in the subjective realm of virtual or implicit knowledge. Indeed, it could not be apprehended objectively unless we already possessed a tacit or virtual grasp of it within ourselves. This reciprocal and dependent relation between the objective manifestation and the virtual knowledge of the *a priori* – first articulated by Plato and revived most recently by Merleau-Ponty, Dufrenne, and Polanyi – has crucial implications for our present purposes. If it is in fact the case that the *a priori* is "immediately recognized [in the object] because the subject already has a virtual knowledge of it",[15] then the objective structure of works of art (so elegantly delineated by New Critics and structuralists) must entail a subjective dimension in and through which it is created and appreciated. This dimension shows significant overlap with the equally subjective realm of imagination. For it is imagination in its *a priori* aspect – i.e., in its structured or schematized character – that makes possible both the *creation* and the *enjoyment* of works of art as objectively structured:

(a) *creation* – Imagination is responsible for the artist's ability to envisage possibilities in his medium which are not yet actualized. It is through imagination in its noetic character of self-inducement, self-containedness, and self-evidence that the artist can surpass the immediate givens of his material by projecting novel ways in which this material can be treated and which represent genuinely new possibilities of realization. (This remains true even in the case of fortuitous 'accidents' in art: the artist must still be able to distinguish the promising from the unpromising accident, and then decide what to do with it in terms of development, exhibition, etc. This necessity of choice and discernment has been evident from Duchamp's *objets trouvés* down to the most recent instances of popart and happenings.)

(b) *appreciation* – The spectator's imagination must be exercised as well. He need not relive and re-enact in imagination the exact act of creation, as Collingwood claims. But he must be able to realize that the work of art is not a static structure, that it always involves several possible modes of access and ways of being experienced. It is primarily his imagination which suggests the available alternatives in this respect. And it is also his imagination which allows him to return to his ordinary life-world after aesthetic experience and to see it in a different way: to see it as having new aspects.[16] More basically still, it is the spectator's imagination in its structured character that allows him to resonate with the work of art; for it is by imagining that he transcends himself toward the work – though only toward the work as a possible, not an actual, world of experience.

In sum, art implies the *a priori* because of the critical role of imagination in aesthetic experience. The *a priori* as objective structure demands a corresponding virtual knowledge in the subject. This virtuality, to be exercised and actualized, requires the intervention of another subjective capacity: imagination. For imagination allows us to project possible actualizations of our virtual knowledge by enabling us to envision what has not yet been realized. Thus we are led to acknowledge an aspect of imagination which is no less important than its *noetico-noematic* structure: its 'possibilizing' function. For the *thetic* modality or belief-character of imagining is that of pure possibility: to imagine is to posit objects as purely possible, as beyond the range of the real. (This status is to be distinguished from that of hypothetical possibility, which is still tied to concrete conditions of realization in experience.) To imagine something is to posit it as possible in a sense that for the moment ignores or transcends reference to the way in which it will be actualized empirically. And it is, strictly speaking, the imagined object that is thus posited as purely possible. It makes no difference whether this object is or is not real in other contexts (e.g., historial or perceptual); as the object of an act of imagining, it is merely and solely possible in its projected ontic status.

Consequently, the notion of possibility itself forms an important intermediary link between art and the *a priori*, thus suggesting another way in which these two disparate terms can be conjoined. On the one hand, possibility, as we have just seen, is an integral aspect of artistic creation and enjoyment, since both of these phases of aesthetic experience involve possibility in an intrinsic way. On the other hand, possibility is intimately related to the *a priori* in terms of both its material type and its virtually known character. (i) As material, the *a priori* is capable of endless multiplication; it is, in Dufrenne's phrase, 'infinitely more diversified than the formal *a priori*.[17] In order to conceive this rich diversity of sub-types, we must be able to think of the material *a priori* in its open range of *possible* meanings. The diversity of the material *a priori* is the diversity of the possible and not of the actual. For the material *a priori* embodies a possible meaning for a particular experience; it is not the only such meaning, and it has none of the rigidity of the formal *a priori* (which is always apprehended as actual). What we grasp immediately is a possible, though nonetheless effectively ingredient, meaning. The joy, for example, that we apprehend as characterizing a Mozart symphony is not the 'objective correlative' (in T. S. Eliot's term) of an emotion which we feel. Rather, the material *a priori* of joy is grasped only as quasi-actual, thus as purely possible. It is a possible meaning of the aesthetic experience, and it remains merely possible even when it is the most probable or likely such meaning.[18] (ii) As virtually known, the *a priori* is also possible in character. For a virtual status is one of possible activation or reactivation. Hence virtual knowledge is knowledge which *can be* actualized: it is at our disposition, and the concept of the possible is thus built into the very idea of the virtual. Further, a virtual knowledge of the *a priori* itself involves a grasp of the *a priori* in its nascent state: i.e., as possible and as not yet concretely embodied in perceptual experience. Such knowledge is implicit in a sense which essentially involves the possible: that is, as *capable* of being made explicit through reflective analysis.

Art, then, is a human phenomenon whose highly structured character is undeniable. Yet it is structured not only by formal *a priori* – as is implicitly maintained by New Critics and structuralists, who seek strictly objective structures having no intrinsic relation to human subjectivity – but also by material *a priori* which solicit the human subject for their creation and apprehension. As we know from Section I, the material *a priori* is a structure that crystallizes an appearance in its immediacy and particularity. Thus it is potentially infinite in its range of possible presentations. But this diversity of appearances – of what we have called its 'sub-types' – does not point merely to an infinite plurality of objective items. For the notion of the material *a priori* cannot be restricted to its appearance in the object alone. We have asserted above that corresponding to each appearance of the material *a priori* there is a virtual knowledge of this *a priori* in the apprehending or creating subject. This virtual knowledge, being implicit and pre-existent, cannot be interpreted as an inductive generalization from successive appearances. It is itself *a priori*, i.e. latent and immanent in the subject. Thus we are forced to speak of an *a priori* or virtual knowledge (in the subject) of the material *a priori* (in the object).[19] Such knowledge is virtual in the sense that it is awakened by the appropriate appearances and not learned from their mere observation or collocation.

It has been Mikel Dufrenne who has provided the most detailed and coherent account of the *a priori* in its bivalent, objective (manifest)-subjective(virtual) nature. Thus it is important before closing to come to terms with Dufrenne's pacemaking efforts, and especially with regard to two controversial issues. The first concerns the ultimate relationship between the objective manifestation of the *a priori* and its virtual knowledge in the subject. In Dufrenne's more cautious utterances, there is said to be an exact correspondence between the 'cosmological' (objective) and the 'existential' (subjective, virtual) *a priori*. But this plausible position rapidly gives way to the more extreme assertion that the cosmological and the existential are only two *aspects* of a more basic *a priori* which is somehow 'anterior' to these aspects. This is to enter into ontology – to "pose the problem of being", as Dufrenne himself says – too hastily for our purposes; and it leads inexorably to the questionable claim that the existential and the cosmological are identical in character.[20] From a phenomenological point of view, however, important differences remain between the existential and the cosmological dimensions of the *a priori*. This is particularly true in aesthetic experience, where it is not descriptively accurate to speak of an equivalence, much less an identity, between the two dimensions. For in this experience, either the subject has the initiative (as in the case of creation) or the object imposes itself on the subject (as in appreciation). Rarely is there a perfect coincidence between the material *a priori* as apprehended in the object and as known implicitly by the subject. Instead, there is typically a dialectical interplay between the two – an interplay which has, however, no necessary final resolution. Differences between the scope of the virtual (which is always general in character) and the particularity of the presented work (which is always singular and unique) remain and cannot be overcome in an ontological syntheses.

A second and more important divergence from Dufrenne's account arises with respect to his conception of the place of feeling and the affective *a priori* in art. In Dufrenne's view, each work of art bears a 'world' which is characterized by certain 'affective qualities'; these qualities are expressed, not represented, by the work. They act as *a priori* structures for the world of the work: "the world of the aesthetic object is structured by an affective quality which is for it an *a priori*".[21] This *a priori* in turn calls for what Dufrenne calls an 'affective category', which is the affective *a priori* as known virtually by the subject.[22] The problem that is posed here is the following: why is the peculiar quality, *a priori*, and category involved in aesthetic experience necessarily *affective* in nature? Within Dufrenne's epistemological theory, it is clear why this is so, for feeling (*le sentiment*) is regarded as the culminating stage of the perceptual experience which is required in the apprehension of art.[23] According to this epistemology, imagination is located at the level of 'representation' and is thus seen as surpassed in aesthetic experience proper, which is expressive in character and for which imagination merely prepares the way. Only feeling possesses the requisite openness and depth that allow the spectator to enter the expressive world of the aesthetic object; thus feeling and world are conceived as reciprocal types of depth: the depth of the expressive world answers to the depth of the subject and vice-versa. The affective *a priori* is what mediates between these two depths:

the affective *a priori* constitutes a consistent and coherent world because it resides in the deepest stratum of the subject, as well as forming the most profound aspect of the aesthetic object.[24]

Yet thus to stress the role of feeling in art at the expense of imagination seems to me mistaken. *Both* factors are required as essential components of aesthetic experience. It is true that if imagination is restricted (as in Kant, whom Dufrenne follows in this regard) to its empirical and transcendental functions, then its role in art will be consequently curtailed and even demoted to a position subordinate to feeling. But if imagination is viewed in a more complete manner, it will be seen to be a central factor in aesthetic experience, and not merely of marginal importance. As we have endeavored to show, this more complete picture involves the recognition of (a) imagination's own *a priori, noetico-noematic* structure and of (b) its capacity to project or entertain pure possibilities. Each of these features of imagination allows it to be essential to art, and thus to the material *a priori* in art – an *a priori* which can no longer be entitled exclusively 'affective', but which is equally 'imaginational'.

To round out our analysis of the role of imagination in art, let us briefly summarize the essential aspects of this role:

(a) In its *noetic* phase, imagination is crucial to both the creation and the appreciation of art: to its creation, because the artist must to some extent be able to self-induce the direction he will take in working with his materials, to be convinced of the self-evident correctness of this direction, and to remain relatively self-enclosed in his ongoing project; and to its appreciation, because the spectator must himself be persuaded to some degree that the aesthetic object which he apprehends is a spontaneous creation and that its structure is self-evident and self-contained. In its *noematic* phase, imagination is crucial insofar as both artist and spectator must be able to envision the aesthetic object as quasi-actual and as set within an indeterminate image-field. This is the case even though the presentation itself, regarded as the bearer of the aesthetic object, is material in character. In attending a play, we witness live actors on a physical stage; but aesthetic appreciation does not consist in positing the actors *as* alive or the stage *as* physically present. Instead, we effect a change in ontic index and transform the aliveness of the actors and the physicality of the stage into immaterial, quasi-actual presences in a surrounding imaginary field.[25]

(b) That is to say, we envisage the characters and the setting in a play – i.e., the world of the aesthetic object – as *possible*. The operation of the imagination in art is not simply negative in effect; it does not consist in merely denying actuality, e.g. aliveness or physicality, in the aesthetic object; instead, it transcends these correlatives of the 'natural attitude' towards a distinctly different ontic status – that of pure possibility in the sense described above. Such possibility is thus essential in the creation and experience of art, which cannot allow itself to be confined to the dimensions of the actual without becoming, say, *Kitsch* or propaganda. This is not to deny that it can still refer to the real – i.e., make use of actual historical figures or objects – but it must do so obliquely and through the lens of the possible: the actual must itself be possibilized, and hence regarded as just one unprivileged possibility among others. To make this reference direct is to lose the essential basis of the expressive world of the aesthetic object: a world which can exist only as possible, never as actual. And it is imagination which provides this basis, for it is imagination which introduces the purely possible into human experience, placing it within the grasp of a consciousness that can project before itself what is not (and may never be) actual.

Therefore, in at least two fundamental ways, imagination can be shown to be essential to the creation and appreciation of art. Whatever the role of feeling may be – and it is unquestionably of considerable importance – the role of imagination in art is at least equally essential. It cannot be considered secondary, as Dufrenne holds, or as obsolete, as much recent Anglo-American aesthetic theory implies. It remains a primary constituent in aesthetic activity, whether this activity be that of the creator, the performer, the spectator, the critic, or the theorist of art. But it assumes this indispensable role without having to be conceived in hyperbolic terms as a supra-personal source of inspiration. Recourse to Romantic rhetoric is not required to shore up the crucial character of imagination in art. Patient phenomenological description reveals it as actively operative and constitutive in aesthetic experience. And it is active there by virtue of its own *a priori* structure, not by virtue of the extraordinary wealth of its possible contents or the bizarreness of its products. This *a priori* structure is continuous on the one hand with the virtual knowledge which we have of all types of *a priori* and on the other hand with the specifically material *a priori* immanent in works of art themselves. The dual continuity between imagination and the virtual and between imagination and the material *a priori* reveals the extent to which imagination is a central and centralizing factor in aesthetic experience. Consequently, it is misleading to give priority to feeling and the affective *a priori* in art. Equally basic, indeed perhaps more basic, are imagination and its own *a priori* aspects. For, as we have been, it is *via* imagination that art implies or involves the *a priori* in the first place. Through imagination's special structure, as well as through its possibilizing function, the material *a priori* in the aesthetic object and its virtual counterpart in the creating or apprehending subject can become effective components in the total aesthetic experience.

In any event, we may conclude that the link between art, imagination, and the *a priori* is closer than is usually admitted or even suspected. When each of these critical terms is properly described – art as creation and appreciation, imagination as *noetico-noematic* in structure and possibilizing in function, the *a priori* as material and virtual as well as formal – we are in a position to affirm a *rapprochement* which is more than merely formal. The three terms not only imply each other in propositional form; they are working partners in the concreteness of aesthetic experience. Within this experience, they are coeval and co-constitutive. Our purpose in this essay has been to underscore this orchestration of factors – factors which have too often been regarded as either opposed or indifferent to one another. We have attempted to overcome mistaken conceptual contrasts by pointing to profound phenomenological continuities: continuities which are best exemplified in aesthetic experience itself. As a result, the apparent paradox with which we began – i.e., the attempt to assimilate two such clashing concepts as imagination and the *a priori*, which seem to clash even more dramatically within the realm of art – is seen to be a pseudo-paradox built upon partial views of the terms at issue. The *a priori*, correctly analyzed as broader in scope than its formal dimension indicates, can no longer be regarded as incompatible with imagination when it is realized that the latter has its own definite structure and that it is not inherently amorphous. The *a priori* as material and as virtually known can merge and actively co-operate with imagination in its distinctive *noetico-noematic* character: one type of structure thus rejoins and reinforces another. And the two are capable of intimate interaction in aesthetic experience because it is there, above all, that the possibilizing role of imagination is most efficacious. Answering to this projective power of imagination is the virtual aspect of the *a priori*, which allows it to be anchored in the creative or appreciative subject. In sum, then, possibilities are projected by imagination in art; they are then embodied in the material *a priori* that permeate and uniquely structure the aesthetic object; and these latter *a priori* are apprehended and clarified by means of the tacit or virtual knowledge residing in the imagining subject (and the same subject can both create and apprehend *by* imagining). It is thus that a profound continuity – a continuity at once conceptual and phenomenological – exists between imagination and the *a priori*, which in the end are conterminous with one another within the domain of art.

NOTES

[1] For this reason, Mikel Dufrenne has written that "the material *a priori* is not [itself] matter, but it structures matter more intimately and specifically than does the formal *a priori*" (*The Notion of the A Priori* (transl. by Edward S. Casey) Northwestern University Press, Evanston 1966, p. 71. Hereafter *NA*).

[2] Here we disagree with Dufrenne, who holds the formal *a priori* to express "the given character of the given" (*ibid.*, p. 71). How can this be, if the formal *a priori* is also considered as expressing "objectivity in general – that is, the most external and general form of the object" (*ibid.*)? But we certainly concur with Dufrenne when he writes that the material *a priori* is "immediately recognized" (*ibid.*, p. 72) and is found only in particular experiences, not in experience in general (*ibid.*).

[3] *Ibid.*, p. 71.

[4] Cf. *ibid.*, pp. 247–48 (index) as well as *Phénoménologie de l'expérience esthétique* II, Presses Universitaires de France, Paris 1953, pp. 543–569. (The latter will be referred to as *PDEE*.)

[5] Cf. *NA*, pp. 108–110.

[6] Cf. Jacques Derrida, *De la grammatologie*, Editions du Minuit, Paris 1967, Part I.

[7] For further analysis, see J. N. Mohanty, *The Concept of Intentionality*, Warren Green, St. Louis 1971, *passim*.

[8] We call 'self-inducement' what a number of philosophers, notably Kant and Sartre, have termed 'spontaneity'. Besides being more descriptive of the character of the act, the notion of self-inducement does not carry the evaluative overtones of creativity which are present in the term 'spontaneity'. Further, the notion of self-inducement has the virtue of illustrating the *noetico-noematic* correlativity in its optimum form; for here noesis and *noema* not only correspond, but coincide.

[9] Cf. Jean-Paul Sartre, *Imagination* (transl. by F. Williams), University of Michigan Press, Ann Arbor, 1960; and *Psychology of Imagination* (transl. by B. Frechtman), Washington Square Press, New York 1966.

[10] Gilbert Ryle, *The Concept of Mind*, Barnes and Noble, New York, 1955, p. 251.

[11] For a fuller treatment of the imagined object and the image, see my article 'Imagination: Imagining and the Image', *Philosophy and Phenomenological Research* (June, 1971).

[12] Cf. R. G. Collingwood, *Principles of Art*, The Clarendon Press, Oxford 1938.

[13] Cf. my articles: 'Meaning in Art', in James Edie (ed.), *New Essays in Phenomenology*, Quadrangle, Chicago 1969; and 'Truth in Art', *Man and World* (Fall-Winter, 1970).

[14] Cf. Jacques Ehrmann (ed.), *Structuralism*, Anchor, New York, 1970.

[15] Mikel Dufrenne, *NA*, p. 72.

[16] On the question of aspect, especially the relation between image and aspect, see Martin Heidegger, *Vorträge und Aufsätze*, Neske, Pfullingen 1954, p. 200; and Ludwig Wittgenstein, *Philosophical Investigations* (transl. by G. E. M. Anscombe), Blackwell, Oxford 1967, §§301, 400; and p. 213.

[17] *NA*, p. 46.

[18] As Dufrenne writes, "the *a priori* of imagination impose themselves... as possible structures of a world" (*NA*, p. 62).

[19] There is also a virtual knowledge in the subject of the formal *a priori*; but this knowl-

edge is itself formal – or more exactly, transcendental – in character and it does not engage the subject in any significant sense. There is nothing, for example, corresponding to creation in the case of the formal *a priori* as virtually known. Thus the subject is not actively *solicited* as he is by the material *a priori*; instead, he simply structures experience (e.g. its spatio-temporal dimensions) in accordance with the universal features of the formal *a priori*.

[20] The quotation is from *PDEE*, p. 562; the identity claim is found in *ibid.*, pp. 561–62.

[21] *Ibid.*, p. 550.

[22] Cf. *ibid.*, pp. 570ff.

[23] Cf. *ibid.*, pp. 462–526.

[24] *Ibid.*, p. 554.

[25] In reference to Proust, Gilles Deleuze writes: "In art, matter is spiritualized, and the context is dematerialized. The work of art is thus a world of signs, but these signs are immaterial and no longer opaque" (*Proust et les signes*, Presses Universitaires de France, Paris 1970, p. 62).